T0319513

Intelligent Surfaces Empowered 6G Wireless Network

Intelligent Surfaces Empowered 6G Wireless Network

Edited by

Qingqing Wu
Shanghai Jiao Tong University
China

Trung Q. Duong
Memorial University of Newfoundland
Canada
Queen's University Belfast
United Kingdom

Derrick Wing Kwan Ng
The University of New South Wales
Australia

Robert Schober
University of Erlangen-Nuremberg
Germany

Rui Zhang
National University of Singapore
Singapore

IEEE PRESS
WILEY

Published by John Wiley & Sons, Inc., Hoboken, New Jersey.
Published simultaneously in Canada.

For general information on our other products and services or for technical support, please contact our Customer Care Department within the United States at (800) 762-2974, outside the United States at (317) 572-3993 or fax (317) 572-4002.

Wiley also publishes its books in a variety of electronic formats. Some content that appears in print may not be available in electronic formats. For more information about Wiley products, visit our web site at www.wiley.com.

Library of Congress Cataloging-in-Publication Data:

Names: Wu, Qingqing (Professor), author.
Title: Intelligent surfaces empowered 6G wireless network / Qingqing Wu
  [and four others].
Description: Hoboken, New Jersey : Wiley, [2024] | Includes index.
Identifiers: LCCN 2023047400 (print) | LCCN 2023047401 (ebook) | ISBN
  9781119913092 (hardback) | ISBN 9781119913108 (adobe pdf) | ISBN
  9781119913115 (epub)
Subjects: LCSH: Surfaces (Technology) | Smart materials. | 6G mobile
  communication systems.
Classification: LCC TA418.7 .W9 2024 (print) | LCC TA418.7 (ebook) | DDC
  621.3845/6–dc23/eng/20231107
LC record available at https://lccn.loc.gov/2023047400
LC ebook record available at https://lccn.loc.gov/2023047401

Cover Design: Wiley
Cover Image: © zf L/Getty Images

Set in 9.5/12.5pt STIXTwoText by Straive, Chennai, India

Contents

About the Editors

Qingqing Wu is an Associate Professor at Shanghai Jiao Tong University. His current research interest includes intelligent reflecting surface (IRS), UAV communications, and reconfigurable MIMO. He is listed as the Clarivate ESI Highly Cited Researcher from 2021 to 2023. He was the recipient of the IEEE Communications Society Asia Pacific Best Young Researcher Award in 2022.

Trung Q. Duong is a Full Professor at Memorial University of Newfoundland, Canada and also a Chair Professor in Telecommunications at Queen's University Belfast, United Kingdom. He has been awarded the Royal Academy of Engineering Research Chair (2021–2025) and Royal Academy of Engineering Research Fellowship (2016–2020). He received the Newton Prize 2017 from the UK government. He is a Fellow of IEEE.

Derrick Wing Kwan Ng is currently a Scientia Associate Professor at the University of New South Wales, Sydney, Australia and an IEEE Fellow. He is the Editor of IEEE Transactions on Communications and the Associate Editor-in-Chief of IEEE Open Journal of the Communications Society. His research interests include global optimization, physical layer security, IRS-assisted communication, UAV-assisted communication, wireless information and power transfer, and green (energy-efficient) wireless communications.

Robert Schober is an Alexander von Humboldt Professor and the Chair for Digital Communication at FAU. His research interests fall into the broad areas of communication theory, wireless and molecular communications, and statistical signal processing. Currently, he serves as Senior Editor of the Proceedings of the IEEE and as ComSoc President-Elect.

Dr. Rui Zhang (Fellow of IEEE, Fellow of the Academy of Engineering Singapore) received the PhD degree from Stanford University in Electrical Engineering in 2007. He is now a Principal's Diligence Chair Professor in School of Science and Engineering, The Chinese University of Hong Kong, Shenzhen. His current research interests include wireless information and power transfer, UAV/satellite communications, IRS, and reconfigurable MIMO.

List of Contributors

Hedieh Ajam
Department of Electrical, Electronics,
and Communication Engineering
Institute for Digital Communications
Friedrich-Alexander-University
Erlangen Nürnberg
Erlangen
Germany

George C. Alexandropoulos
Department of Informatics and
Telecommunications
National and Kapodistrian University
of Athens,
Athens
Greece

and

Technology Innovation Institute
Masdar City
Abu Dhabi
United Arab Emirates

Stefano Buzzi
Department of Electric and
Information Engineering (DIEI)
University of Cassino and Southern
Latium,
Cassino (FR)
Italy

and

Consorzio Nazionale
Interuniversitario per le
Telecomunicazioni (CNIT)
Parma
Italy

and

Dipartimento di Elettronica
Informazione e Bioingegneria
Politecnico di Milano
Milano
Italy

Zhi Chen
National Key Laboratory of
Wireless Communications
University of Electronic Science
and Technology of China
Chengdu
Sichuan
China

Yuhao Chen
Department of Electronic Engineering
Tsinghua University
Beijing
China

Qiang Cheng
State Key Laboratory of Millimeter
Waves
Southeast University
Nanjing
China

Bruno Clerckx
Department of Electrical and
Electronics Engineering
Imperial College London
London
UK

Tie Jun Cui
State Key Laboratory of Millimeter
Waves
Southeast University
Nanjing
China

Jun Yan Dai
State Key Laboratory of Millimeter
Waves
Southeast University
Nanjing
China

Linglong Dai
Department of Electronic Engineering
Tsinghua University
Beijing
China

Carmen D'Andrea
Department of Electric and
Information Engineering (DIEI)
University of Cassino and Southern
Latium
Cassino (FR)
Italy

and

Consorzio Nazionale
Interuniversitario per le
Telecomunicazioni (CNIT)
Parma
Italy

Zhiguo Ding
School of Electrical and Electronic
Engineering
The University of Manchester
Manchester
UK

Marco Di Renzo
Université Paris-Saclay
CNRS, CentraleSupélec, Laboratoire
des Signaux et Systèmes
Gif-sur-Yvette
France

Trung Q. Duong
Memorial University of Newfoundland
Canada

and

Queen's University Belfast
UK

Fang Fang
Department of Electrical and
Computer Engineering and the
Department of Computer Science
Western University
London
Canada

Jiguang He
Technology Innovation Institute
Abu Dhabi
United Arab Emirates

Shaokang Hu
School of Electrical Engineering and
Telecommunications
University of New South Wales
Sydney
NSW
Australia

Xiaoyan Hu
School of Information and
Communications Engineering
Xi'an Jiaotong University
Xi'an
China

Giovanni Interdonato
Department of Electric and
Information Engineering (DIEI)
University of Cassino and Southern
Latium
Cassino (FR)
Italy

and

Consorzio Nazionale
Interuniversitario per le
Telecomunicazioni (CNIT)
Parma
Italy

Shi Jin
National Mobile Communications
Research Laboratory
Southeast University
Nanjing
China

and

School of Information Science and
Engineering
Southeast University
Nanjing
China

Tao Jiang
The Edward S. Rogers Sr. Department
of Electrical and Computer
Engineering
University of Toronto
Toronto
Ontario
Canada

Konstantinos D. Katsanos
Department of Informatics and
Telecommunications
National and Kapodistrian University
of Athens
Athens
Greece

Hyowon Kim
Department of Electronics
Engineering
Chungnam National University
Daejeon
South Korea

Ruiqi (Richie) Liu
State Key Laboratory of Mobile
Network and Mobile Multimedia
Technology
Shenzhen
China

and

Wireless and Computing Research
Institute
ZTE Corporation
Beijing
China

Zhi-Quan Luo
Shenzhen Research Institute of
Big Data
Shenzhen
China

Christos Masouros
Department of Electronic and
Electrical Engineering
University College London
London
UK

Kaitao Meng
State Key Laboratory of Internet
of Things for Smart City
University of Macau
Macau
China

Derrick Wing Kwan Ng
School of Electrical Engineering
and Telecommunications
University of New South Wales
Sydney
NSW
Australia

Boyu Ning
National Key Laboratory of Wireless
Communications
University of Electronic Science and
Technology of China
Chengdu
Sichuan
China

Robert Schober
Department of Electrical, Electronics
and Communication Engineering
Institute for Digital Communications
Friedrich-Alexander-University
Erlangen Nürnberg
Erlangen
Germany

Kaiming Shen
School of Science and Engineering
The Chinese University of Hong Kong
(Shenzhen)
Shenzhen
China

Foad Sohrabi
Nokia Bell Labs
Murray Hill
NJ
USA

Wankai Tang
National Mobile Communications
Research Laboratory
Southeast University
Nanjing
China

Jinghe Wang
National Mobile Communications
Research Laboratory
Southeast University
Nanjing
China

Kai-Kit Wong
Department of Electronic and
Electrical Engineering
University College London
London
UK

Qingqing Wu
Department of Electronic Engineering
Shanghai Jiaotong University
Shanghai
China

Henk Wymeersch
Department of Electrical Engineering
Chalmers University of Technology
Gothenburg
Sweden

Ximing Xie
School of Electrical and Electronic
Engineering
The University of Manchester
Manchester
UK

Qiumo Yu
Department of Electronic Engineering
Tsinghua University
Beijing
China

Wei Yu
The Edward S. Rogers Sr. Department
of Electrical and Computer
Engineering
University of Toronto
Toronto
Ontario
Canada

Rui Zhang
Department of Electrical and
Computer Engineering
National University of Singapore
Singapore

Zijian Zhang
Department of Electronic Engineering
Tsinghua University
Beijing
China

Yang Zhao
Department of Electrical and
Electronics Engineering
Imperial College London
London
UK

Preface

The next generation of wireless technology (6G) promises to transform wireless communication and human interconnectivity like never before. Intelligent surface, which adopts significant numbers of small reflective surfaces to reconfigure wireless connections and improve network performance, has recently come to be recognized as a critical component for enabling the future 6G. The next phase of wireless technology demands engineers and researchers familiar with this technology and able to cope with the challenges.

Intelligent Surfaces Empowered 6G Wireless Network provides a thorough overview of intelligent surface technologies and their applications in wireless networks and 6G. It includes an introduction to the fundamentals of intelligent surfaces, before moving to more advanced content for engineers who understand them and look to apply them in the 6G realm. Its detailed discussion of the challenges and opportunities posed by intelligent surfaces empowered wireless networks makes it the first work of its kind.

Intelligent Surfaces Empowered 6G Wireless Network readers will also find:

- An editorial team including the original pioneers of intelligent surface technology.
- Detailed coverage of subjects including MIMO, terahertz, NOMA, energy harvesting, physical layer security, computing, sensing, machine learning, and more.
- Discussion of hardware design, signal processing techniques, and other critical aspects of IRS engineering.

Intelligent Surfaces Empowered 6G Wireless Network is a must for students, researchers, and working engineers looking to understand this vital aspect of the coming 6G revolution.

10 October 1991	*Qingqing Wu*
29 October 1979	*Trung Q. Duong*
5 November 1984	*Derrick Wing Kwan Ng*
10 June 1971	*Robert Schober*
14 September 1976	*Rui Zhang*

Acknowledgement

Q. Wu's work is supported by National Key R&D Program of China (2023YFB2905000), NSFC 62371289, NSFC 62331022, FDCT under Grant 0119/2020/A3.

Part I

Fundamentals of IRS

1

Introduction to Intelligent Surfaces

Kaitao Meng[1], Qingqing Wu[2], Trung Q. Duong[3,4], Derrick Wing Kwan Ng[5], Robert Schober[6], and Rui Zhang[7]

[1] *State Key Laboratory of Internet of Things for Smart City, University of Macau, Macau, China*
[2] *Department of Electronic Engineering, Shanghai Jiaotong University, Shanghai, Country*
[3] *Queen's University Belfast, United Kingdom*
[4] *Memorial University of Newfoundland, Canada*
[5] *School of Electrical Engineering and Telecommunications, University of New South Wales, Sydney, NSW, Australia*
[6] *Institute for Digital Communications, Friedrich-Alexander University of Erlangen-Nuremberg, Erlangen, Germany*
[7] *Department of Electrical and Computer Engineering, National University of Singapore, Singapore*

1.1 Background

In the forthcoming era of Internet of Everything (IoE), worldwide mobile data traffic is expected to grow at an annual rate of roughly 55% between 2020 and 2030, eventually reaching 5016 exabytes by 2030 (Andrews et al. 2014). According to a recent report, by 2025, the number of connected devices will increase to more than 30 billion globally. Also, considering the rapid emergence of new wireless applications such as smart cities, intelligent transportation, and augmented/virtual reality (Nikitas et al. 2020), it is foreseen that the fifth generation (5G) may encounter capacity and performance limitations in supporting the accommodating these low-latency, high-capacity, ultra-reliable, and massive-connectivity wireless communication services. In addition to supporting these high-quality wireless communications, next-generation wireless networks are required to provide several other heterogeneous services, e.g., extremely high-accuracy sensing (Meng et al. 2023) and low-latency computing capabilities. Specifically, the representative key performance indicators advocated by the sixth generation (6G) are summarized as follows (Letaief et al. 2019; Tataria et al. 2021; Jiang et al. 2021):

Intelligent Surfaces Empowered 6G Wireless Network, First Edition.
Edited by Qingqing Wu, Trung Q. Duong, Derrick Wing Kwan Ng, Robert Schober, and Rui Zhang.
© 2024 The Institute of Electrical and Electronics Engineers, Inc. Published 2024 by John Wiley & Sons, Inc.

- The **peak data rates** under ideal wireless propagation conditions are higher than terabits per second for both indoor and outdoor connections, which is 100–1000 times that of the state-of-the-art 5G;
- The **energy efficiency** of 6G is 10–100 times that of 5G to achieve green communications;
- Five times more the **spectral efficiency** of 5G is pursued by utilizing the limited frequency spectrum more efficiently;
- **Connection density** could be 10 times that of 5G, about 10^7 devices/km^2 to satisfy the high demand for massive connectivity in IoE and enhanced mMTC;
- **Reliability** is larger than 99.99999% to support more enhanced ultra-reliable and low-latency communication (URLLC) compared to 5G;
- Shorter than 100 μs **latency** is required to support numerous enhanced URLLC;
- Centimeter (cm)-level **positioning accuracy** in three-dimensional (3D) space is required to fulfill the harsh demands of various vertical and industrial applications, instead of requiring meter (m)-level positioning accuracy in two-dimensional space (2D) space as in 5G;

However, even with the existing technologies such as massive multiple-input-multiple-output (M-MIMO) and millimeter wave (mmWave)/terahertz (THz), the abovementioned performance requirements for IoE services may not be fully realized due to the following reasons:

- First, dense deployment of active nodes such as access points (APs), base stations (BSs), and relays can shorten the communication distance, thereby enhancing network coverage and capacity, which, however, incurs higher energy consumption and backhaul/deployment/maintenance costs.
- Second, installing substantially more antennas at APs/BSs/relays to take advantage of the huge M-MIMO gains inevitably results in increased hardware/energy costs and signal processing complexity, as well as exacerbates more severe and complicated network interference issues (Lu et al. 2014).
- Third, migrating to higher frequency bands, such as mmWave and THz frequencies, is able to harness their larger available unlicensed bandwidth (Niu et al. 2015). Yet, it requires the deployment of more active nodes and more antennas to compensate for the associated severer propagation attenuation over distance.
- Fourth, the diffraction and scattering effects of high-frequency radio are weakened, such that propagating electromagnetic waves can be easily blocked by obstacles such as urban buildings. As a result, the effective coverage radius of APs/BSs decreases while the potential number of blind spots increase. Thus, it will be difficult to ensure universal coverage and wireless services exploiting traditional cellular technologies.

Taking into account the above limitations and issues, it is highly imperative to develop disruptively new and innovative technologies to realize spectrum- and energy-efficient and cost-effective capacity growth of future wireless networks.

1.2 Concept of Intelligent Surfaces

The fundamental challenge to achieve high-throughput and ultra-reliable wireless communication arises from time-varying wireless channels caused by user mobility (Neely 2006). The conventional approaches to this challenge are mainly to utilize various modulation, coding, and diversity techniques to counteract for channel fading, or to adapt to the channel through adaptive power/rate control and beamforming techniques (Chen and Laneman 2006). Traditionally, the fading wireless channel is treated as uncontrollable block-box and becomes one of the main limiting factors for performance improvement.

Motivated by the above, the advancing radio environment reconfiguration technique has recently emerged as a promising new paradigm to achieve smart and highly controllable ratio propagation channels for next-generation wireless communication systems. This has been achieved by the recently proposed intelligent surfaces, such as intelligent reflecting/refracting surfaces (IRSs) (Wu and Zhang 2019; Zheng et al. 2022; Huang et al. 2022), also called reconfigurable intelligent surfaces (RISs) (ElMossallamy et al. 2020; Liu et al. 2021), or large intelligent surfaces (LISs) (Hu et al. 2018; Jung et al. 2021). Generally speaking, an intelligent surface is a planar surface comprising a large number of passive reflecting/refracting elements, each of which is able to induce a controllable amplitude and/or phase change to the incident signal independently. More specifically, intelligent surfaces can be realized by adopting metamaterial or patch-array-based technologies (Liu et al. 2021). With a dense deployment of smart surfaces in a wireless network and smartly coordinating their reflection/refraction, the signal propagation/radio between transmitters and receivers can be flexibly reconfigured to achieve the desired realizations and/or distributions. This paradigm serves as a new approach to fundamentally address the wireless channel fading impairment and interference issues, and it is possible to achieve a dramatic improvement in wireless communication capacity and reliability.

As shown in Fig. 1.1, a typical architecture of an intelligent surface consists of three layers and a smart controller. On the right-most layer, a large number of metal patches are printed on a dielectric substrate to directly interact with the incident signal. Behind this layer, a copper plate is adopted to avoid signal energy leakage. Finally, the left-most layer is the control board responsible for adjusting

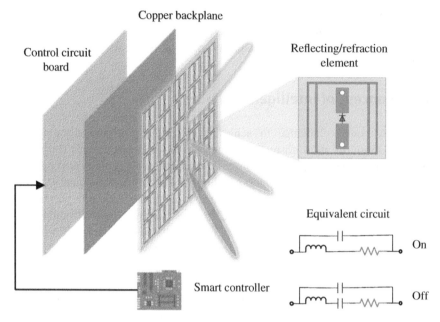

Figure 1.1 Architecture of intelligent surfaces. Source: Adapted from (Wu et al. 2021).

the reflection/refraction amplitude/phase shift of each element, triggered by a smart controller connected to the intelligent surface (Jian et al. 2022). Practically, a field-programmable gate array (FPGA) can be implemented as a controller, which also acts as a gateway, communicating and coordinating with other network components (such as BS, AP, and user terminals) through an out-of-band wireless link to achieve low power consumption rate information is exchanged with them.

In the following, we further highlight the main differences between intelligent surfaces and other related technologies such as active relaying and backscatter communication. First, compared to active wireless relays that assist source–destination communication through signal regeneration and retransmission, intelligent surfaces neither demodulate nor generate any information sources but act as passive array (or weakly active array, such as active intelligent surfaces; Long et al. 2021) to reflect/refract the received signals. Additionally, active relays generally operates with a half-duplex (HD) protocol and are therefore generally less spectral efficient than that of intelligent surfaces operating in full-duplex (FD) mode. Second, a tag reader in backscatter communication requires to perform interference cancellation at its receiver to decode the radio frequency identification tags' message (Wang et al. 2016). Actually, intelligent surfaces can also modulate its information during reflection, but which are primarily designed to facilitate existing communication links.

1.3   Advantages of Intelligence Surface

Benefiting from the appealing ability to reconfigure wireless channels, it is envisioned that intelligent surfaces are suitable to be massively deployed in wireless networks and possess various practical advantages for implementation as follows:

- First, in general, intelligent surfaces mainly reflect/refract incident signals in a passive manner without requiring any transmit radio-frequency (RF) chains. Thus, they can be implemented with orders-of-magnitude lower hardware/ energy costs as compared to traditional active antenna arrays or the recently proposed active surfaces (Hu et al. 2018).
- Second, intelligent surfaces are not only able to reconfigure the wireless propagation environment by compensating for the power loss over long distances but also optimize channel rank for improving the potential spatial multiplexing gain and spectral efficiency by introducing more controllable signal paths between the transmitters and receivers in multi-antenna communications.
- Third, intelligent surfaces operate in full-duplex mode and are free of any antenna noise amplification as well as self-interference (Abdullah et al. 2020), which thus offers competitive advantages over traditional active relays, e.g., the half-duplex relay that suffers from low spectral efficiency as well as the FD relay that requires sophisticated techniques for self-interference cancellation.
- Fourth, interference suppression becomes effective by utilizing intelligent surfaces which results in a better signal quality for the cell-edge users (Ma et al. 2021). For multi-user wireless networks, the resources of intelligent surfaces can be allocated over the time/spatial domains to assist data transmissions of different users. In this case, better quality-of-service (QoS) provisioning can be provided to improve the sum-rate performance or max-min fairness among different users.
- Fifth, the effective network coverage can be extended by utilizing intelligent surfaces to establish virtual line-of-sight (LoS) links and bypass signal blockages between the transceivers. Based on this, such virtual LoS links establishment can also offer new opportunities to achieve localization in blind spots (Aubry et al. 2021). In this case, intelligent surfaces can be seamlessly integrated into existing wireless systems to support ubiquitous connectivity and provide great flexibility for sensing and communication.
- Finally, the intelligent surface is made of low-cost passive scattering elements, which is easier to be deployed and more sustainable to be operated compared

to active nodes (e.g., APs, BSs, and relays) since the intelligent surfaces can be battery-less and wirelessly powered by RF-based energy harvesting. Moreover, it can be easily attached to and removed from the facades of buildings, indoor walls, ceilings, and even mobile vehicles/trains.

Motivated by the above advantages, it is foreseen that intelligent surfaces will bring fundamental paradigm shifts in wireless network design in the future. For example, as compared to employing an M-MIMO antenna array with an extremely large size, it is more sustainable and energy-efficient for the new small/moderate-scale MIMO networks assisted by intelligent surfaces. As such, different from M-MIMO leveraging tens and even hundreds of active antennas to generate sharp beams steering toward users' direction, intelligent surface-assisted MIMO systems can create fine-grained reflect/refract beams via smart passive reflection/refraction design by exploiting the large aperture of intelligent surfaces, which only requires substantially fewer antennas while satisfying the users' QoS requirements. Thus, this significantly reduces the system hardware cost and energy consumption, especially for wireless systems migrating to higher frequency bands in next-generation wireless networks. On the other hand, while existing wireless networks rely on a heterogeneous multi-layer architecture consisting of macros and small BSs/APs, they are active nodes that generate new signals. Therefore, complex coordination and interference management between these nodes are required to achieve the premise of network space capacity enhancement. However, this approach inevitably increases network operating overhead, and it may not be cost-effective to sustain the growth of wireless network capacity in the future. Differently, integrating intelligent surfaces into wireless networks creates a new hybrid architecture comprising active and passive components intelligently. Since smart surfaces are much lower cost than their active counterparts, they can be deployed more densely in wireless networks at a lower cost without the need for sophisticated interference management. By optimally designing the ratio between active BSs and passive intelligent surfaces deployed in a hybrid network, sustainable network capacity scaling with cost can be achieved (Lyu and Zhang 2021).

1.4 Potential Applications

Figure 1.2 illustrates an envisioned future wireless network assisted by intelligent surfaces with a variety of promising applications, including service enhancement in the dead zone, smart office, smart industry, and smart transportation. Specifically, in outdoor environments, the intelligent surface can be coated on building facades, lampposts, billboards, and even the surfaces of high-speed

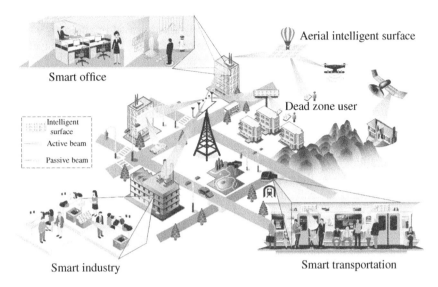

Aerial intelligent surface

Smart office

Dead zone user

Intelligent surface
Active beam
Passive beam

Smart industry

Smart transportation

Figure 1.2 Smart city empowered by intelligent surfaces.

vehicles. By effectively compensating for the Doppler effect (Basar 2021), intelligent surfaces can support smart cities, intelligent transportation, and other applications. On the other hand, in indoor environments, intelligent surfaces can also be attached to the ceilings, walls, furniture, and even behind the paintings/decorations to effectively tackle the issues of limited or unavailable coverage caused by occlusions and to achieve high-capacity hot-spot for eMBB and mMTC applications,[1] thereby providing low-cost and easy-to-implement coverage enhancement solutions for manufacturing and remote control. Overall, intelligent surfaces are a disruptive technology that can make our current passive environment intelligent, active, and controllable, and it may benefit a wide range of 5G/6G vertical industries. Furthermore, there is a high interest in implementing and commercializing intelligent surface-like technologies to create new valuable industry chains, and several pilot projects about intelligent surfaces have been launched to advance research in this new area. For example, NTT DOCOMO and Metawave company demonstrated a 28 GHz band based on the constructed meta-structure reflectarray in 2018 (Chen et al. 2021). Moreover, Greenerwave company developed physics-inspired algorithms for reconfigurable metasurfaces. In September 2020, ZTE Company joined hands with more than ten domestic and foreign enterprises and universities, including China Unicom, to establish

---

1 Note that such surfaces can actually be fabricated as mirrors or lenses for signal reflection and refraction, respectively, depending on the application scenario where the wireless transmitter and receiver are located on the same or opposite sides of the surface.

the "RIS Research Project" in CCSA TC 5-WG 6 (Liu et al. 2022). In September 2021, the IMT-2030 (6G) Promotion Group officially released the industry's first research report on smart metasurface technology at the 6G seminar (Wang et al. 2022).

Aiming at the main applications of intelligent surfaces in future wireless networks, several typical scenarios are shown in Fig. 1.3. For example, for a user located in a service shadow zone in Fig. 1.3(a), intelligent surfaces can be deployed in a proper location to create a virtual LoS link between the user and its serving BS. This is especially useful for coverage extension of mmWave and THz communications that are highly susceptible to obstruction. In addition, deploying intelligent surfaces at the cell edge not only helps increase the expected signal power of users at the cell edge but also helps suppress co-channel interference

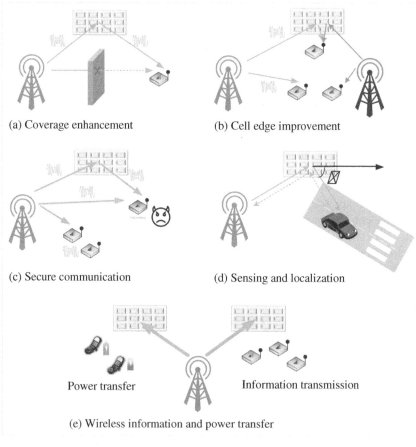

(a) Coverage enhancement   (b) Cell edge improvement

(c) Secure communication   (d) Sensing and localization

Power transfer   Information transmission

(e) Wireless information and power transfer

Figure 1.3 Illustration of intelligent surface applications.

from adjacent cells to users, as shown in Fig. 1.3(b). Similarly, intelligent surfaces play an important role in secure communications by controlling the potential signal leakage energy at the eavesdropper's location, c.f. Fig. 1.3(c). More specifically, when the link distance from the BS to the eavesdropper is smaller than that to the legitimate user, or the eavesdropper lies in the same direction as the legitimate user, the achievable secrecy communication rates are highly limited (even by employing transmit beamforming at the BS in the latter case). However, if an intelligent surface is deployed in the vicinity of the eavesdropper, the reflected signal from the intelligent surface can be adapted to cancel the signal from the BS at the eavesdropper, thus effectively reducing the information leakage. Besides improving communication performance, multi-antenna radar can also utilize intelligent surfaces to increase the signal power reflected from the target in Fig. 1.3(d). In particular, intelligent surfaces can also help to obtain the target information by establishing artificial virtual LoS links from the radar to the targets. Furthermore, to improve the efficiency of simultaneous wireless information and power transfer (SWIPT) from BSs/APs to wireless devices (Wu and Zhang 2020), the large aperture of the intelligent surface can be exploited to compensate for significant power loss at long distances by reflection/refraction to its nearby counterparts, as shown in Fig. 1.3(e).

In addition to the basic applications mentioned above, several new trends in intelligent surface applications are being investigated, including the transition from single surfaces to network-level surfaces, from passive surfaces to active and hybrid surfaces, and from reflective/refractive surfaces to omni-surfaces, from ground surfaces to integrated air-ground surfaces, from fixed surfaces to mobile surfaces, from simple connection surface to beyond diagonal surfaces, and so on.

The rest of this book is organized as follows. Part I includes another three chapters to illustrate several key aspects of the system design of intelligent surfaces in detail, such as the intelligent surface architecture and hardware design in Chapter 2, the channel modeling methods for intelligent surfaces in Chapter 3, as well as the associated main challenges and solutions in Chapter 4. Furthermore, Part II focuses on several promising designs for intelligent surface empowered 6G wireless systems. First, Chapter 5 provides an overview of intelligent surfaces for 6G and industry advance, followed by typical combing with conventional techniques, such as IRS for massive MIMO/OFDM in Chapter 6; IRS design for URLLC and sensing/localization in Chapters 7 and 8, respectively; besides these conventional techniques, IRS-aided mmWave/THz communications, IRS-aided NOMA, interference nulling assisted by IRS are investigated in Chapters 9, 10, and 12, respectively; except for communication, intelligent surfaces can also be designed for assisting edge computing and machine learning, which are studied in Chapters 11 and 13, respectively. Moreover, in Part III, we discuss other relevant topics on intelligent surfaces for broadening their scopes, e.g.,

Chapter 14 discusses the key problem of IRS-aided wireless power transfer and energy harvesting; Chapter 15 presents the advantages of IRS for physical layer security design; finally, the potential for wireless optical communication are also investigated.

1.5 Conclusion

In this chapter, a comprehensive introduction to the new intelligent surface technology is provided. A detailed concept of intelligent surfaces and the corresponding promising applications are presented to overview existing related works and inspire future direction. Finally, the organizational structure of the book is briefly explained.

Bibliography

Abdullah, Z., Chen, G., Lambotharan, S., and Chambers, J.A. (2020). Optimization of intelligent reflecting surface assisted full-duplex relay networks. *IEEE Wireless Communications Letters* 10 (2): 363–367.

Andrews, J.G., Buzzi, S., Choi, W. et al. (2014). What will 5G be? *IEEE Journal on Selected Areas in Communications* 32 (6): 1065–1082. https://doi.org/10.1109/JSAC .2014.2328098.

Aubry, A., De Maio, A., and Rosamilia, M. (2021). Reconfigurable intelligent surfaces for N-LOS radar surveillance. *IEEE Transactions on Vehicular Technology* 70 (10): 10735–10749.

Basar, E. (2021). Reconfigurable intelligent surfaces for Doppler effect and multipath fading mitigation. *Frontiers in Communications and Networks* 2: 672857. https://doi .org/10.3389/frcmn.2021.672857.

Chen, D. and Laneman, J.N. (2006). Modulation and demodulation for cooperative diversity in wireless systems. *IEEE Transactions on Wireless Communications* 5 (7): 1785–1794. https://doi.org/10.1109/TWC.2006.1673090.

Chen, Z., Ma, X., Han, C., and Wen, Q. (2021). Towards intelligent reflecting surface empowered 6G terahertz communications: a survey. *China Communications* 18 (5): 93–119.

ElMossallamy, M.A., Zhang, H., Song, L. et al. (2020). Reconfigurable intelligent surfaces for wireless communications: principles, challenges, and opportunities. *IEEE Transactions on Cognitive Communications and Networking* 6 (3): 990–1002. https://doi.org/10.1109/TCCN.2020.2992604.

Hu, S., Rusek, F., and Edfors, O. (2018). Beyond massive MIMO: the potential of positioning with large intelligent surfaces. *IEEE Transactions on Signal Processing* 66 (7): 1761–1774. https://doi.org/10.1109/TSP.2018.2795547.

Huang, Z., Zheng, B., and Zhang, R. (2022). Transforming fading channel from fast to slow: intelligent refracting surface aided high-mobility communication. *IEEE Transactions on Wireless Communications* 21 (7): 4989–5003. https://doi.org/10.1109/TWC.2021.3135685.

Jian, M., Alexandropoulos, G.C., Basar, E. et al. (2022). Reconfigurable intelligent surfaces for wireless communications: overview of hardware designs, channel models, and estimation techniques. *Intelligent and Converged Networks* 3 (1): 1–32.

Jiang, W., Han, B., Habibi, M.A., and Schotten, H.D. (2021). The road towards 6G: a comprehensive survey. *IEEE Open Journal of the Communications Society* 2: 334–366. https://doi.org/10.1109/OJCOMS.2021.3057679.

Jung, M., Saad, W., and Kong, G. (2021). Performance analysis of active large intelligent surfaces (LISs): uplink spectral efficiency and pilot training. *IEEE Transactions on Communications* 69 (5): 3379–3394. https://doi.org/10.1109/TCOMM.2021.3056532.

Letaief, K.B., Chen, W., Shi, Y. et al. (2019). The roadmap to 6G: AI empowered wireless networks. *IEEE Communications Magazine* 57 (8): 84–90. https://doi.org/10.1109/MCOM.2019.1900271.

Liu, R., Wu, Q., Di Renzo, M., and Yuan, Y. (2022). A path to smart radio environments: an industrial viewpoint on reconfigurable intelligent surfaces. *IEEE Wireless Communications* 29 (1): 202–208.

Liu, Y., Liu, X., Mu, X. et al. (2021). Reconfigurable intelligent surfaces: principles and opportunities. *IEEE Communications Surveys & Tutorials* 23 (3): 1546–1577. https://doi.org/10.1109/COMST.2021.3077737.

Long, R., Liang, Y.-C., Pei, Y., and Larsson, E.G. (2021). Active reconfigurable intelligent surface-aided wireless communications. *IEEE Transactions on Wireless Communications* 20 (8): 4962–4975.

Lu, L., Li, G.Y., Swindlehurst, A.L. et al. (2014). An overview of massive MIMO: benefits and challenges. *IEEE Journal of Selected Topics in Signal Processing* 8 (5): 742–758.

Lyu, J. and Zhang, R. (2021). Hybrid active/passive wireless network aided by intelligent reflecting surface: system modeling and performance analysis. *IEEE Transactions on Wireless Communications* 20 (11): 7196–7212. https://doi.org/10.1109/TWC.2021.3081447.

Ma, Z., Wu, Y., Xiao, M. et al. (2021). Interference suppression for railway wireless communication systems: a reconfigurable intelligent surface approach. *IEEE Transactions on Vehicular Technology* 70 (11): 11593–11603.

Meng, K., Wu, Q., Ma, S. et al. (2023). Throughput maximization for UAV-enabled integrated periodic sensing and communication. *IEEE Transactions on Wireless Communications* 22 (1): 671–687. https://doi.org/10.1109/TWC.2022.3197623.

Neely, M.J. (2006). Energy optimal control for time-varying wireless networks. *IEEE Transactions on Information Theory* 52 (7): 2915–2934.

Nikitas, A., Michalakopoulou, K., Njoya, E.T., and Karampatzakis, D. (2020). Artificial intelligence, transport and the smart city: definitions and dimensions of a new mobility era. *Sustainability* 12 (7).

Niu, Y., Li, Y., Jin, D. et al. (2015). A survey of millimeter wave communications (mmWave) for 5G: opportunities and challenges. *Wireless Networks* 21 (8): 2657–2676.

Tataria, H., Shafi, M., Molisch, A.F. et al. (2021). 6G wireless systems: vision, requirements, challenges, insights, and opportunities. *Proceedings of the IEEE* 109 (7): 1166–1199. https://doi.org/10.1109/JPROC.2021.3061701.

Wang, G., Gao, F., Fan, R., and Tellambura, C. (2016). Ambient backscatter communication systems: detection and performance analysis. *IEEE Transactions on Communications* 64 (11): 4836–4846. https://doi.org/10.1109/TCOMM.2016.2602341.

Wang, Z., Du, Y., Wei, K. et al. (2022). Vision, application scenarios, and key technology trends for 6G mobile communications. *Science China Information Sciences* 65 (5): 151301.

Wu, Q. and Zhang, R. (2019). Intelligent reflecting surface enhanced wireless network via joint active and passive beamforming. *IEEE Transactions on Wireless Communications* 18 (11): 5394–5409. https://doi.org/10.1109/TWC.2019.2936025.

Wu, Q. and Zhang, R. (2020). Joint active and passive beamforming optimization for intelligent reflecting surface assisted SWIPT under QoS constraints. *IEEE Journal on Selected Areas in Communications* 38 (8): 1735–1748. https://doi.org/10.1109/JSAC.2020.3000807.

Wu, Q., Zhang, S., Zheng, B. et al. (2021). Intelligent reflecting surface-aided wireless communications: a tutorial. *IEEE Transactions on Communications* 69 (5): 3313–3351. https://doi.org/10.1109/TCOMM.2021.3051897.

Zheng, B., You, C., Mei, W., and Zhang, R. (2022). A survey on channel estimation and practical passive beamforming design for intelligent reflecting surface aided wireless communications. *IEEE Communications Surveys & Tutorials* 24 (2): 1035–1071. https://doi.org/10.1109/COMST.2022.3155305.

2

IRS Architecture and Hardware Design

Zijian Zhang, Yuhao Chen, Qiumo Yu, and Linglong Dai

Department of Electronic Engineering, Tsinghua University, Beijing, China

2.1 Metamaterials: Basics of IRS

In a broad sense, IRS is a branch of electromagnetic metamaterials. Specifically, metamaterials can be divided into three-dimensional metamaterials and two-dimensional metasurfaces (Di Renzo et al. 2020). The widely studied metasurfaces can be divided into the metasurfaces with fixed physical parameters and those with adjustable electromagnetic properties.

Generally, IRS is a kind of dynamically adjustable metasurface. Metamaterials were first known as "left-hand materials" and "double-negative electromagnetic media." In 1967, Viktor Veselago published a Russian paper on metamaterials (Veselago 1967), which was translated into English in 1968. In this paper, Veselago first proposed the concept of left-hand materials, which refer to the materials with negative dielectric constant $\varepsilon < 0$ and negative permeability $\mu < 0$. He also systematically analyzes the propagation characteristics of electromagnetic waves in double-negative media, and predicts several novel anomalous electromagnetic phenomena in theory. In 1996, John B. Pendry verified the existence of negative permittivity (Pendry et al. 1996), and he further proposed the realization of periodic arrangement and verified the negative permeability (Pendry et al. 1999). The earliest research on artificial metasurfaces was the mushroom-shaped high-impedance surface (HIS) proposed by the team leaded by Daniel F. Sievenpiper in Sievenpiper et al. (1999).

The electromagnetic properties of three-dimensional metamaterials can be described by using traditional equivalent medium parameters (e.g., dielectric constant and permeability), but these parameters are no longer applicable to

Intelligent Surfaces Empowered 6G Wireless Network, First Edition.
Edited by Qingqing Wu, Trung Q. Duong, Derrick Wing Kwan Ng, Robert Schober, and Rui Zhang.

the analysis of metasurfaces (Cui et al. 2014). Regarding the two-dimensional structure of metasurfaces, researchers have successively proposed a variety of theories for analysis and modeling, among which the most representative is the generalized Snell's law proposed by Federico Capasso's team in 2011 (Yu et al. 2011). It is proved that the generalized Snell's law can well describe the physical properties of electromagnetic metasurfaces, which has become a common method for analysis and modeling for a long time.

2.2 Programmable Metasurfaces

The early metasurfaces are designed and fabricated based on the metamaterials. When the physical structure of these metasurfaces is determined, their functional performances are fixed. Since these metasurfaces do not support dynamic adjustments, the flexibility of their applications is limited. To solve this problem, the programmable metasurface has become a mainstream of researches. Various tunable metamaterials have been used to implement programmable metasurfaces, as shown in Fig. 2.1. Particularly, as a representative realization, active components (such as switching diodes and varactors), or adjustable materials (such as graphene) are widely integrated into the programmable metasurfaces. Through adjusting the external excitation, the fixed physical structure of the metasurfaces shows reconfigurable electromagnetic characteristics, which provides a possibility to implement a practical IRS.

In the early research, electromagnetic metasurfaces usually use continuous or quasi-continuous polarizability, impedance, amplitude, phase, and other parameters to characterize the electromagnetic characteristics on their interfaces. These characterization methods are used to design metasurfaces from a perspective of physics; thus, these metasurfaces can be called "analog metasurfaces," In 2014, Tiejun Cui of Southeast University put forward the concept of "digital coding and programmable metasurfaces," and used the form of binary coding to characterize the electromagnetic characteristics of metasurfaces (Cui et al. 2014). After modeling the adjustable physical properties, the mature coding algorithm and software in computer can be used to optimize the physical parameters of metasurfaces. Besides, it is also convenient to better use artificial intelligence (AI) to intelligently reconfigure the electromagnetic responses of metasurfaces (Ma et al. 2019). In 2017, Cui's team published a paper to summarize the existing research and propose the concept of "information metasurfaces," which introduces well-known IRS technology (Cui et al. 2017).

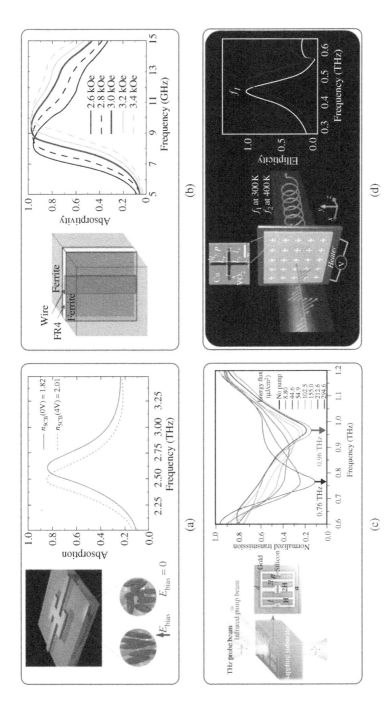

Figure 2.1 Programmable metasurfaces made of different stimuli-sensitive metamaterials. (a) Electric-sensitive liquid crystal-based metasurface. (b) Magnetic sensitive ferrite-based metasurface. (c) Light-sensitive semiconductor-based metasurface. (d) Thermal-sensitive VO$_2$-based metasurface. Source: Fu Liu et al. 2018/Reproduced from IEEE.

2.3 IRS Hardware Design

2.3.1 IRS System Architecture

Whether it is used as a terminal antenna or a control device for a channel, the hardware architecture of a smart metasurface consists of three main parts: a reconfigurable electromagnetic surface, a feeding system, and a control system.

The reconfigurable electromagnetic surface is the main part of the system to control the space wave. Its structure is an array composed of periodic or quasi-periodic surface elements. Each element integrates nonlinear devices such as PIN diodes (Mamedes et al. 2020), varactors (Rains et al. 2022) or MEMS switches (Baghchehsaraei and Oberhammer 2013). Generally speaking, these nonlinear devices respond to low-frequency control signals given by the control system, changing the electromagnetic properties of the local elements, thereby regulating the high-frequency signals from the feeding system (Fig. 2.2).

The feeding system provides energy and controls the communication signals of the entire system. According to the way of feeding and output, it can be divided into reflection type (Dai et al. 2018), transmission type (Asadchy et al. 2016), active and passive integrated type (Venkatesh et al. 2022), and near-field transmission type. Its function is to input the electromagnetic wave to be modulated onto the electromagnetic surface. Among them, the far-field reflective and transmissive feeding systems can be actively transmitted by the feed antenna in the same system, or can passively receive long-range electromagnetic waves from other

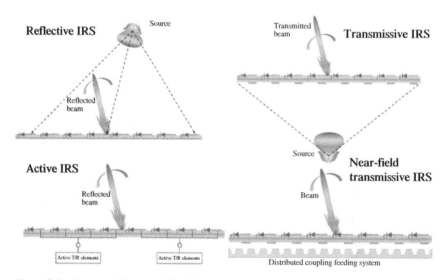

Figure 2.2 Four typical types of feeding system.

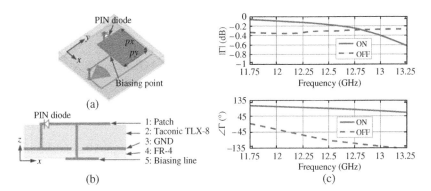

Figure 2.3 1-bit IRS element. (a) Perspective view. (b) Side view. (c) Simulated reflection magnitude and phase.

signal sources. At this time, there will be no physical entities in the feed system, but still an important part of the overall system.

The control system is usually integrated on the field programmable gate array (FPGA) or other programmable platforms to control the electromagnetic property of the surface. Control signal on low frequency is generated on it based on control decision of higher-level system, adapting the voltage on the nonlinear devices on the electromagnetic surface, so that the property of the surface can be controlled in real time (Fig. 2.3).

2.3.2 IRS Element Design

IRS element design is the core of IRS design. The design goals need to be determined based on the actual application requirements. The main part of the element and bias circuit are then carefully designed and optimized.

The procedure of IRS element design is as follows. First, a main model should be established in electromagnetic simulation software (such as HFSS and CST), and periodic boundary conditions, Floquet port excitation, and equivalent RLC model of nonlinear components are set up. Then the appropriate element geometry is selected for design optimization so as to meet the preset design requirements in the required frequency band, such as the 1-bit reflection element requires that the reflection amplitude is close to 0 dB and the reflection phase difference is 180°. Finally, structures such as bias circuit used to connect to the control system is considered to verify their impact on element performance.

Yang et al. (2016) gave a design example of a 1-bit digital phase-controlled element, which achieves 1-bit digital phase modulation around 14.5 GHz. In the specific implementation, a classic rectangular resonant element is selected, and the resonant property of the element is changed by turning on and off the PIN diode, thereby changing the resonant frequency point, so that there is a 180° phase difference between the two resonant frequency points. Finally, a bias line for controlling the PIN diode and a fan-shaped stub for AC–DC isolation were added, verifying that it has little effect on the electrical characteristics within the element's operating frequency band.

2.3.3 IRS Array Design

Based on the well-designed IRS element, the IRS array is designed and fabricated using printed circuit board (PCB) technology. Traditionally, the elements are arranged periodically with distance of half-wavelength, while some irregular located design are proposed to obtain narrow main lobe width and low side lobe levels with fewer elements (Zhang et al. 2022). Furthermore, the circuit should be designed to connect the control system and IRS elements.

Due to the low cost and quasi-passive property of IRS element, IRS can be designed as an array with extremely large size to realize higher beamforming gain. Figure 2.4 shows an example of IRS array with 2304 elements (Cui et al. 2022). The array consists of 9 PCB boards, each of which consists of a subarray with 16 × 16 IRS elements and an FPGA chip as controller. The subarrays work independently of each other, so the array can be extended to reach larger size by simply increasing the number of subarrays. Each IRS element is connected to an output pin of the FPGA so that the phase of IRS elements can be switched at the

Figure 2.4 An example of IRS array with 2304 elements. Source: Linglong Dai, IEEE Fellow and an Associate Professor at the Department of Electronic Engineering, Tsinghua University.

same time and the switching frequency can reach over 100kHz ideally. The FPGA can be replaced by microcontroller unit (MCU) to reduce the energy consumption and the system cost. However, the limited number of output pins of MCU will make it hard to realize large-scale IRS.

2.3.4 IRS Controller Design

In a communication system, a typical application scenario of IRS is the direct path from the transmitter to the receiver and the reflection path through the IRS. Obstacles in the direct path often lead to the deterioration of the direct path channel, while the reflection path can provide a better communication environment with the aid of IRS. In order to realize the auxiliary capability in the communication system, IRS needs to adjust the phase of each element of itself so that the reflected signals of each element received by the receiving end can be superimposed in phase. Generally, IRS will be equipped with a control module. After the control module performs phase compensation on the IRS front element, its reflection pattern forms a focused beam in the target direction, thereby establishing a stable channel to the receiving end.

The main function of the control module is to provide different phase distributions for the IRS element according to the exit direction of the IRS beam. The quantized phase distribution of each element is also called the code table. Using the 0° beam code table to phase the IRS can make the output beam move toward 0° direction. The control module is mainly composed of a host computer, a control chip, and a drive circuit, as shown in Fig. 2.5. The host computer can be connected to the communication system to provide the output direction to the control chip in real time. The control chip, taking the FPGA chip as an example, assigns the stopwatch to the output pin according to the output direction, and the pin is connected

Figure 2.5 An example of IRS control system. Source: SICSTOCK.COM/https://www.sicstock.com/products/xc6slx9-2tqg144c-xc6-6s?_pos=4&_sid=ffcb9a8ef&_ss=r/last accessed September 20, 2023.

Host computer

FPGA

Control board

to the IRS through the drive circuit and cable. Since the IRS elements are usually based on PIN, HEMT or varactor diode and other components to realize the phase reconfigurable function, the drive circuit of the control board only needs to output the corresponding voltage or current according to the digital logic of the pin to complete the IRS phase layout to complete the control of the IRS.

2.3.5 Full-Wave Simulation and Field Test

Theoretically, after finishing the antenna element design, the structure of the whole IRS surface can be obtained by periodically arranging the element structure. However, due to higher-order modes and the existence of mutual coupling among units, the actual performance of the IRS surface may suffer from a measure of degradation. In order to guarantee the required performance, full-wave simulation for the whole IRS surface is of vital importance.

Specifically, during full-wave simulations, the whole IRS surface is firstly modeled in simulation software. Then, the practical source is utilized to send electromagnetic waves. The response of the IRS is recorded finally to analyze the performance of the whole IRS surface. Full-wave simulations need a large amount of computing resource, which often takes several hours, even several days. The results of full-wave simulations are more reliable than the results derived from single units but still cannot replace the results of field test.

The field test, as shown in Fig. 2.6, includes waveguide test, near-field test, and far-field test. The waveguide test needs a specially designed waveguide, in which two designed units are placed to measure the practical performance. The near-field test is usually carried out when the beam gain is high. The units on IRS surface are firstly all set in the same state. After obtaining the near-field distribution, the performance of the whole IRS surface can be derived. The far-field test is carried out in a practical scenario or far-field anechoic chamber to directly acquire the performance of IRS surface.

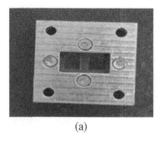
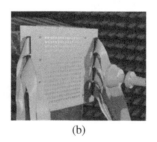

(a)   (b)   (c)

Figure 2.6   (a) Waveguide test, (b) Near-field test, (c) Far-field test.

2.4 State-of-the-Art IRS Prototype

2.4.1 Passive IRS Prototype by Tsinghua

The IRS-based prototype developed by the researchers from Tsinghua University (THU) in China integrates real-time electronic equipment and tunable metamaterials (Dai et al. 2020). The researchers use PIN diode to design reconfigurable reflecting elements with different phase-shift resolutions and working frequency bands, as shown in Fig. 2.7.

The developed IRS consists of 16×16 elements, and the reflection mode of each element is determined by the on-off state of its integrated PIN diode. The on-off state of PIN diode is equivalent to a high-frequency single-port feedback circuit with different impedance, and the reconfigurable phase shift of the reflected signals can be realized. By reconfiguring the control state of each element, the IRS can adjust the outgoing angle of the incident signals in a desired manner. Particularly, the diode status on each IRS element is independently controlled by the logic circuit deployed behind the patch. The control signal of the entire 256 element array is sent from the FPGA board to each component. Based on this design idea, Tsinghua University has designed and fabricated IRSs working in 2.3 and 28.5 GHz, respectively.

In the software simulations, the high-performance server HP Xeon E5-2697 and MATLAB are used to carry out the system-level verification of RIS-assisted wireless communications, which provides important references and benchmarks for the design of hardware platform. In the online hardware system, first, the host computer can realize the RF setting, data receiving and transmitting, and real-time baseband signal processing, then USRP is used to achieve analog signal processing and transmission, and the designed IRS is deployed to reflect the RF signals. The components of the receiver are the same as those of the transmitter. The RF signals received by the receiver antenna are amplified, down converted, quantized, and desampled (using ADC) by the USRP. Then, the processed signals are converged to the control module through the PCI switch box, and the baseband signals are analyzed and processed by the FPGA and the embedded processor in the control

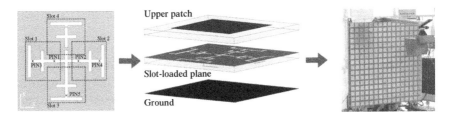

Figure 2.7 PIN diode-based IRS. Source: Tsinghua University Press Ltd.

Figure 2.8 Two IRS prototypes (256 elements). Source: Tsinghua University Press Ltd.

Figure 2.9 The IRS prototypes (2304 elements). Source: Tsinghua University Press Ltd.

module, so that the decoded symbols can be displayed as a high-definition video. Following the above idea, two IRS-based wireless communication prototypes are developed, as shown in Fig. 2.8. Then, in 2022, a low-power IRS prototype with a larger number of elements (2304 elements), which also combines an end-to-end neural network for decoding, is designed and made by (Cui et al. 2022), as shown in Fig. 2.9.

Based on the prototypes in Fig. 2.8, Tsinghua University has then finished the transmission verification in the indoor and outdoor environment, respectively. For example, the measured array gain of the 28.5 GHz 2304-element IRS is about 30 dB, and the working bandwidth of the communication system is 800 MHz. The communication prototype can support wireless transmission over a distance of 5 m in the indoor environment, with a transmission rate of 1.683 Gbps.

2.4.2 Active IRS Prototype by Tsinghua

To overcome the "multiplicative fading" effect introduced by IRSs, the researchers at Tsinghua University proposed active IRS. Unlike the traditional passive IRSs which passively reflect signals, the active IRS can amplify the reflected signals. As shown in Fig. 2.10, in addition to the phase-shifting circuit, each active IRS element also integrates a reflection-type amplifier, which can enhance the reflected signals and improve the ability of IRSs to actively control the wireless environment.

Based on the above antenna principle of active IRSs, the researchers designed and fabricated a 64-element active IRS, as shown in Fig. 2.11. The designed active IRS center frequency point is 3.5 GHz, and each element can realize 2-bit phase shifting. The phase shifting part of each antenna unit uses two low-power PIN

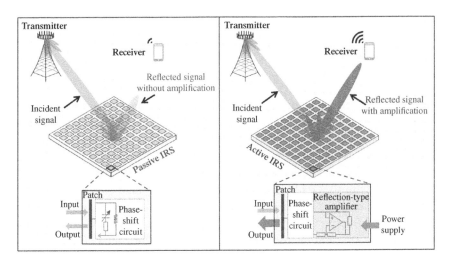

Figure 2.10 A comparison between passive IRSs and active IRSs.

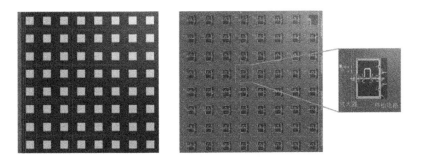

Figure 2.11 A 64-element active IRS. Source: Tsinghua University Press Ltd.

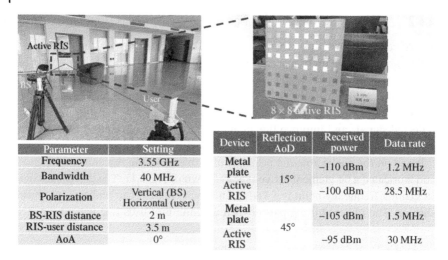

Parameter	Setting
Frequency	3.55 GHz
Bandwidth	40 MHz
Polarization	Vertical (BS) Horizontal (user)
BS-RIS distance	2 m
RIS-user distance	3.5 m
AoA	0°

Device	Reflection AoD	Received power	Data rate
Metal plate	15°	−110 dBm	1.2 MHz
Active RIS		−100 dBm	28.5 MHz
Metal plate	45°	−105 dBm	1.5 MHz
Active RIS		−95 dBm	30 MHz

Figure 2.12 Active IRS prototype and experimental measurement results. Source: Tsinghua University Press Ltd.

diodes and two SPDTs to replace the traditional high-precision phase shifters, which can greatly reduce the hardware power consumption. Then, the researchers provide an active IRS-aided communication prototype, as shown in Fig. 2.12.

Based on the developed prototype, an experimental environment is established for further validation. The operating frequency of the active IRS to $f = 3.5$ GHz and the bandwidth to 40 MHz by adjusting the circuit impedance of active elements. The polarization of the antenna at the BS and that at the user is selected as vertical and horizontal, respectively. The transmit power is set to −10 mW. The heights of the BS, the IRS, and the user are fixed as 1 m. The horizontal distance of the BS-IRS link and that of the IRS-user link are set to 2 and 3.5 m, respectively. The angle of arrival (AoA) at the active IRS is fixed as zero, and the angle of departure (AoD) will be specified to evaluate the performance gain of active IRSs at different orientations. To observe the reflection gain of the active IRS, a metal plate with the same aperture size is used as the active IRS for performance comparison.

By moving the user at different AoDs and configuring the phase shift of the active IRS with discrete Fourier transform (DFT) codebook, the experimental results are shown in Fig. 2.12. One can observe that, compared with the received power for the metal plate, the active IRS can always achieve a gain of about 10 dB. The data rate for the active IRS can hold at about 30 Mbps, while that for the metal plate only ranges from 1 to 2 Mbps. The reason is that, the beamforming at the active IRS can make the reflected beam with high array gain and reflection gain, while the metal plate can only reflect the incident signals randomly without in-phase combination or amplification, which validates the significant gain of active IRSs.

2.4.3 IRS Modulation Prototype by SEU

Researchers from Southeast University (SEU) in China built a prototype based on IRS and validated the capacity of IRS in wireless communication systems (Tang et al. 2019). This prototype utilized the control signal in the baseband to configure the phase shift of each element on IRS and realized a data rate of 2.048 Mbps. This result was encouraging since the cost of IRS is far lower than traditional antennas. The specific design of this prototype is elaborated as follows.

2.4.3.1 Transmitter Design

The diagram of the transmitter is illustrated in Fig. 2.13. In the transmitter of this prototype, software-defined radio (SDR) firstly maps the bit stream to the corresponding IRS configuration. Then, the control signal is sent to IRS and sets the IRS dynamically, thus realizing quadrature phase shift keying (QPSK) modulation by IRS.

Specifically, the relationship between the IRS reflecting coefficient and the bit stream waiting to be modulated can be presented as:

$$\Gamma(t) = \sum_{n=1}^{N} \Gamma_n h(t - nT), \Gamma_n \in \{P_1, P_2, P_3, P_4\}, \tag{2.1}$$

where $\Gamma(t)$ denotes the whole reflecting coefficient of IRS at time t, N denotes the number of units on IRS, Γ_n denotes the reflecting coefficient of the n-th element, $h(t)$ denotes the sampling function, and $\{P_1, P_2, P_3, P_4\}$ denotes four different IRS reflecting coefficient, which can be mapped to four constellation points of QPSK.

2.4.3.2 Frame Structure Design

The frame structure used in this prototype is illustrated in Fig. 2.14. One frame is composed of 1 synchronization subframe, 1 pilot subframe, and 9 data subframes.

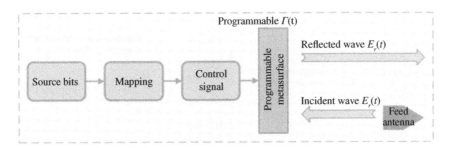

Figure 2.13 Designing diagram of the transmitter.

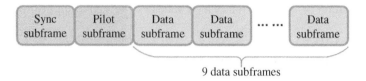

9 data subframes

Figure 2.14 Frame structure.

The synchronization subframe is composed of a synchronization sequence with a length of 420. The pilot subframe is composed of 2048 pilot symbols and 160 cyclic prefix (CP) symbols. Similarly, every data subframe is composed of 2048 data symbols and 160 CP symbols. Every frame can transmit 36,864 bits. The data rate of the transmitter mainly depends on the sampling rate. In this prototype, a data rate of 2.048 Mbps and a sampling rate of 1.25 MSaps have been realized.

2.4.3.3 Receiver Design

The diagram of the receiver is illustrated in Fig. 2.15. In this prototype, the receiver utilizes traditional orthogonal sampling zero IF structure, which realizes synchronization, channel estimation, channel equalization, and QPSK demodulation by processing the IQ signal in baseband.

In this diagram, the purpose of frame synchronization is to find the beginning position of each frame. It is worth noting that the IRS in this prototype can only achieve a phase shift ranging from 0° to 255°, so regular synchronization sequence such as Zadoff–Chu (ZC) Squence cannot be realized. In order to realize frame synchronization, an extended Barker Code Sequence is utilized in this prototype. The receiver conducts the autocorrelation in the searching window and realizes the frame synchronization by Barker Code. As for the phase retrieval, the receiver utilizes CP to achieve the joint maximum likelihood estimation of carrier frequency offset and eliminates the rotation of the constellation. After the synchronization, the channel estimation, channel equalization, and QPSK demodulation are carried out sequentially.

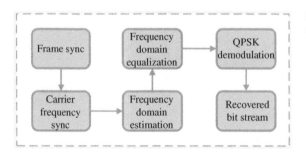

Figure 2.15 Designing diagram of the receiver.

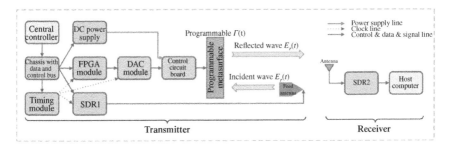

Figure 2.16 Designing diagram of the system.

2.4.3.4 System Design

The diagram of the whole system is illustrated in Fig. 2.16, where the left part represents the transmitter, which is composed of IRS, control circuits, and the PXIe system. In the PXIe system, the mapping of bit stream to IRS configuration is realized on the central controller. Then, the sequence after mapping is sent to the FPGA module and DAC module to generate the control signal sequence, which is then utilized to configure the IRS. Finally, the incident signal with single carrier of 4 GHz is generated by SDR1 and reflected by the IRS to realize the data transmission.

The receiver in this prototype is composed of receiving antenna, SDR, and the host computer. The host computer acquires the IQ data in the baseband through regular RF chain, and the synchronization and demodulation are completed thereafter.

2.4.4 Transmissive IRS Prototype by MIT

Researchers from the Massachusetts Institute of Technology (MIT) built an IRS prototype that can simultaneously transmit and reflect signals (Arun and Balakrishnan 2020). In this prototype, each reconfigurable element is divided into several sub-units and is connected to a 1-bit RF switch. The RF switches can control the connection state of different sub-units, thus controlling the transmission and reflection state of each element. The IRS is configured and integrated thereafter, as illustrated in Fig. 2.17.

It is worth noting that the IRS in the MIT prototype is different from regular ones since the units cannot realize phase shift. Instead, it can only switch between transmitting signals and reflecting signals. Specifically, each element is divided into several independent sub-units. When the RF switch is on, the element can be viewed as a metal plate and can reflect the incident signals. When the RF switch is off, the element can be viewed as several metal wires, which cannot reflect the incident signals and the signal can completely transmit through the IRS (Fig. 2.18).

Figure 2.17 The IRS prototype developed by MIT. Source: Jason Dorfman, MIT CSAIL.

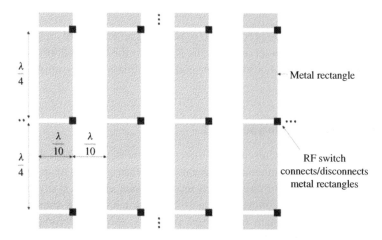

Figure 2.18 The element structure of the IRS developed by MIT.

Furthermore, by integrating a large number of units, the IRS can optimize both the transmitting signals and the reflecting signals to serve users at both sides. Specifically, unlike traditional reflective phase-shifting RIS, the design principle of the radiation coefficient of this IRS is shown in Fig. 2.19. The dotted line represents the IRS, the white and gray areas represent the two spaces separated by the IRS, and the black arrow represents the amplitude and phase of the incident

Figure 2.19 The design principle of the IRS radiation coefficient proposed by MIT.

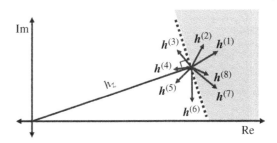

signals at the receiver. The goal is to design the permeability of each IRS element so that as many signals as possible can be reflected and coherently added at the receiver, while those signals harmful to the received signal strength can pass through the IRS.

Researchers from MIT also carried out some system tests in communication systems. The central frequency was 2.4 GHz and the test was carried out in the indoor office scenario. The test results showed that the IRS can improve the power of communication signals by a range of 3.8–20.0 dB.

2.4.5 IRS Prototype by China Mobile

Researchers from China Mobile proposed the three steps of the development of IRS. The first step is realizing the passive IRS, which can be deployed effectively and extend the coverage of the communication signals, especially in the locations where the original signals are weak or even cannot cover. The second step is to develop the semi-static IRS, which can realize the beam selection by different configurations of the IRS elements. With the deployment of semi-static IRS, the sum-rate of the cells can be further improved. In the third step, the dynamic IRS is realized, where the location of users can be tracked by the algorithms designed for IRS-assisted systems (Fig. 2.20).

To validate the performance of IRS, researchers from China Mobile also built a prototype based on the industrial base station and IRS and conducted several system tests (CMCC 2022).

In the first test, three fixed points were selected to evaluate the data rate performance, as illustrated in Fig. 2.21 (a). The results showed that IRS can improve the data rate of the selected points. Specifically, the reference signal receiving power (RSRP) was improved by 27 dB and the data rate of user was improved 5.4 times after deploying the IRS in the communication system.

In the second test, researchers evaluated the coverage performance of moving users, as illustrated in Fig. 2.21 (b). The results showed that the IRS can improve the RSRP by 23 dB and the data rate of user was improved 5 times in the whole test field.

Figure 2.20 The scenario where IRS can extend the coverage of signals. Source: courtesy of Min.

Figure 2.21 The test scenario of China Mobile: fixed points.

2.4.6 IRS Prototype by DOCOMO

Researchers from NTT DOCOMO in Japan built a prototype based on the IRS composed of a large number of sub-wavelength element elements, which were placed in a periodic arrangement on a two-dimensional surface covered with a glass substrate (DOCOMO 2018). The IRS units had three modes, which can be controlled by moving the glass substrate slightly: (i) full penetration of the incident signals; (ii) partial reflection of the incident signals; and (iii) full reflection of all signals. This IRS was highly transparent and did not interfere aesthetically or physically with the surrounding environment or with the line-of-sight of people, so it was suitable for use within buildings or on vehicles. In practical communication systems, the designed IRS can manipulate the radio waves according to the specific installation environment. For example, in locations that are not suited for installing base stations, such as in built-up areas or in indoor areas where the reception of signals needs to be blocked selectively, the proposed IRS is highly adaptive (Fig. 2.22).

Researchers from DOCOMO also conducted a few system tests (Fig. 2.23). Specifically, the 5G data communication using 28 GHz band was measured between the 5G base station located on the rooftop and the 5G mobile station running experimental vehicle. The meta-structure-based IRS used in the test, the reflection direction and the beam shape of the reflected wave are determined so that the area of the 5G site was expanded using these tiny structures to form the beam arranged in the plane of the reflect array. By deploying a meta-structure-based IRS, the communication quality was greatly improved in the area where 5G data communication was previously impossible with range extension of about 35 m. By using the IRS, the communication speed improved by 500 Mbps for the vehicle equipped with the 5G mobile station.

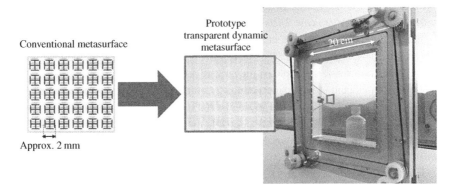

Figure 2.22 The prototype. Source: NTT DOCOMO.

Figure 2.23 The test scenario. Source: Metawave Corporation.

Bibliography

Arun, V. and Balakrishnan, H. (2020). RFocus: beamforming using thousands of passive antennas. *17th USENIX Symposium on Networked Systems Design and Implementation (NSDI 20)*, 1047–1061.

Asadchy, V.S., Albooyeh, M., Tcvetkova, S.N. et al. (2016). Perfect control of reflection and refraction using spatially dispersive metasurfaces. *Physical Review B* 94: 075142. https://doi.org/10.1103/PhysRevB.94.075142.

Baghchehsaraei, Z. and Oberhammer, J. (2013). Parameter analysis of millimeter-wave waveguide switch based on a MEMS-reconfigurable surface. *IEEE Transactions on Microwave Theory and Techniques* 61 (12): 4396–4406. https://doi.org/10.1109/TMTT.2013.2287682.

CMCC (2022). New progress in 5G of 6G technology: China mobile research institute and industrial partners completed the industry's first prototype verification of dynamic intelligent metasurface technology, August 2022. http://www.cww.net.cn/article?id=567155 (accessed 16 August 2023).

Cui, T.J., Qi, M.Q., Wan, X. et al. (2014). Coding metamaterials, digital metamaterials and programmable metamaterials. *Light: Science & Applications* 3 (10): e218–e218.

Cui, T.J., Liu, S., and Zhang, L. (2017). Information metamaterials and metasurfaces. *Journal of Materials Chemistry C* 5 (15): 3644–3668.

Cui, M., Wu, Z., Chen, Y. et al. (2022). Demo: low-power communications based on RIS and AI for 6G. *2022 IEEE International Conference on Communications*

Workshops (ICC Workshops), 1–2. https://doi.org/10.1109/ICCWorkshops53468 .2022.9915019.

Dai, J.Y., Zhao, J., Cheng, Q., and Cui, T.J. (2018). Independent control of harmonic amplitudes and phases via a time-domain digital coding metasurface. *Light: Science & Applications* 7 (1): 1–10.

Dai, L., Wang, B., Wang, M. et al. (2020). Reconfigurable intelligent surface-based wireless communications: antenna design, prototyping, and experimental results. *IEEE Access* 8: 45913–45923.

Di Renzo, M., Zappone, A., Debbah, M. et al. (2020). Smart radio environments empowered by reconfigurable intelligent surfaces: how it works, state of research, and the road ahead. *IEEE Journal on Selected Areas in Communications* 38 (11): 2450–2525.

DOCOMO (2018). NTT DOCOMO and metawave announce successful demonstration of 28GHz-band 5G using world's first meta-structure technology, November 2018. https://www.businesswire.com/news/home/20181204005253/en/ NTT-DOCOMO-Metawave-Announce-Successful-Demonstration-28GHz-Band.

Liu, F., Pitilakis, A., Mirmoosa, M.S. et al. (2018). Programmable metasurfaces: state of the art and prospects. *Proceedings of the 2018 IEEE International Symposium on Circuits and Systems (ISCAS)*, 1–5. IEEE.

Ma, Q., Bai, G.D., Jing, H.B. et al. (2019). Smart metasurface with self-adaptively reprogrammable functions. *Light: Science & Applications* 8 (1): 1–12.

Mamedes, D.F., Neto, A.G., and Bornemann, J. (2020). Reconfigurable corner reflector using PIN-diode-switched frequency selective surfaces. *2020 IEEE International Symposium on Antennas and Propagation and North American Radio Science Meeting*, 127–128. https://doi.org/10.1109/IEEECONF35879.2020.9329791.

Pendry, J.B., Holden, A.J., Stewart, W.J., and Youngs, I. (1996). Extremely low frequency plasmons in metallic mesostructures. *Physical Review Letters* 76 (25): 4773.

Pendry, J.B., Holden, A.J., Robbins, D.J., and Stewart, W.J. (1999). Magnetism from conductors and enhanced nonlinear phenomena. *IEEE Transactions on Microwave Theory and Techniques* 47 (11): 2075–2084.

Rains, J., Kazim, J.U.R., Tukmanov, A. et al. (2022). Varactor-based reconfigurable intelligent surface with dual linear polarisation at K-band. *2022 IEEE International Symposium on Antennas and Propagation and USNC-URSI Radio Science Meeting (AP-S/URSI)*, 673–674. https://doi.org/10.1109/AP-S/USNC-URSI47032.2022 .9887367.

Sievenpiper, D., Zhang, L., Broas, R.F.J. et al. (1999). High-impedance electromagnetic surfaces with a forbidden frequency band. *IEEE Transactions on Microwave Theory and Techniques* 47 (11): 2059–2074.

Tang, W., Li, X., Dai, J.Y. et al. (2019). Wireless communications with programmable metasurface: transceiver design and experimental results. *China Communications* 16 (5): 46–61.

Venkatesh, S., Saeidi, H., Lu, X., and Sengupta, K. (2022). Active tunable millimeter-wave reflective surface across 57-64GHz for blockage mitigation and physical layer security. *2022 IEEE Radio Frequency Integrated Circuits Symposium (RFIC)*, 63–66. https://doi.org/10.1109/RFIC54546.2022.9863076.

Veselago, V.G. (1967). Electrodynamics of substances with simultaneously negative and. *Uspekhi Fizicheskikh Nauk* 92 (7): 517.

Yang, H., Yang, F., Xu, S. et al. (2016). A 1-bit 10 × 10 reconfigurable reflectarray antenna: design, optimization, and experiment. *IEEE Transactions on Antennas and Propagation* 64 (6): 2246–2254. https://doi.org/10.1109/TAP.2016.2550178.

Yu, N., Genevet, P., Kats, M.A. et al. (2011). Light propagation with phase discontinuities: generalized laws of reflection and refraction. *Science* 334 (6054): 333–337.

Zhang, H., Wu, W., Cheng, Q. et al. (2022). Reconfigurable reflectarray antenna based on hyperuniform disordered distribution. *IEEE Transactions on Antennas and Propagation* 70 (9): 7513–7523. https://doi.org/10.1109/TAP.2022.3193230.

3

On Path Loss and Channel Reciprocity of RIS-Assisted Wireless Communications

Wankai Tang[1], Jinghe Wang[1], Jun Yan Dai[2], Marco Di Renzo[3], Shi Jin[1], Qiang Cheng[2], and Tie Jun Cui[2]

[1] *National Mobile Communications Research Laboratory, Southeast University, Nanjing, China*
[2] *State Key Laboratory of Millimeter Waves, Southeast University, Nanjing, China*
[3] *Université Paris-Saclay, CNRS, CentraleSupélec, Laboratoire des Signaux et Systèmes, Gif-sur-Yvette, France*

3.1 Introduction

In recent years, reconfigurable intelligent surfaces (RISs) have been under active discussion and investigation as a promising technology to enable future 6G networks (Di Renzo et al. 2019; Basar et al. 2019; Huang et al. 2019; Wu and Zhang 2019; Han et al. 2019). An RIS is an artificial electromagnetic surface comprised of a large number of sub-wavelength unit cells (Cui et al. 2014). They can flexibly control the parameters of wireless signals, such as phase, amplitude, and polarization (Cui et al. 2017), thus enabling the emerging concept known as "smart radio environment" (Di Renzo et al. 2020b). The unique programmable feature makes RISs especially appealing for wireless applications, including coverage enhancement (Wu and Zhang 2020), wireless power transfer (Yu et al. 2019; Pan et al. 2020), secure communication (Chen et al. 2019), wireless sensing and positioning (Hu et al. 2020), direct modulation (Cui et al. 2019; Tang et al. 2020), etc.

Path loss and channel reciprocity are two basic characteristics of wireless channels (Goldsmith 2005). The former represents the basic relation between the wireless signal power and the transmission distance, which can provide information on how far a wireless signal can be successfully transmitted. The latter determines whether the uplink channel state information can be directly used for designing the downlink transmission scheme. Due to the importance of characterizing path loss and channel reciprocity characteristics of RIS-assisted

Intelligent Surfaces Empowered 6G Wireless Network, First Edition.
Edited by Qingqing Wu, Trung Q. Duong, Derrick Wing Kwan Ng, Robert Schober, and Rui Zhang.
© 2024 The Institute of Electrical and Electronics Engineers, Inc. Published 2024 by John Wiley & Sons, Inc.

wireless communications, researchers have recently conducted related studies that are based on different analytical assumptions and approaches.

Di Renzo et al. (2020a) derived asymptotic scaling laws of the path loss as a function of the transmission distances and the size of the RIS in the far-field and near-field cases. The results are obtained by leveraging the scalar Huygens–Fresnel principle in a two-dimensional space. Özdogan et al. (2020) studied the path loss model of the RIS in the far-field region by using a metal plate as the benchmark. Garcia et al. (2020) calculated the radiation density of the scattered field in the near-field and far field under the assumption of dipole antennas and discussed the scaling laws as a function of the transmission distances numerically. Ellingson (2021) proposed a physical model for the path loss of an RIS-assisted wireless link under the assumption that the antenna gain of the transmitter/receiver is constant over the RIS. Najafi et al. (2021) developed a physics-based RIS path loss model, in which the impact of grouped unit cells on the wireless channel is obtained by solving the integral equations for electromagnetic vector fields under the far-field assumption. Danufane et al. (2021) generalized the model in Di Renzo et al. (2020a) to a three-dimensional space by using the vector generalization of Green's theorem, and characterized the scaling laws of the path loss as a function of the transmission distances and the size of the RIS based on scattering theory. Wang et al. (2021) proposed a radar cross-section-based path loss model that introduces an angle-dependent reflection phase behavior of RIS unit cells. Gradoni and Di Renzo (2021) developed a path loss model that is based on the theory of mutual impedances of thin dipole antennas. The end-to-end channel model resembles a multiple input multiple output (MIMO) communication system and considers the mutual coupling among RIS unit cells. Recently, Di Renzo et al. (2022) proposed a path loss model, based on scattering theory, assuming that the incident signals are not constant over the RIS elements. A summary of the abovementioned research works on RIS path loss modeling is available in Table I of (Degli-Esposti et al. 2022). As for the aspect of channel reciprocity, Kord et al. (2020) summarized design methods of nonreciprocal metasurfaces and metamaterials. However, the influence of RIS technology on channel reciprocity was not discussed from the perspective of wireless communications.

In this chapter, we will introduce a path loss model for RIS-assisted communication systems that is based on antenna theory. The proposed RIS path loss model is easy to use yet sufficiently accurate and has been validated by measurement results (Tang et al. 2021a, 2022). Moreover, the channel reciprocity of RIS-assisted wireless links will be discussed and experimentally demonstrated in this chapter (Tang et al. 2021b). These aspects are helpful to understand the basic characteristics of large-scale fading and channel reciprocity in RIS-assisted wireless networks.

3.2 Path Loss Modeling and Channel Reciprocity Analysis

This section describes the considered RIS-assisted wireless communication system, introduces free-space path loss models for RIS-assisted wireless communications, and discusses the channel reciprocity of RIS.

3.2.1 System Description

Considering the RIS-assisted wireless communication system shown in Fig. 3.1, this chapter focuses on the free-space path loss of the reflected path provided by the RIS, and therefore assumes that the direct path between the transmitter and the receiver is completely blocked. As shown in Fig. 3.1, the RIS is deployed in the x–o–y plane of a Cartesian coordinate system and its geometric center coincides with the origin of the coordinate system. The RIS consists of a regular arrangement of unit cells with configurable electromagnetic response in N rows and M columns. The length of each unit cell is d_x along the x-axis and d_y along the y-axis, and its size is typically on the sub-wavelength scale. The normalized power radiation pattern of the unit cell is denoted by $F(\theta, \varphi)$, and characterizes the induced and scattered signal power intensity as a function of the angles for a single unit cell, and G is the scattering gain of the unit cell. $U_{n,m}$ denotes the unit cell located in the nth row and mth column with $n \in [1, N]$ and $m \in [1, M]$. The coordinate position of $U_{n,m}$ is $(x_{n,m}, y_{n,m}, 0)$, its distance to the center of the RIS is $d_{n,m}$, and its corresponding programmable reflection coefficient is $\Gamma_{n,m}$.

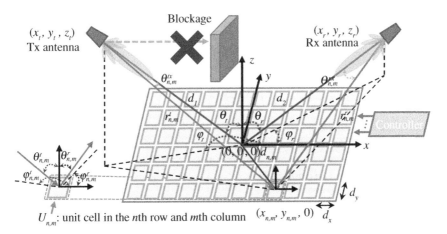

Figure 3.1 A typical RIS-assisted wireless communication system.

In addition, d_1, d_2, θ_t, φ_t, θ_r, and φ_r denote the distance from the transmitter to the center of the RIS, the distance from the receiver to the center of the RIS, the elevation angle and the azimuth angle from the center of the RIS to the transmitter, and the elevation angle and the azimuth angle from the center of the RIS to the receiver, respectively. For the unit cell $U_{n,m}$, $r^t_{n,m}$, $r^r_{n,m}$, $\theta^t_{n,m}$, $\varphi^t_{n,m}$, $\theta^r_{n,m}$, and $\varphi^r_{n,m}$ denote the distance from the transmitter to $U_{n,m}$, the distance from the receiver to $U_{n,m}$, the elevation and azimuth angles from $U_{n,m}$ to the transmitter, and the elevation and azimuth angles from $U_{n,m}$ to the receiver, respectively.

As shown in Fig. 3.1, the transmitter at (x_t, y_t, z_t) transmits a wireless signal with power P_t and wavelength λ to the RIS. The normalized power radiation pattern of the transmit antenna is $F^{tx}(\theta, \varphi)$ and the gain is G_t. The signal is then reflected by the RIS and received by the receiver at (x_r, y_r, z_r) and the normalized power radiation pattern of the receive antenna is $F^{rx}(\theta, \varphi)$ and the gain is G_r. The parameters $\theta^{tx}_{n,m}$, $\varphi^{tx}_{n,m}$, $\theta^{rx}_{n,m}$, and $\varphi^{rx}_{n,m}$ denote the elevation angle and the azimuth angle of the transmit antenna to $U_{n,m}$, the elevation angle and the azimuth angle of the receive antenna to $U_{n,m}$, respectively. The symbols and definitions of the parameters in the considered RIS-assisted wireless communication system are listed in Table 3.1.

3.2.2 General Path Loss Model

By explicitly considering the effects of the above physical and electromagnetic parameters on the path loss of the RIS, the following theorem gives the path loss of the reflected wireless link provided by the unit cell $U_{n,m}$ as a function of key design parameters.

Theorem 3.1 *The free-space path loss model of the cascaded sub-channel corresponding to a single unit cell $U_{n,m}$ can be expressed as*

$$\mathrm{PL}_{U_{n,m}} = \frac{P_t}{P^r_{n,m}} = \frac{16\pi^2 \left(r^t_{n,m} r^r_{n,m}\right)^2}{G_t G_r \left(d_x d_y\right)^2 F^{\text{combine}}_{n,m} |\Gamma_{n,m}|^2}, \tag{3.1}$$

where $P^r_{n,m}$ is the signal power at the receiver that is reflected from the unit cell $U_{n,m}$. $F^{\text{combine}}_{n,m}$ is the joint normalized power radiation pattern of the transmit antenna, the unit cell, and the receive antenna. It is an angle-dependent factor with a maximum value of 1, which can be expressed as

$$F^{\text{combine}}_{n,m} = F^{tx}\left(\theta^{tx}_{n,m}, \varphi^{tx}_{n,m}\right) F\left(\theta^t_{n,m}, \varphi^t_{n,m}\right) F\left(\theta^r_{n,m}, \varphi^r_{n,m}\right) F^{rx}\left(\theta^{rx}_{n,m}, \varphi^{rx}_{n,m}\right). \tag{3.2}$$

Proof: The proof of Theorem 3.1 is given in Appendix 3.A.1. ☐

Table 3.1 Symbols and definitions.

Symbol	Definition	Symbol	Definition
N	Number of rows of unit cells	M	Number of columns of unit cells
$U_{n,m}$	Unit cell in nth row, mth column	$\Gamma_{n,m}$	Reflection coefficient of $U_{n,m}$
d_x	Length of each unit cell along x-axis	d_y	Length of each unit cell along y-axis
$(x_{n,m}, y_{n,m}, 0)$	Location of the unit cell $U_{n,m}$	$d_{n,m}$	Distance between $U_{n,m}$ and the center of the RIS $d_{n,m} = \sqrt{x_{n,m}^2 + y_{n,m}^2}$
(x_t, y_t, z_t)	Location of the transmitter	(x_r, y_r, z_r)	Location of the receiver
d_1	Distance between the transmitter and the center of the RIS $d_1 = \sqrt{x_t^2 + y_t^2 + z_t^2}$	d_2	Distance between the receiver and the center of the RIS $d_2 = \sqrt{x_r^2 + y_r^2 + z_r^2}$
(θ_t, φ_t)	Elevation angle and azimuth angle from the center of the RIS to the transmitter	$r_{n,m}^t$	Distance between the transmitter and the unit cell $U_{n,m}$, $r_{n,m}^t = \sqrt{(x_t - x_{n,m})^2 + (y_t - y_{n,m})^2 + z_t^2}$
(θ_r, φ_r)	Elevation angle and azimuth angle from the center of the RIS to the receiver	$r_{n,m}^r$	Distance between the receiver and the unit cell $U_{n,m}$, $r_{n,m}^r = \sqrt{(x_r - x_{n,m})^2 + (y_r - y_{n,m})^2 + z_r^2}$
$(\theta_{n,m}^t, \varphi_{n,m}^t)$	Elevation angle and azimuth angle from $U_{n,m}$ to the transmitter	$F(\theta, \varphi)$	Normalized power radiation pattern of each unit cell
$(\theta_{n,m}^r, \varphi_{n,m}^r)$	Elevation angle and azimuth angle from $U_{n,m}$ to the receiver	G	Scattering gain of a single unit cell
$F^{tx}(\theta, \varphi)$	Normalized power radiation pattern of the transmit antenna	$F^{rx}(\theta, \varphi)$	Normalized power radiation pattern of the receive antenna
G_t	Gain of the transmit antenna	G_r	Gain of the receive antenna
$(\theta_{n,m}^{tx}, \varphi_{n,m}^{tx})$	Elevation angle and azimuth angle from the transmit antenna to $U_{n,m}$	λ	Wavelength
$(\theta_{n,m}^{rx}, \varphi_{n,m}^{rx})$	Elevation angle and azimuth angle from the receive antenna to $U_{n,m}$	P_t	Transmitted signal power

Theorem 3.1 reveals that the free-space path loss of the cascaded sub-channel formed by a single unit cell $U_{n,m}$ is proportional to the square of the product of its two distances to the transmitter and the receiver, and inversely proportional to the square of the area of the unit cell $U_{n,m}$ (i.e., proportional to the fourth power of the operating frequency of the wireless signal) and inversely proportional to the angle-dependent loss factor $F^{\text{combine}}_{n,m}$.

Equation (3.1) is a general model of the free-space path loss for a single unit cell. The path loss model for the entire RIS is given in the following corollary.

Corollary 3.1 *The free-space path loss model of the cascaded channel corresponding to the entire RIS can be expressed as*

$$
PL_{\text{general}} = \frac{16\pi^2}{G_t G_r \left(d_x d_y\right)^2 \left| \sum_{m=1}^{M} \sum_{n=1}^{N} \frac{\sqrt{F^{\text{combine}}_{n,m}} \, \Gamma_{n,m}}{r^l_{n,m} r^r_{n,m}} e^{\frac{-j2\pi(r^l_{n,m}+r^r_{n,m})}{\lambda}} \right|^2 }, \tag{3.3}
$$

where $F^{\text{combine}}_{n,m}$ is defined in (3.2).

Proof: The proof of Corollary 3.1 follows directly from Theorem 3.1 by using the superposition principle. □

Corollary 3.1 reveals that the free-space path loss of the reflected cascaded wireless link provided by the entire RIS is inversely proportional to the square of the area of the unit cell (i.e., proportional to the fourth power of the operating frequency). It also shows that the path loss PL_{general} is related to the joint normalized power radiation pattern of the transmit/receive antenna and the unit cell, the number of unit cells, the distance between the transmitter/receiver and each unit cell and, in particular, to the design of the reflection coefficients of unit cells of the RIS. However, the relation between them is not that straightforward and it will be further analyzed and discussed in this chapter under several typical scenarios.

Furthermore, Theorem 3.1 and Corollary 3.1 indicate that, as long as the reflection coefficient $\Gamma_{n,m}$ of the unit cell is transmit-receive reciprocal, the uplink and downlink wireless channels of an RIS-assisted time-division duplexing (TDD) wireless communication system are reciprocal, which is an important property in TDD wireless communication systems.

The proposed RIS path loss model contains an angle-dependent loss factor $F^{\text{combine}}_{n,m}$ associated with the power radiation pattern of the transmit antenna, the unit cell, and the receive antenna, which is generally expressed as in (3.2). A case study is given as follows.

If the directions of peak radiation of both the transmit and receive antennas are steered to the center of the RIS, the loss factor $F^{\text{combine}}_{n,m}$ in (3.2) can be further expressed as

$$F_{n,m}^{\text{combine}} \overset{(a)}{=} \left(\cos\theta_{n,m}^{tx}\right)^{\left(\frac{G_t}{2}-1\right)} \left(\cos\theta_{n,m}^{t}\right)^{\alpha} \left(\cos\theta_{n,m}^{r}\right)^{\alpha} \left(\cos\theta_{n,m}^{rx}\right)^{\left(\frac{G_r}{2}-1\right)}$$

$$= \left(\frac{\left(d_1\right)^2 + \left(r_{n,m}^t\right)^2 - \left(d_{n,m}\right)^2}{2d_1 r_{n,m}^t}\right)^{\left(\frac{G_t}{2}-1\right)} \left(\frac{z_t}{r_{n,m}^t}\right)^{\alpha} \tag{3.4}$$

$$\times \left(\frac{z_r}{r_{n,m}^r}\right)^{\alpha} \left(\frac{\left(d_2\right)^2 + \left(r_{n,m}^r\right)^2 - \left(d_{n,m}\right)^2}{2d_2 r_{n,m}^r}\right)^{\left(\frac{G_r}{2}-1\right)},$$

where step (a) follows by using the power form of $\cos\theta$ to describe the normalized radiation pattern of the transmit antenna, the unit cell $U_{n,m}$ and the receive antenna in a simplified but effective way (Stutzman and Thiele 2012). The coefficient α is employed to fit the actual radiation pattern of the unit cell, which is relevant to the specific hardware design of the unit cell. Generally and efficiently, we can use $\alpha = 1$ for numerical simulations and theoretical analyses. In addition, the distances d_1, d_2, $r_{n,m}^t$, $r_{n,m}^r$, and $d_{n,m}$ in (3.4) can be calculated based on the relative positions of the transmitter, the unit cell $U_{n,m}$ and the receiver. The detailed calculation formulations are given in Table 3.1.

3.2.3 Path Loss Models for Typical Scenarios

As shown in Fig. 3.2, there are two typical wireless communication scenarios for RISs, including RIS-assisted beamforming and RIS-assisted broadcasting. In the

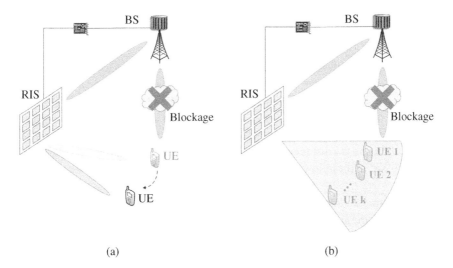

(a)   (b)

Figure 3.2 Two typical application scenarios of RIS-assisted wireless communications. (a) Beamforming. (b) Broadcasting.

RIS-assisted beamforming application, the signals reflected by the unit cells are aligned to maximize the received signal power of a specific user. In the RIS-assisted broadcasting application, the signals reflected by the unit cells cover all the users in a specific area. The realization of these applications is determined by the design of the reflection coefficients of the RIS unit cells. Therefore, the total path loss of the cascaded channel provided by the entire RIS under each typical scenario can be analyzed based on the general path loss model given in Corollary 3.1 to obtain intuitive insights.

In the beamforming scenario, by appropriately designing the reflection phase $\phi_{n,m}$ of each unit cell, the phases of the reflected signals from all the unit cells to the receiver can be aligned to enhance the received signal power.

Proposition 3.1 *Assume that the reflection coefficients of all the unit cells of the RIS have the same amplitude component A and have adjustable phase shift values $\phi_{n,m}$ (i.e., $\Gamma_{n,m} = Ae^{j\phi_{n,m}}$). By beamforming the reflected signals to the receiver, the total path loss of the cascaded channel corresponding to the entire RIS can be expressed as*

$$
PL^{beamform} = \frac{16\pi^2}{G_t G_r (d_x d_y)^2 A^2 \left| \sum_{m=1}^{M} \sum_{n=1}^{N} \frac{\sqrt{F_{n,m}^{combine}}}{r_{n,m}^t r_{n,m}^r} \right|^2}, \tag{3.5}
$$

where the phase shift $\phi_{n,m}$ is designed as

$$
\phi_{n,m} = \mathrm{mod}\left(\frac{2\pi \left(r_{n,m}^t + r_{n,m}^r \right)}{\lambda}, 2\pi \right) \tag{3.6}
$$

Proof: It is proved by substituting $\Gamma_{n,m} = Ae^{j\phi_{n,m}}$ and (3.6) into (3.3). □

Proposition 3.1 provides a phase shift design method for the RIS unit cells that maximize the received signal power in the near-field and far-field regions of the entire RIS, but the direct relation between the entire path loss of the RIS-assisted wireless link and the distance parameters d_1 and d_2 cannot be intuitively obtained. It is discussed in the following special case, i.e., the far-field beamforming scenario.

Proposition 3.2 *Assume that both the transmit antenna and the receive antenna are located in the far field of the RIS and their peak radiation directions are directed toward the center of the RIS. The reflection coefficients of all the unit cells of the RIS have the same amplitude component A and have adjustable phase shift values $\phi_{n,m}$ (i.e., $\Gamma_{n,m} = Ae^{j\phi_{n,m}}$). By performing far-field beamforming to the receiver through the RIS, the total path loss of the cascaded channel provided by the entire RIS can be* *expressed as*

$$
PL_{\text{farfield}}^{\text{beamform}} = \frac{16\pi^2 \left(d_1 d_2\right)^2}{G_t G_r \left(MN d_x d_y\right)^2 A^2 F\left(\theta_t, \varphi_t\right) F\left(\theta_r, \varphi_r\right)}
\tag{3.7}
$$

Proof: Under the far-field condition, the joint normalized power radiation pattern $F_{n,m}^{\text{combine}}$ degenerates to $F\left(\theta_t, \varphi_t\right) F\left(\theta_r, \varphi_r\right)$, and the distance parameters $r_{n,m}^t$ and $r_{n,m}^r$ degenerate to d_1 and d_2, respectively. Substituting them into (3.5), the proof is completed. □

Compared to the general total path loss model of the entire RIS given in (3.3), the total path loss model given in (3.7) for the RIS-assisted far-field beamforming scenario is more intuitive. Proposition 3.2 reveals that the entire path loss in the RIS-assisted far-field beamforming scenario is proportional to the square of the product of the two distances (i.e., $\left(d_1 d_2\right)^2$), is inversely proportional to the square of the total area of the RIS (i.e., $\left(MN d_x d_y\right)^2$), and is related to the power radiation pattern $F(\theta, \varphi)$ of the unit cell.

In addition to beamforming applications, broadcast applications can be considered for an RIS, so that the reflected signals cover a specific area. The following proposition gives a simple example.

Proposition 3.3 *Assume that the transmitter is very close to the RIS, the electrical size of the RIS is large enough, and all the unit cells of the RIS have the same reflection coefficient $\Gamma_{n,m} = Ae^{j\phi}$. Thus, the RIS performs specular reflection and signal broadcasting. The total path loss of the cascaded channel corresponding to the entire RIS can be approximately expressed as*

$$
PL_{\text{nearfield}}^{\text{broadcast}} \approx \frac{16\pi^2 \left(d_1 + d_2\right)^2}{G_t G_r \lambda^2 A^2}
\tag{3.8}
$$

Proof: It can be proved according to the principle of geometric optics. The signal transmission process is equivalent to that of a wireless signal that is transmitted by a virtual transmitter symmetrically located with respect to the RIS, and is received after traveling a distance $(d_1 + d_2)$. □

Proposition 3.3 indicates that, in the RIS-assisted near-field broadcasting scenario, the total path loss of the cascaded channel provided by the entire RIS is proportional to $\left(d_1 + d_2\right)^2$, rather than $\left(d_1 d_2\right)^2$ as for the RIS-assisted far-field beamforming scenario.

3.2.4 Discussion on RIS Path Loss and Channel Reciprocity

The path loss model in (3.3) is the total path loss of the wireless link provided by the entire RIS and is directly related to the configuration of the reflection coefficients of the RIS unit cells. Specifically, the reason why the specific path loss models given in the preceding three propositions are different is due to the different methods to configure the reflection coefficient $\Gamma_{n,m}$ in each scenario.

On the other hand, the path loss model (3.1) is for the cascaded sub-channel provided by a single unit cell, which is often more suitable for theoretical analysis of RIS-assisted wireless communication systems. According to (3.1) and ignoring the noise at the receiver, the received signal corresponding to the reflection from the unit cell $U_{n,m}$ can be written as

$$
s_{n,m}^r = s_t \frac{\overbrace{\sqrt{G_t G_r}(d_x d_y)}^{\text{Size}} \overbrace{\sqrt{F_{n,m}^{\text{combine}}}}^{\text{Angle-dependent factor}} \overbrace{|\Gamma_{n,m}|}^{\text{Efficiency}}}{4\pi \underbrace{\left(r_{n,m}^t r_{n,m}^r\right)}_{\text{Product distance}}} e^{-j\left(\frac{2\pi}{\lambda}\left(r_{n,m}^t + r_{n,m}^r\right) - \angle\Gamma_{n,m}\right)} \quad (3.9)
$$

where s_t is the transmitted signal. Equation (3.9) can be used to characterize the fundamental properties of large-scale fading of the cascaded sub-channel provided by each unit cell, which is proportional to the size $d_x d_y$ and reflection efficiency (amplitude) $|\Gamma_{n,m}|$ of the unit cell, and is inversely proportional to the product of the distances $r_{n,m}^t$ and $r_{n,m}^r$. In addition, $s_{n,m}^r$ is proportional to the square root of the joint normalized power radiation pattern $F_{n,m}^{\text{combine}}$, which is an angle-dependent loss factor as defined in (3.2).

As far as channel reciprocity is concerned, it can be observed from (3.9) that, as long as the reflection coefficient $\Gamma_{n,m}$ of the unit cell is transmit-receive reciprocal, the wireless channel assisted by the RIS is reciprocal. It is worth noting that, this condition can be fulfilled even if $\Gamma_{n,m}$ is angle-sensitive, as long as $\Gamma_{n,m}$ is reciprocal for the exchange of the angle of incidence $(\theta_{n,m}^t, \varphi_{n,m}^t)$ and the angle of observation $(\theta_{n,m}^r, \varphi_{n,m}^r)$ shown in Fig. 3.1 (Tang et al. 2021b; Pei et al. 2021). In other words, if $\Gamma_{n,m}$ is not only sensitive to the angle of incidence but also sensitive to the angle of observation, angle-sensitive regulation and electromagnetic reciprocity can coexist for reciprocal RISs. In fact, channel reciprocity is common, while channel nonreciprocity is rare and requires special RIS hardware designs, such as using active nonreciprocal radio-frequency circuits and nonreciprocal materials in RIS implementations (Kord et al. 2020; Asadchy et al. 2020). Therefore, the reciprocity of the reflection coefficient $\Gamma_{n,m}$ holds for commonly designed and fabricated RISs, which ensures that the uplink and the downlink channels of RIS-assisted wireless communication systems are often reciprocal.

Channel reciprocity means that the downlink channel is equal to the uplink channel. However, channel reciprocity may not ensure beam reciprocity in RIS-assisted wireless communication systems. Yue et al. (2022) pointed out that the channel reciprocity holds, whereas the beam reciprocity does not for angle-sensitive RISs. This phenomenon means that due to the angle-sensitive properties of the RIS, the beam patterns of the reflected signals in the uplink timeslot and in the downlink timeslot are nonreciprocal, which indicates that the angle dependency of the reflection coefficient $\Gamma_{n,m}$ should be considered for RIS-assisted beamforming designs.

3.3 Path Loss Measurement and Channel Reciprocity Validation

In this section, we conduct a measurement campaign to validate the proposed RIS path loss models and the channel reciprocity of RIS-assisted wireless communication systems.

3.3.1 Two Fabricated RISs

The two fabricated RISs employed in the measurement campaign are shown in Fig. 3.3, which have different operating frequencies and different sizes. The detailed parameters of the RISs and antennas are illustrated in Table 3.2.

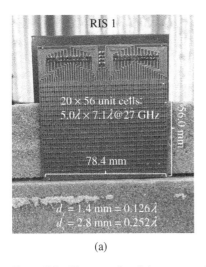

(a) (b)

Figure 3.3 Photographs of the two employed RISs. (a) RIS 1. (b) RIS 2.

Table 3.2 Parameters of the RISs and antennas employed in the measurement.

Name	Parameters
RIS 1	$N = 20, M = 56, d_x = 1.4$ mm, $d_y = 2.8$ mm, the operating frequency is 27 GHz, 1-bit phase control. $\angle \Gamma_{n,m} = 165°$ and $|\Gamma_{n,m}| = 0.9$ when code "0" is applied to $U_{n,m}$, $\angle \Gamma_{n,m} = 0°$ and $|\Gamma_{n,m}| = 0.7$ when code "1" is applied to $U_{n,m}$
RIS 2	$N = 40, M = 40, d_x = 3.8$ mm, $d_y = 3.8$ mm, the operating frequency is 33 GHz, 1-bit phase control. $\angle \Gamma_{n,m} = 150°$ and $|\Gamma_{n,m}| = 0.8$ when code "0" is applied to $U_{n,m}$, $\angle \Gamma_{n,m} = 0°$ and $|\Gamma_{n,m}| = 0.8$ when code "1" is applied to $U_{n,m}$
Antennas	When the operating frequency is $f = 27$ GHz, $G_t = G_r = 109.6 = 20.4$ dB. According to (3.4), $F^{tx}(\theta, \varphi) = F^{rx}(\theta, \varphi) = (\cos \theta)^{53.8}$ when $\theta \in \left[0, \frac{\pi}{2}\right]$, $F^{tx}(\theta, \varphi) = F^{rx}(\theta, \varphi) = 0$ when $\theta \in \left(\frac{\pi}{2}, \pi\right]$. When the operating frequency is $f = 33$ GHz, $G_t = G_r = 128.8 = 21.1$ dB. According to (3.4), $F^{tx}(\theta, \varphi) = F^{rx}(\theta, \varphi) = (\cos \theta)^{63.4}$ when $\theta \in \left[0, \frac{\pi}{2}\right]$, $F^{tx}(\theta, \varphi) = F^{rx}(\theta, \varphi) = 0$ when $\theta \in \left(\frac{\pi}{2}, \pi\right]$

3.3.2 Two Measurement Systems

As shown in Fig. 3.4(a), the constructed RIS path loss measurement system consists of an radio frequency (RF) signal generator, an RF signal analyzer, an RIS, two antennas, and a pair of RF cables. During the measurement process, the RF signal generator provides the transmit signal with power P_t and operating frequency f, and the RF signal analyzer displays the information of the received signal power. The RIS is placed on a stable tripod. The transmit antenna and the receive antenna can be flexibly moved in order to perform path loss measurements under different configurations of d_1, d_2, θ_t, and θ_r. The calibration method proposed by Tang et al. (2022) is employed for the RIS path loss measurement and data analysis.

The same experimental system architecture can be used to explore whether RIS-assisted wireless channels remain reciprocal. As shown in Fig. 3.4(b), when performing the uplink measurement, antenna 2 is connected to the RF signal generator and antenna 1 is connected to the RF signal analyzer. On the contrary, when performing the downlink measurement, antenna 1 is connected to the RF signal generator and antenna 2 is connected to the RF signal analyzer. Therefore, the channel reciprocity measurement system can be switched between the uplink mode and the downlink mode, simply and conveniently, by swapping the RF connection of the RF signal generator and the RF signal analyzer. Therefore, the

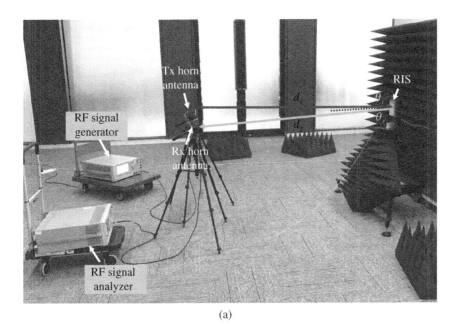

(a)

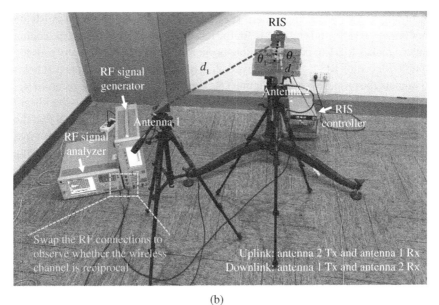

(b)

Figure 3.4 Photographs of the measurement systems. (a) Path loss measurement system. (b) Channel reciprocity measurement system.

reciprocity of RIS-assisted uplink and downlink channels can be validated by observing whether the received signals in the uplink and downlink modes are consistent.

3.3.3 Validation of RIS Path Loss Models

The path loss measurement is first carried out with RIS 1 in the experimental system shown in Fig. 3.4(a). The RF signal generator transmits a wireless signal with a power of 20 dBm and an operating frequency of 27 GHz. We set $d_1 = 2$ m, $\theta_t = \theta_r = 10°$, and vary d_2 in the interval [1 m, 5 m], and measure the path loss versus d_2. The measurement setup satisfies the far-field condition for RIS 1. We configure all the unit cells of RIS 1 to the coding state "0," thus RIS 1 performs far-field specular beamforming. The experimental measurement is also performed for the case $\theta_t = \theta_r = 45°$. The measurement results are shown in Fig. 3.5. It can be observed that the measurement results are in good agreement with the general path loss model (3.3) and the specific path loss model (3.7) for the far-field beamforming scenario. In addition, the path loss increases as the angles of incidence and reception increase.

Furthermore, the received signal power versus the angle of reception is measured. We set $\theta_t = 0°$, $d_1 = 1.3$ m, $d_2 = 2.6$ m, and configure the unit cells of RIS 1 according to a stripe coding pattern (i.e., coding "0" for the unit cells satisfying $\mod(m, 14) \in [1, 7]$ and coding "1" for the rest of the unit cells). This stripe coding pattern makes RIS 1 to perform double-beam reflection. There are many possible options for coding RIS 1, and the above coding method is just an example that helps to validate the proposed model. Figure 3.5(c) illustrates the measured received signal power versus the angle of reception θ_r. RIS 1 successfully achieves dual-beam reflection and the two beams are oriented to $|\theta_r| = 34°$. It can be observed that the measured curve matches well with the theoretical curve of the general path loss model in (3.3). Figure 3.5(d) shows the measured path loss versus d_2 when using the above stripe coding method. The measurement is performed by setting $\theta_t = 0°$, $\theta_r = 34°$, and $d_1 = 0.5$ m. It can be observed that the measurement result is in good agreement with the general path loss model given in (3.3).

The path loss measurement is then carried out with RIS 2 in the same experimental system. The RF signal generator provides a transmit signal with a power of 20 dBm and an operating frequency of 33 GHz. We set $d_1 = 5$ or 0.25 m, $\theta_t = \theta_r = 10°$, and measure the path loss versus d_2. We configure all the unit cells of RIS 2 to the coding state "0." When $d_1 = 5$ m, d_2 varies in the interval [5 m, 8 m], which satisfies the far-field condition for RIS 2, and RIS 2 performs far-field beamforming. As shown in Fig. 3.6(a), the measurement results are in good agreement with the general path loss model in (3.3) and the specific path loss model in (3.7) for the far-field beamforming scenario. When $d_1 = 0.25$ m, d_2 varies

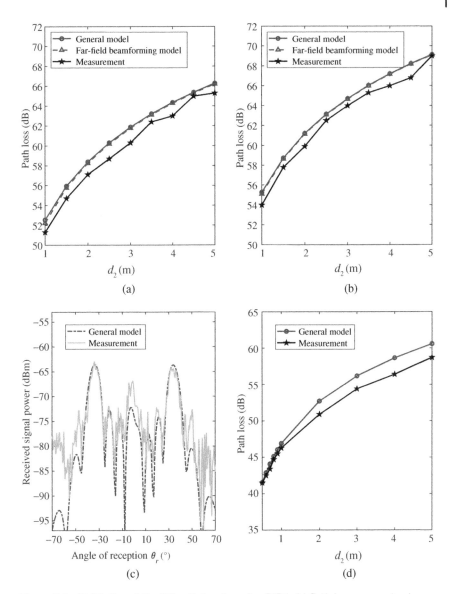

Figure 3.5 Validation of the RIS path loss by using RIS 1. (a) Path loss versus d_2 when performing far-field beamforming with $d_1 = 2$ m and $\theta_t = \theta_r = 10°$. (b) Path loss versus d_2 when performing far-field beamforming with $d_1 = 2$ m and $\theta_t = \theta_r = 45°$. (c) Received signal power versus θ_r when performing dual-beam reflection with $\theta_t = 0°$, $d_1 = 1.3$ m and $d_2 = 2.6$ m. (d) Path loss versus d_2 when performing dual-beam reflection with $\theta_t = 0°$, $\theta_r = 34°$, and $d_1 = 0.5$ m.

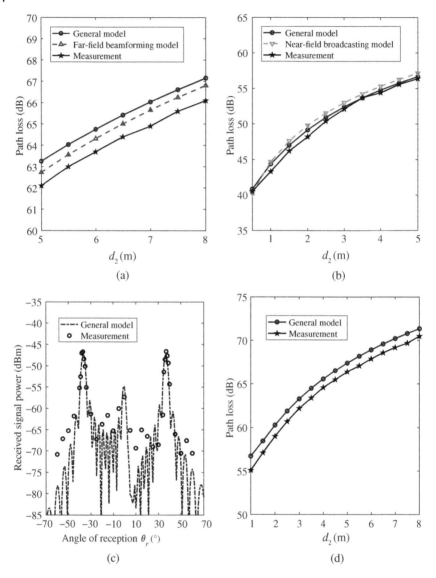

Figure 3.6 Validation of the RIS path loss by using RIS 2. (a) Path loss versus d_2 when performing far-field beamforming with $d_1 = 5\,$m and $\theta_t = \theta_r = 10°$. (b) Path loss versus d_2 when performing near-field broadcasting with $d_1 = 0.25\,$m and $\theta_t = \theta_r = 10°$. (c) Received signal power versus θ_r when performing dual-beam reflection with $\theta_t = 0°$, $d_1 = 5\,$m and $d_2 = 5\,$m. (d) Path loss versus d_2 when performing dual-beam reflection with $\theta_t = 0°$, $\theta_r = 37°$, and $d_1 = 5\,$m.

in the interval [0.5 m, 5 m], which satisfies the near-field condition of RIS 2, and RIS 2 performs near-field broadcasting. As shown in Fig. 3.6(b), the measurement results are in good agreement with the general path loss model in (3.3) and the specific path loss model in (3.8) for the near-field broadcasting scenario.

In addition, the received signal power versus the angle of reception is measured. We set $\theta_t = 0°$, $d_1 = 5$ m, $d_2 = 5$ m, and configure the unit cells of RIS 2 with a stripe coding pattern (i.e., coding "0" for the unit cells satisfying mod $(m, 4) \in$ [1, 2] and coding "1" for the rest of the unit cells). Figure 3.6(c) presents the measured received signal power versus the angle of reception θ_r. RIS 2 successfully achieves dual-beam reflection and the two beams are oriented to $|\theta_r| = 37°$. It can be observed that the measurement results match well with the theoretical curve of the general path loss model in (3.3). Figure 3.6(d) shows the measured path loss versus d_2 when using the stripe coding configuration. The measurement is performed by setting $\theta_t = 0°$, $\theta_r = 37°$, and $d_1 = 5$ m. It can be observed that the measured curve is in good agreement with the proposed general path loss model given in (3.3).

3.3.4   Validation of RIS Channel Reciprocity

The channel reciprocity measurement is performed with RIS 1 in the experimental system shown in Fig. 3.4(b). The received signal powers in the uplink and downlink transmission modes (the phase of the received signal is not measured due to the limitation of the experimental equipment) are measured and compared to validate whether the channel reciprocity holds in RIS-assisted wireless channels. The unit cells of RIS 1 are configured with an identical coding pattern and a stripe coding pattern, respectively. The corresponding measurement results are shown in Table 3.3. It can be observed that the received signal powers (amplitudes) in the uplink and downlink setups agree well with each other, which validates that the reciprocity of RIS-assisted wireless channels still holds for the considered hardware platforms.

The above experimental results verify that channel reciprocity holds in RIS-assisted TDD wireless communication systems, as long as the employed RISs are commonly designed and fabricated, and conform to the tensor identity of permittivity and permeability (Asadchy et al. 2020). In addition, Table 3.3 indicates that an appropriate coding pattern can significantly increase the received signal power, which is an important function of RIS-assisted wireless communications. If the channel reciprocity holds, the improvement of the received signal power is the same in the uplink and downlink channels.

Last but not least, it is worth noting that, even though it is often necessary to design efficient wireless communication protocols based on channel reciprocity, there are special cases that need to break channel reciprocity, such as secure

Table 3.3 Experimental results for validating RIS channel reciprocity.

Measurement setting	Coding pattern	Received signal power (uplink/downlink, dBm)
$P_t = 0\,\text{dBm}$, $f = 27\,\text{GHz}$, $d_1 = 1\,\text{m}, \theta_1 = 35°$, $d_2 = 0.5\,\text{m}, \theta_2 = 0°$	Identical coding	−58.0/−57.8
	Stripe coding	−49.4/−49.2
$P_t = 0\,\text{dBm}$, $f = 27\,\text{GHz}$, $d_1 = 1\,\text{m}, \theta_1 = 5°$, $d_2 = 0.5\,\text{m}, \theta_2 = 0°$	Identical coding	−49.3/−49.2
	Stripe coding	−58.2/−58.2

wireless communication and wireless power transfer. Active non-reciprocal RF circuits (Ma et al. 2019), time-varying controls (Zhang et al. 2019), and nonlinearities and structural asymmetries (Mahmoud et al. 2015) can be used in RIS hardware designs to break channel reciprocity in RIS-assisted wireless networks.

3.4 Conclusion

In this chapter, a general free-space path loss model for RIS-assisted wireless communication systems is introduced, which is simple to use yet sufficiently accurate. Based on the proposed general model, specific path loss models for typical scenarios are derived to characterize the overall path loss characteristics of RIS-assisted beamforming and RIS-assisted broadcasting channels, respectively. These specific models reveal the explicit relation between the path loss and physical parameters, including the transmission distances and the size of the RIS. Two fabricated RISs are employed to conduct path loss measurements under different configurations and scenarios. The experimental results are in good agreement with the proposed path loss models for RISs. In addition, it is pointed out that the uplink and downlink channels of RIS-assisted TDD wireless communication systems can be either reciprocal or non-reciprocal, depending on the design and manufacture of the employed RISs. The measurement results validate that channel reciprocity holds in common RIS-assisted TDD wireless communication systems. This chapter may help researchers understand the basic characteristics of large-scale fading and channel reciprocity in RIS-assisted wireless communication systems, which can be used for link budget calculation and system performance analysis.

3.A Appendix

3.A.1 Proof of Theorem 3.1

The signal power incident into the unit cell $U_{n,m}$ can be written as

$$P_{n,m}^{\text{in}} = P_t \frac{G_t G \lambda^2}{\left(4\pi r_{n,m}^t\right)^2} F^{tx}\left(\theta_{n,m}^{tx}, \varphi_{n,m}^{tx}\right) F\left(\theta_{n,m}^t, \varphi_{n,m}^t\right). \tag{3.A.10}$$

According to the law of energy conservation, for the unit cell $U_{n,m}$, the reflected signal power is equal to the incident signal power multiplied by the square of the absolute value of the reflection coefficient. Thus, we have

$$P_{n,m}^{\text{reflect}} = P_{n,m}^{\text{in}} |\Gamma_{n,m}|^2, \tag{3.A.11}$$

where the reflection coefficient can be expressed as

$$\Gamma_{n,m} = A_{n,m} e^{j\phi_{n,m}}, \tag{3.A.12}$$

where $A_{n,m}$ and $\phi_{n,m}$ denote the amplitude component and the phase component of the reflection coefficient of the unit cell $U_{n,m}$, respectively.

The signal power at the receiver that is reflected from the unit cell $U_{n,m}$ can be expressed as

$$P_{n,m}^r = P_{n,m}^{\text{reflect}} \frac{G G_r \lambda^2}{\left(4\pi r_{n,m}^r\right)^2} F\left(\theta_{n,m}^r, \varphi_{n,m}^r\right) F^{rx}\left(\theta_{n,m}^{rx}, \varphi_{n,m}^{rx}\right). \tag{3.A.13}$$

The aperture of the unit cell $U_{n,m}$ is $d_x d_y$, thus its scattering gain is

$$G = \frac{4\pi d_x d_y}{\lambda^2}. \tag{3.A.14}$$

The proof of Theorem 3.1 is obtained by substituting (3.A.10), (3.A.11), and (3.A.14) into (3.A.13).

Bibliography

Asadchy, V.S., Mirmoosa, M.S., Díaz-Rubio, A. et al. (2020). Tutorial on electromagnetic nonreciprocity and its origins. *Proceedings of the IEEE* 108 (10): 1684–1727.

Basar, E., Di Renzo, M., De Rosny, J. et al. (2019). Wireless communications through reconfigurable intelligent surfaces. *IEEE Access* 7: 116753–116773.

Chen, J., Liang, Y.-C., Pei, Y., and Guo, H. (2019). Intelligent reflecting surface: a programmable wireless environment for physical layer security. *IEEE Access* 7: 82599–82612.

Cui, T.J., Qi, M.Q., Wan, X. et al. (2014). Coding metamaterials, digital metamaterials and programmable metamaterials. *Light: Science & Applications* 3: 1–9.

Cui, T.J., Liu, S., and Zhang, L. (2017). Information metamaterials and metasurfaces. *Journal of Materials Chemistry C* 5: 3644–3668.

Cui, T.J., Liu, S., Bai, G.D., and Ma, Q. (2019). Direct transmission of digital message via programmable coding metasurface. *Research* 2019: 2584509.

Danufane, F.H., Di Renzo, M., de Rosny, J., and Tretyakov, S. (2021). On the path-loss of reconfigurable intelligent surfaces: an approach based on green's theorem applied to vector fields. *IEEE Transactions on Communications* 69 (8): 5573–5592.

Degli-Esposti, V., Vitucci, E.M., Di Renzo, M., and Tretyakov, S. (2022). Reradiation and scattering from a reconfigurable intelligent surface: a general macroscopic model. *IEEE Transactions on Antennas and Propagation* 70 (10): 8691–8706.

Di Renzo, M., Debbah, M., Phan-Huy, D.-T. et al. (2019). Smart radio environments empowered by reconfigurable AI meta-surfaces: an idea whose time has come. *EURASIP Journal on Wireless Communications and Networking* 2019: 1–20.

Di Renzo, M., Danufane, F.H., Xi, X. et al. (2020a). Analytical modeling of the path-loss for reconfigurable intelligent surfaces - anomalous mirror or scatterer? *2020 IEEE 21st International Workshop on Signal Processing Advances in Wireless Communications (SPAWC)*.

Di Renzo, M., Zappone, A., Debbah, M. et al. (2020b). Smart radio environments empowered by reconfigurable intelligent surfaces: how it works, state of research, and the road ahead. *IEEE Journal on Selected Areas in Communications* 38 (11): 2450–2525.

Di Renzo, M., Ahmed, A., Zappone, A. et al. (2022). Digital reconfigurable intelligent surfaces: on the impact of realistic reradiation models. https://arxiv.org/abs/2205 .09799.

Ellingson, S.W. (2021). Path loss in reconfigurable intelligent surface-enabled channels. *2021 IEEE 32nd Annual International Symposium on Personal, Indoor and Mobile Radio Communications (PIMRC)*.

Garcia, J.C.B., Sibille, A., and Kamoun, M. (2020). Reconfigurable intelligent surfaces: bridging the gap between scattering and reflection. *IEEE Journal on Selected Areas in Communications* 38 (11): 2538–2547.

Goldsmith, A. (2005). *Wireless Communications*. Cambridge: Cambridge University Press.

Gradoni, G. and Di Renzo, M. (2021). End-to-end mutual coupling aware communication model for reconfigurable intelligent surfaces: an electromagnetic-compliant approach based on mutual impedances. *IEEE Wireless Communications Letters* 10 (5): 938–942.

Han, Y., Tang, W., Jin, S. et al. (2019). Large intelligent surface-assisted wireless communication exploiting statistical CSI. *IEEE Transactions on Vehicular Technology* 68 (8): 8238–8242.

Hu, J., Zhang, H., Di, B. et al. (2020). Reconfigurable intelligent surface based RF sensing: design, optimization, and implementation. *IEEE Journal on Selected Areas in Communications* 38 (11): 2700–2716.

Huang, C., Zappone, A., Alexandropoulos, G.C. et al. (2019). Reconfigurable intelligent surfaces for energy efficiency in wireless communication. *IEEE Transactions on Wireless Communications* 18 (8): 4157–4170.

Kord, A., Sounas, D.L., and Alù, A. (2020). Microwave nonreciprocity. *Proceedings of the IEEE* 108 (10): 1728–1758.

Ma, Q., Chen, L., Jing, H.B. et al. (2019). Controllable and programmable nonreciprocity based on detachable digital coding metasurface. *Advanced Optical Materials* 7 (24): 1901285.

Mahmoud, A.M., Davoyan, A.R., and Engheta, N. (2015). All-passive nonreciprocal metastructure. *Nature Communications* 6 (1): 1–7.

Najafi, M., Jamali, V., Schober, R., and Poor, H.V. (2021). Physics-based modeling and scalable optimization of large intelligent reflecting surfaces. *IEEE Transactions on Communications* 69 (4): 2673–2691.

Özdogan, Ö., Björnson, E., and Larsson, E.G. (2020). Intelligent reflecting surfaces: physics, propagation, and pathloss modeling. *IEEE Wireless Communications Letters* 9 (5): 581–585.

Pan, C., Ren, H., Wang, K. et al. (2020). Intelligent reflecting surface aided MIMO broadcasting for simultaneous wireless information and power transfer. *IEEE Journal on Selected Areas in Communications* 38 (8): 1719–1734.

Pei, X., Yin, H., Tan, L. et al. (2021). RIS-aided wireless communications: prototyping, adaptive beamforming, and indoor/outdoor field trials. *IEEE Transactions on Communications* 69 (12): 8627–8640.

Stutzman, W.L. and Thiele, G.A. (2012). *Antenna Theory and Design*. New York: Wiley.

Tang, W., Dai, J.Y., Chen, M.Z. et al. (2020). Mimo transmission through reconfigurable intelligent surface: system design, analysis, and implementation. *IEEE Journal on Selected Areas in Communications* 38 (11): 2683–2699.

Tang, W., Chen, M.Z., Chen, X. et al. (2021a). Wireless communications with reconfigurable intelligent surface: path loss modeling and experimental measurement. *IEEE Transactions on Wireless Communications* 20 (1): 421–439.

Tang, W., Chen, X., Chen, M.Z. et al. (2021b). On channel reciprocity in reconfigurable intelligent surface assisted wireless networks. *IEEE Wireless Communications* 28 (6): 94–101.

Tang, W., Chen, X., Chen, M.Z. et al. (2022). Path loss modeling and measurements for reconfigurable intelligent surfaces in the millimeter-wave frequency band. *IEEE Transactions on Communications* 70 (9): 6259–6276.

Wang, Z., Tan, L., Yin, H. et al. (2021). A received power model for reconfigurable intelligent surface and measurement-based validations. *2021 IEEE 22nd*

International Workshop on Signal Processing Advances in Wireless Communications (SPAWC).

Wu, Q. and Zhang, R. (2019). Intelligent reflecting surface enhanced wireless network via joint active and passive beamforming. *IEEE Transactions on Wireless Communications* 18 (11): 5394–5409.

Wu, Q. and Zhang, R. (2020). Towards smart and reconfigurable environment: intelligent reflecting surface aided wireless network. *IEEE Communications Magazine* 58 (1): 106–112.

Yu, S., Liu, H., and Li, L. (2019). Design of near-field focused metasurface for high-efficient wireless power transfer with multifocus characteristics. *IEEE Transactions on Industrial Electronics* 66 (5): 3993–4002.

Yue, S., Zeng, S., Zhang, H. et al. (2022). Intelligent omni-surfaces aided wireless communications: does the reciprocity hold? https://arxiv.org/abs/2211.03059.

Zhang, L., Chen, X.Q., Shao, R.W. et al. (2019). Breaking reciprocity with space-time-coding digital metasurfaces. *Advanced Materials* 31 (41): 1904069.

4

Intelligent Surface Communication Design: Main Challenges and Solutions

Kaitao Meng[1], Qingqing Wu[2], and Rui Zhang[3]

[1] *State Key Laboratory of Internet of Things for Smart City, University of Macau, Macau, China*
[2] *Department of Electronic Engineering, Shanghai Jiaotong University, Shanghai, China*
[3] *Department of Electrical and Computer Engineering, National University of Singapore, Singapore*

4.1 Introduction

Despite many benefits mentioned in Chapter 1, there are several new challenges in the intelligent surface-aided wireless communication design due to significantly differing from the traditional network comprising active components, such as BSs, APs, and terminals. Thus, in this chapter, the main challenges and solutions are presented for intelligent surface-assisted communication system design, including channel estimation in Section 4.2, passive beamforming design in Section 4.3, and deployment strategy in Section 4.4.

4.2 Channel Estimation

4.2.1 Problem Description and Challenges

In order to fully realize the various performance improvements brought by intelligent surfaces, the acquisition of accurate channel state information (CSI) is fundamentally crucial. However, accurate CSI in intelligent surface-aided networks is practically challenging due to the plethora of additional channels involved between the intelligent surfaces and their associated transceivers, as well as the lack of transmit/receive capabilities of the intelligent surface elements. As a typical intelligent surface, the channel estimation issue of IRSs is investigated in this section.

In particular, an IRS-assisted MIMO system is considered to provide uplink services for K users equipped with M_u antennas, where the signal received at the BS

Intelligent Surfaces Empowered 6G Wireless Network, First Edition.
Edited by Qingqing Wu, Trung Q. Duong, Derrick Wing Kwan Ng, Robert Schober, and Rui Zhang.
© 2024 The Institute of Electrical and Electronics Engineers, Inc. Published 2024 by John Wiley & Sons, Inc.

with M_B antennas can be given by

$$y = \sum_{k=1}^{K} \left(G^T \Theta H_{r,k} + H_{d,k} \right) x_k + z, \tag{4.1}$$

where $G \in \mathbb{C}^{N \times M_B}$, $H_{r,k} \in \mathbb{C}^{N \times M_u}$ and $H_{d,k} \in \mathbb{C}^{M_B \times M_u}$ respectively denote the BS-IRS, user k-IRS, and user k-BS links, with $k \in \mathcal{K} = \{1, \dots, K\}$; $x_k \in \mathbb{C}^{M_u \times 1}$ represents the signal sent by user k and $z \in \mathbb{C}^{M_B \times 1}$ is the additive white Gaussian noise (AWGN) vector at the BS. In (4.1), $\Theta = \text{diag}\left(e^{j\theta_1}, e^{j\theta_2}, \dots, e^{j\theta_N} \right)$ denotes the reflection-coefficient matrix of one IRS with N reflecting elements. Here, the reflection amplitude of each element is set to one for simplicity. In this case, to design the IRS phase-shifting matrix Θ for achieving full passive beamforming gains of IRS, accurate CSI of G, $H_{r,k}$, and $H_{d,k}$ is required, and thus the total number of channel coefficients consists of two parts: $K \times N M_u + M_B N$ for channels $\{H_{r,k}\}_{k=1}^{K}$ and G, as well as $K \times M_B M_u$ for the direct channel $\{H_{d,k}\}_{k=1}^{K}$, where the former is newly introduced as compared to the conventional wireless communication systems. As such, a large number of additional channels are involved between the IRS and its associated BSs/users, which leads to an excessive amount of channel parameters to be estimated. Note that it is also different in the total number of channel coefficients for time-division duplexing (TDD) and frequency-division duplexing (FDD) systems. In particular, the FDD system needs to estimate twice the number of channel coefficients in (4.1) due to the general uplink and downlink channel asymmetry; in contrast, the TDD system may only need to use the uplink and downlink channel reciprocity to obtain uplink or downlink channel coefficient. Besides the significant increase in IRS-induced channel coefficients, another challenge for IRS channel estimation arises from the lack of active radio frequency (RF) chains and thus cannot transmit pilot signals to facilitate channel estimation. This is in stark contrast to an active BS/AP in traditional wireless systems.

In the literature, there are two main IRS channel estimation methods based on two different IRS configurations, depending on whether it is installed with sensing devices (receiving RF chain), called semi-passive IRS and fully passive IRS (Zheng et al. 2020), as shown in Fig. 4.1. Moreover, a hybrid semi- and fully-passive channel estimation can be achieved by combing these two configurations. In the following, we introduce these IRS configurations in detail, discuss state-of-the-art results on IRS channel estimation based on them respectively, and finally highlight the remaining important issues to be addressed in future work.

4.2.2 Semi-Passive IRS Channel Estimation

Although IRS dispenses with the transmit RF chain for cost and power reduction, some IRS reflective elements can still be equipped with a low-power/cost receive RF chain (namely channel sensors) to endow the IRS with sensing capability for

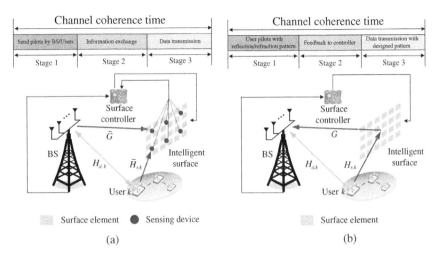

Figure 4.1 Two practical intelligent surface configurations and their respective transmission protocols. (a) Semi-passive intelligent surfaces. (b) Fully-passive intelligent surfaces.

channel estimation. As shown in Fig. 4.1(a), a small number of reflective elements equipped with channel sensors are distributed over a large IRS. Accordingly, the semi-passive IRS generally operates in two modes alternately over time: channel sensing mode and reflection mode. In the form mode, sensors receive pilot signals from BSs/users to estimate their respective channels to the IRS, while all reflective elements are turned OFF; in the latter mode, the IRS elements are turned on to reflect signals from the BSs/users for enhancing communication and the receive sensors are deactivated.

In particular, a general transmission protocol for semi-passive IRS is illustrated in Fig. 4.1(a), where each channel coherence interval comprises three stages. In the first stage, by adopting a similar approach to channel estimation in conventional networks without IRS, the BS/users transmit pilot signals to estimate the direct link between the BS and users, while the IRS simultaneously obtains the CSI from the BS/users by analyzing the received signal at its equipped sensors. Then, in the second stage, the acquired CSI is exchanged between the BS and the IRS controller for transmit beamforming and IRS phase shift design. Finally, in the third stage, the IRS operates in the reflection mode to enhance data transmission between the BS and users with the designed phase shifts. One should keep in mind that for semi-passive IRSs, only the CSI of BS/users→IRS channels can be estimated by the IRS sensors due to the lack of transmit RF chains, while the CSI of IRS→BS/users is unavailable in the FDD system for semi-passive IRS. Their corresponding reverse links can only be exploited in TDD systems by leveraging channel reciprocity.

To reduce the number of receive RF chains, the subarray-based method can be applied at the IRS, i.e., each subarray consists of a group of adjacent elements arranged vertically and/or horizontally, and each cluster is equipped with a receive RF chain for channel estimation. As shown in Fig. 4.1(a), $\bar{G} \in \mathbb{C}^{N_s \times M_B}$ and $\bar{H}_{r,k} \in \mathbb{C}^{N_s \times M_u}$ denote the BS-user k and user k-IRS sensors channels, respectively, with $N_s < N$. Importantly, the channel between the BS/user and the IRS sensors is different from that between the BS/user and the IRS elements given in (4.1), while they are generally correlated due to the close proximity. As such, an essential challenge for semi-passive IRS channel estimation is how to construct the channels G and $H_{r,k}$ in (4.1) based on the estimated CSI of the relatively low-dimensional channels $\bar{G}$ and $\bar{H}_{r,k}$. To tackle this issue, efficient signal processing methods, such as compressed sensing, data interpolation, and machine learning (Wu et al. 2021), can be applied to construct the CSI of BS/users→IRS links by exploiting certain IRS channel properties (such as low-rank, sparsity, and spatial correlation) among the reflecting elements. Moreover, the accuracy of the sampled CSI at the channel sensors highly depends on the number of sensors as well the resolution of analog-to-digital converters (ADCs), which also further affects the CSI accuracy for constructing the BS/users→IRS channels. Intuitively, mounting more sensors can provide more channel measurements to reduce IRS CSI construction errors, and applying higher-resolution ADCs can reduce quantization errors. However, a systematic investigation of the practical algorithms and the cost-performance trade-offs for semi-passive IRS channel estimation is still lacking in the literature, although there are a handful of recent preliminary works (Alexandropoulos and Vlachos 2020; Taha et al. 2021; Hu et al. 2022) addressing some of these aspects. For instance, Alexandropoulos and Vlachos (2020) proposed an alternating optimization (AO) method using random spatial sampling and analog combination techniques for semi-passive IRS channel estimation under a channel model based on beam space, where its mean square error (MSE) performance was evaluated in terms of channel sensing time. Another important issue is the information exchange between IRS and BS, which requires a more efficient feedback mechanism with less delay under certain link constraints. In Taha et al. (2021), semi-passive IRS channel estimation algorithms based on compressed sensing and deep learning techniques were proposed for the IRS-aided single-input-single-output (SISO) system, where the impact of both the number of IRS sensors and channel sensing time on the achievable rate was evaluated by simulations.

4.2.3 Fully-Passive IRS Channel Estimation

Different from semi-passive IRS, for fully-passive IRS, there are no sensors mounted on the IRS for low-cost implementation, and the CSI between the IRS and BS/users cannot directly be obtained in general. In this case, an alternative approach is to estimate the concatenated user-IRS-BS channel simultaneously. Notice that the CSI on the cascaded links is generally sufficient for the optimal passive/active beamforming design. Specifically, the IRS-aided uplink multi-user MIMO system is considered (c.f. (4.1)), where the cascaded user-IRS-BS channel is the transposed Khatri-Rao product of the user-IRS and IRS-BS channels, with each channel coefficient given by $[G]_{n,m_B}[H_{r,k}]_{n,m_u}, \forall n \in \mathcal{N} = \{1,\dots,N\}, m_B \in \mathcal{M}_B = \{1,\dots,M_B\}, m_u \in \mathcal{M}_u = \{1,\dots,M_u\}$, and $k \in \mathcal{K}$. Thus, the total number of channel coefficients for these concatenated user-IRS-BS channels is $K \times NM_BM_u$, and the corresponding training overhead is usually significantly larger than that for separate channel estimation of semi-passive IRS. This suggests that there is an inherent redundancy in the concatenated channels due to the fact that all users' uplink/downlink channels involve the common IRS-BS channel G^T, however, this is difficult to estimate from any user's concatenated channel analysis.

In Fig. 4.1(b), the uplink transmission protocol with fully-passive IRS is illustrated, where each channel coherence interval is divided into three stages. In the first stage, users send orthogonal pilots to the BS, and the IRS changes its reflection coefficient according to the predesigned reflection pattern, and the BS estimates the user-BS direct channel and the concatenated user-IRS-BS channel based on this. Then, in the second stage, the IRS reflection coefficients for data transmission are designed at the BS together with its receive beamforming, and then sent to the IRS controller via the backhaul link. Finally, in the third stage, the reflection coefficients are set accordingly to facilitate independent data transmissions from the users to the BS. For the TDD system, the estimated uplink CSI can also be used to design the IRS reflection for the downlink transmission from the BS to users; while for the FDD system, the transmission protocol in Fig. 4.1(b) is still applicable for the downlink communication, the only modification is to swap the roles of BS and users in the first stage.

For fully-passive IRS channel estimation, the key issue is to accurately estimate direct user-BS and concatenated user-IRS-BS channels with minimal training overhead by jointly designing pilot sequences, IRS reflection patterns, and signal processing algorithms at the receiver. For exposition purposes, a simple IRS-assisted single-user system is considered first (i.e., $K = 1$), where both BS and user are equipped with one antenna (i.e., $M_B = M_u = 1$). In this case, a practical approach for IRS channel estimation is to employ an ON/OFF based IRS reflection mode (Mishra and Johansson 2019), i.e., each IRS element is turned on in sequence, while other elements are set to OFF each time, so that the user-BS direct channels and the concatenated channels associated with different IRS elements are estimated separately. Note that this approach requires at least $N + 1$ pilot symbols to estimate the total $N + 1$ channel coefficients in this system. Although simple to implement, such IRS reflective mode inevitably results in a significant reflecting power loss since only one element is turned on at a time. To overcome this problem, an all-ON IRS reflection method can be used with time-varying orthogonal reflection coefficients, e.g., those drawn from different columns of an $(N + 1) \times (N + 1)$ discrete Fourier transform (DFT) matrix (Zheng and Zhang 2020), whereby $N + 1$ channel coefficients can be estimated together within $N + 1$ pilot symbol durations. Another important extension is to explore channel estimation in the multi-antenna case, where each receive antenna can estimate its associated channel in parallel, and certain channel properties (e.g., low rank, sparsity, spatial correlation, etc.) can be exploited to further improve estimation efficiency (Gao et al. 2021; Liu et al. 2023). Moreover, to reduce the training overhead of the IRS with a large number of elements and simplify the IRS reflective design for data transmission, an effective method is to group adjacent IRS elements into the subsurface. Accordingly, only the concatenated channel associated with each subsurface needs to be estimated, thus greatly reducing the training overhead. This element grouping strategy provides a flexible trade-off between training/design overhead/complexity and IRS passive beamforming gain in practice. Actually, for IRS channel estimation under different settings, there is a general trade-off between the training overhead for channel estimation and the passive beamforming gain for data transmission. Specifically, too little training leads to inaccurate/insufficient CSI, which leads to inefficient reflective design and reduces passive beamforming gain for data transfer; while too much training leads to reduced data transfer time, which leads to a lower achievable rate.

For channel estimation using passive IRS to serve multiple users (i.e., $K > 1$), a straightforward approach is to adopt a single-user channel estimation design, estimating the channels of K users separately in successive times, however, this increases the total training overhead by a factor of K compared to the single-user case, so it is effectively prohibitive if K is very large. Recall that all users actually share the common IRS-BS channel G^T in (4.1) in their cascaded user IRS-BS channels, based on this fact, the training overhead for IRS channel estimation in the multiuser case can be significantly reduced. For example, a user may be selected as a reference user whose concatenated channel is first estimated. Then, based on this reference CSI, the concatenated channels of the remaining $K - 1$ users can be efficiently estimated (Wang et al. 2020). Note that this result exploits the redundancy of BS receive antennas to reduce the training overhead, which is in sharp contrast to conventional multiuser channel estimation without IRS.

For broadband systems with frequency-selective fading channels, more channel coefficients need to be estimated due to the multipath delay spread. Furthermore, although the channel is frequency selective, the IRS reflection coefficient is frequency-flat, and thus cannot be flexibly designed for different frequencies. For the above reasons, IRS channel estimation for broadband frequency selective fading channels is more challenging than for narrowband flat fading channels.

Fortunately, since the number of orthogonal frequency-division multiplexing (OFDM) subcarriers is usually much larger than the maximum number of delay paths in practical systems, there is large redundancy for channel estimation, which can be used to design OFDM-based pilot symbols to efficiently estimate the channel for multiple users simultaneously (Yang et al. 2020). Moreover, it can also make full use of the redundancy of receive antennas and OFDM subcarriers, and further improve the training efficiency of the general IRS-assisted broadband MIMO system.

As compared to semi-passive IRS systems, the higher estimation accuracy of the cascaded channel estimation can be achieved with less hardware cost and energy consumption in fully-passive IRS systems. However, it is worth noting that the CSI error and cost-performance trade-offs of individual and cascaded channel estimation methods strongly depend on the employed signal processing technique, channel model, training design, and hardware constraints, which deserve a more in-depth future comparison. Moreover, how to design an effective hybrid/combined channel estimation method to achieve the potential of reducing the real-time training overhead is still an open problem.

4.3 Passive Beamforming Optimization

In this section, we study the passive beamforming optimization of IRS-assisted wireless communications under various system settings, including single/multi-user, single/multi-antenna, narrow/broad-band, and single/multi-cell communication. In particular, a single-cell IRS-assisted multiuser system is illustrated in Fig. 4.2, where an IRS with N elements is deployed to improve the downlink communication between the BS and K users. It is assumed that the BS is equipped with M_t antennas, and each user is equipped with M_r antennas.

4.3.1 IRS-aided SISO System: Passive Beamforming Basics and Power Scaling Order

First, the SISO case is considered, i.e., $K = 1$ and $M_t = M_r = 1$. The baseband equivalent channels from the BS to the IRS, from the IRS to the user, and from the BS to the user are denoted by $g \in \mathbb{C}^{N \times 1}$, $h_r^H \in \mathbb{C}^{1 \times N}$, and $h_d \in \mathbb{C}$, respectively. The information signal to the user is an independent and identically distributed (i.i.d.) random variable with zero mean and unit variance, denoted by x. Then, the signal received by the user is expressed as

$$y = (h_r^H g + h_d)\sqrt{P_t}x + z. \tag{4.2}$$

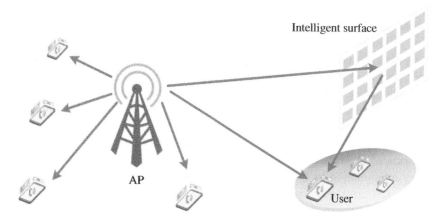

Figure 4.2 IRS-aided multiuser communication system.

In (4.2), P_t is the transmit power at the BS, and z denotes AWGN at the user receiver modeled as circularly symmetric complex Gaussian (CSCG) with zero mean and variance σ^2. Accordingly, the receive SNR at the user can be given by

$$\gamma = \frac{P_t|\mathbf{h}_r^H \mathbf{g} + h_d|^2}{\sigma^2} = \frac{P_t|\sum_{n=1}^{Nx} h_{r,n}\beta_n e^{j\theta_n} g_n + h_d|^2}{\sigma^2}. \tag{4.3}$$

Thus, the maximum achievable rate of the considered IRS-aided point-to-point SISO link can be expressed in bits per second per Hertz (bps/Hz), i.e., $r = \log_2(1 + \gamma)$.

Our objective is to maximize the achievable rate r by optimizing the IRS phase shifts (or namely passive reflect beamforming). After ignoring the constant terms, the optimization problem is formulated as

$$(\text{P1}) : \max_{\theta,\beta} \left| \sum_{n=1}^{N} h_{r,n} g_n \beta_n e^{j\theta_n} + h_d \right|^2 \tag{4.4}$$

$$\text{s.t.} \quad 0 \le \theta_n < 2\pi, n = 1, \dots, N, \tag{4.5}$$

$$0 \le \beta_n \le 1, n = 1, \dots, N, \tag{4.6}$$

where $\theta = [\theta_1, \dots, \theta_N]^T$ and $\beta = [\beta_1, \dots, \beta_N]^T$. In (P1), continuous reflection amplitude and phase shift are assumed. It is not difficult to verify that for any given $\beta \ge 0$, the optimal phase-shift solutions to (P1) are given by (Wu and Zhang 2019a)

$$\theta_n^\star = \text{mod}[\zeta - (\phi_n + \psi_n), 2\pi], n = 1, \dots, N, \tag{4.7}$$

where ϕ_n, ψ_n, and ζ denote the phases of $h_{r,n}$, g_n, and h_d, respectively. Actually, the solutions in (4.7) do not depend on the values in β, which implies that they are indeed optimal to (P1). Also, if the direct link between the user and the BS is negligible compared to the IRS reflective link (e.g., when the former is heavily blocked), it thus can be ignored, i.e., $h_d = 0$, the optimal solution in (4.7) can be multiplied by any common phase shift without changing the optimal value of (P1). By substituting (4.7) into (P1), this problem is generally reduced to

$$(P1.1): \max_{\beta} \left| \sum_{n=1}^{N} |h_{r,n}||g_n|\beta_n + |h_d| \right|^2 \tag{4.8}$$

$$\text{s.t.} \quad 0 \leq \beta_n \leq 1, n = 1, \dots, N. \tag{4.9}$$

It can be readily verified that the optimal reflection amplitude solutions are given by $\beta_n^\star = 1, \forall n$ since the reflection amplitude maximization can facilitate to improve the receive power at the user due to coherent combining. Interestingly, it is observed that the optimal IRS reflection design depends on the cascaded channels via the IRS only, i.e., $\text{diag}(h_r^H)g$, without the need for the individual channels, h_r^H and g.

A fundamental question about the passive beamforming performance of IRS is how the maximum received SNR grows with the number of IRS elements N. It was shown in Wu and Zhang (2019a) that as $N \to \infty$, the asymptotic receive power is approximately given by

$$P_r \approx N^2 \frac{P_t \pi^2 \varrho_h^2 \varrho_g^2}{16}. \tag{4.10}$$

It can be seen that when N is sufficiently large, the received power of the user increases quadratically with N (Wu and Zhang 2019a). Alternatively, the transmitted power of BS can be reduced by $1/N^2$ without affecting the received SNR. This is because IRS gets not only the reflected beamforming gain of $\mathcal{O}(N)$ on the IRS-User link but also the additional aperture gain of $\mathcal{O}(N)$ on the BS-IRS link.

4.3.2 IRS-aided MISO System: Joint Active and Passive Beamforming

In this subsection, a single-user system is considered with multiple antennas at the BS, i.e., $M_t > 1$. In this case, the BS transmit/receive beamforming needs to be optimized jointly with the IRS passive beamforming. Here, we consider the multiple-input-single-output (MISO) case while the results are also applicable to the single-input-multiple-output (SIMO) case in the reverse user-BS link. Let G, h_r^H, and h_d^H denote the BS-IRS channel matrix, the IRS-user channel vector, and the BS-user channel vector, respectively. Similar to (P1), the rate/SNR maximization problem in the MISO case is equivalently formulated as

$$(P2): \max_{w,\theta} \quad |(h_r^H G + h_d^H)w|^2 \tag{4.11}$$

$$\text{s.t.} \quad \|w\|^2 \leq P_t, \tag{4.12}$$

$$0 \leq \theta_n < 2\pi, n = 1, \dots, N, \tag{4.13}$$

where w and P_t are the transmit beamforming vector at the BS and the maximum transmit power, respectively. However, (P2) is a non-convex optimization problem because its objective function is non-concave with respect to (w.r.t.) w and θ. It can be observed that with a fixed transmit beamforming vector w, (P2) reduces to the corresponding SISO, i.e., (P1). This thus motivates AO methods to sub-optimally solve (P2) (Wu and Zhang 2019a). Specifically, when w is fixed, the optimal phase shift can be obtained by solving the resulting (P1); when θ is fixed, the optimal transmission beamforming solution of (P2) is given by the maximum-ratio transmission (MRT), i.e., $w_{\text{MRT}} = \sqrt{P_t} \frac{(h_r^H G + h_d^H)^H}{\|h_r^H G + h_d^H\|}$. The above process is repeated until convergence is reached, which guarantees at least a local optimal solution of (P2).

Another way to solve (P2) is to first reduce it to an optimization problem w.r.t. the phase shift only. Specifically, by replacing w_{MRT} with (4.11), (P2) reduces to the following problem involving only θ,

$$(P3): \max_{\theta} \quad \|h_r^H G + h_d^H\|^2 \tag{4.14}$$

$$\text{s.t.} \quad 0 \leq \theta_n \leq 2\pi, n = 1, \dots, N. \tag{4.15}$$

In fact, it is a non-deterministic polynomial-time hard (NP-hard) problem w.r.t. N, so it is impossible to obtain an optimal solution of (P3) with polynomial complexity N. For ease of analysis, let $v = [v_1, \dots, v_N]^H$ where $v_n = e^{j\theta_n}$, $\forall n$, then, the constraints in (4.15) are equivalent to the unit-modulus constraints, i.e., $|v_n|^2 = 1, \forall n$. By applying the variable transformation $h_r^H G = v^H \Phi$ where $\Phi = \text{diag}(h_r^H) G \in \mathbb{C}^{N \times M_t}$, (P3) is equivalently rewritten as (P4), i.e.,

$$(P4): \max_{v} \quad v^H \Phi \Phi^H v + v^H \Phi h_d + h_d^H \Phi^H v + \|h_d^H\|^2 \tag{4.16}$$

$$\text{s.t.} \quad |v_n|^2 = 1, n = 1, \dots, N. \tag{4.17}$$

(P4) is a non-convex quadratically constrained quadratic program (QCQP), which is generally NP-hard (So et al. 2007). Therefore, various approaches have been presented to obtain high-quality sub-optimal solutions to (P4), such as (i) semidefinite relaxation (SDR) and Gaussian randomization (Wu and Zhang 2019a), (ii) AO (also known as block coordinate descent method), where the phase shift of each IRS element is optimized iteratively (Wu and Zhang 2019b) (iii) penalty-based method (Dong et al. 2021), (iv) manifold optimization (Zhang et al. 2022a), (v) alternating direction method of multipliers (ADMM) (Yao et al. 2020), (vi) deep learning-based method (Jiang et al. 2021).

4.3.3 IRS-Aided MIMO System

Unlike SISO/MISO/SIMO systems which can only support the transmission of a single data stream, the capacity of MIMO systems (i.e., $M_t > 1$ and $M_r > 1$) is usually achieved by transmitting multiple data streams in parallel. Specifically, let $H_d \in \mathbb{C}^{M_r \times M_t}$, $G \in \mathbb{C}^{N \times M_t}$, and $H_r \in \mathbb{C}^{M_r \times N}$ denote the BS-user direct channel, the BS-IRS channel, and the IRS-user channel, respectively. Here, the equivalent IRS-aided MIMO channel can be given by $\tilde{H} = H_d + H_r \Theta G$. Then, the capacity optimization problem can be formulated as

$$(P5): \quad \max_{\Theta, Q} \quad \log_2 \det \left(I_{M_r} + \frac{1}{\sigma^2} \tilde{H} Q \tilde{H}^H \right) \tag{4.18}$$

$$\text{s.t.} \quad 0 \le \theta_n < 2\pi, \quad n = 1, \dots, N, \tag{4.19}$$

$$\beta_n = 1, \quad n = 1, \dots, N, \tag{4.20}$$

$$\text{tr}(Q) \le P_t, \tag{4.21}$$

$$Q \succeq 0. \tag{4.22}$$

In (P5), $Q \in \mathbb{C}^{M_t \times M_t}$ denotes the transmit covariance matrix. Due to the non-concavity of its objective function w.r.t Θ and Q, (P5) is a non-convex optimization problem. Zhang and Zhang (2020) proposed an efficient AO-based algorithm to solve (P5), where one IRS element's phase shift and the transmit covariance matrix are optimized in an iterative manner. Specifically, with a given IRS reflection matrix Θ, the optimal transmit covariance matrix is given by the well-known eigenmode transmission with water-filling-based spatial power allocation (Tse and Viswanath 2005). The above AO-based algorithm is guaranteed to converge at least to a locally optimal solution of (P5), with a polynomial complexity of only N, M_t, or M_r (Zhang and Zhang 2020). Note that various key parameters of IRS-enhanced MIMO channels, including channel total power, condition number, and rank. (Zhang and Zhang 2020) can be significantly improved by properly designing the IRS reflection coefficient. Especially, for practical scenarios where the BS-user direct channel is low-rank, AO-based algorithms produce higher-rank MIMO channels that are comparable to those without IRS. Compared with the MIMO channel, it has a greater spatial multiplexing gain, which greatly improves the channel capacity in the high-SNR regime.

4.3.4 IRS-Aided OFDM System

To illustrate the corresponding signal and channel model, both the BS and the user are considered to be equipped with a single antenna. Specifically, the numbers of delay taps in the BS-user direct channel, the BS-IRS channel, and the IRS-user channel are denoted by L_0, L_1, and L_2, respectively. Let $h^d = [h_0^d, \dots, h_{L_0-1}^d]^T \in \mathbb{C}^{L_0 \times 1}$ denote the baseband equivalent time-domain direct channel between the BS and the user; $h_n^{r_1} = [h_{n0}^{r_1}, \dots, h_{n(L_1-1)}^{r_1}]^T \in \mathbb{C}^{L_1 \times 1}$ and $h_n^{r_2} = [h_{n0}^{r_2}, \dots, h_{n(L_2-1)}^{r_2}]^T \in \mathbb{C}^{L_2 \times 1}$ denote that for the BS-IRS and IRS-user channels associated with the nth IRS reflecting element, respectively. The passband signal impinging on IRS element n

can be expressed as

$$y_{in,n}(t) = \text{Re}\left\{ \left[\sum_{l=0}^{L_1-1} h_{nl}^{r_1} x(t - \tau_l(t)) \right] e^{j2\pi f_c t} \right\}, \tag{4.23}$$

where $x(t)$ is the baseband transmit signal at the BS and $\tau_l(t) = l/B$ is the time delay of the lth tap. The reflected signal by the nth IRS element can be given by

$$y_{out,n}(t) = \beta_n y_{in,n}(t - t_n) = \text{Re}\left\{ \beta_n \left[\sum_{l=0}^{L_1-1} h_{nl}^{r_1} x(t - t_n - \tau_l(t)) \right] e^{j2\pi f_c(t-t_n)} \right\}$$

$$\approx \text{Re}\left\{ \beta_n e^{j\theta_n} \left[\sum_{l=0}^{L_1-1} h_{nl}^{r_1} x(t - \tau_l(t)) \right] e^{j2\pi f_c t} \right\}, \tag{4.24}$$

where $t_n \in [0, 1/f_c]$ denotes the time delay at the nth IRS element; $x(t - t_n - \tau_l(t)) \approx x(t - \tau_l(t))$ holds since $t_n \leq 1/f_c \ll 1/B$; and $\theta_n = 2\pi - 2\pi f_c t_n \in [0, 2\pi)$ denotes the phase shift at the nth IRS element. After undergoing the IRS-user channel, the receive passband signal at the user via the reflection of the nth IRS element is expressed as

$$y_{r,n}(t) = \text{Re}\left\{ \left[\sum_{l'=0}^{L_2-1} h_{nl'}^{r_2} \beta_n e^{j\theta_n} \sum_{l=0}^{L_1-1} h_{nl}^{r_1} x(t - \tau_l(t) - \tau_{l'}(t)) \right] e^{j2\pi f_c t} \right\}. \tag{4.25}$$

Thus, the baseband signal model of (4.25) is given by

$$y_n(t) = (h_n^{r_2} \circledast \beta_n e^{j\theta_n} \circledast h_n^{r_1}) x(t) \overset{\Delta}{=} h_n^r x(t). \tag{4.26}$$

In (4.26), $h_n^r = h_n^{r_2} \circledast \beta_n e^{j\theta_n} \circledast h_n^{r_1} = \beta_n e^{j\theta_n}(h_n^{r_2} \circledast h_n^{r_1}) \in \mathbb{C}^{L^r \times 1}$ denotes the cascaded BS-IRS-user channel via the nth IRS element, with $L^r = L_1 + L_2 - 1$, where $\circledast$ denotes the convolution operation. It is worth noting that the concatenated channel is the convolution of the BS-IRS multipath channel, the (single-tap) IRS reflection coefficient, and the IRS-user multipath channel.

Here, the total bandwidth of the considered OFDM system is equally divided into $Q \geq 1$ orthogonal sub-bands. For the purpose of exposition, let $\tilde{g}_n^r = [(h_n^{r_2} \circledast h_n^{r_1})^T, 0, \ldots, 0]^T \in \mathbb{C}^{Q \times 1}$ denote the zero-padded convolved BS-IRS and IRS-user time-domain channel via the nth IRS element, and define $\tilde{G}^r = [\tilde{g}_1^r, \ldots, \tilde{g}_N^r] \in \mathbb{C}^{Q \times N}$. Thus, the cascaded BS-IRS-user channel can be given by $\tilde{h}^r = \tilde{G}^r \theta$, where $\theta = [\beta_1 e^{j\theta_1}, \ldots, \beta_N e^{j\theta_N}]^T$. Let $\tilde{h}^d = [h^{d^T}, 0, \ldots, 0]^T \in \mathbb{C}^{Q \times 1}$ denote the zero-padded time-domain channel from the BS to the user. Then, the superposed effective channel impulse response (CIR) is expressed as $\tilde{h} = \tilde{h}^d + \tilde{h}^r = \tilde{h}^d + \tilde{G}^r \theta$. Note that the number of delay taps in the effective BS-user channel is $L = \max(L_0, L_1 + L_2 - 1)$. By ignoring the rate loss due to the cyclic prefix insertion, the achievable rate of the IRS-aided OFDM system in bps/Hz is given by

$$r = \frac{1}{Q} \sum_{q=1}^{Q} \log_2 \left(1 + \frac{p_q |\boldsymbol{f}_q^H \tilde{\boldsymbol{h}}^d + \boldsymbol{f}_q^H \tilde{\boldsymbol{G}}^r \boldsymbol{\theta}|^2}{\bar{\sigma}^2} \right), \tag{4.27}$$

where p_q denotes the transmit power allocated to sub-carrier q with $\sum_{q=1}^{Q} p_q \leq P_t$, and $\bar{\sigma}^2$ denotes the average receiver noise power at each sub-carrier.

To maximize the achievable rate in (4.27), the IRS phase shift in $\boldsymbol{\theta}$ needs to cater to the frequency-varying channel at different subcarriers, or equivalently, the time-domain channels at different delayed taps. In addition, $\boldsymbol{\theta}$ needs to be jointly optimized with the transmit power allocation on the Q subcarriers. Thus, compared to (P1), the problem is more difficult to solve. To handle this problem, Yang et al. (2020) proposed an efficient successive convex approximation (SCA) based algorithm by approximating the objective function in (4.27) as its concave lower bound. Zhang and Zhang (2020) proposed an efficient AO-based algorithm by extending the algorithm in the narrowband MIMO case and utilizing convex relaxation techniques.

4.3.5 Passive Beamforming with Discrete Reflection Amplitude and Phase Shift

The continuous phase shifts of the IRS with the maximum reflection magnitude are mainly considered in the previous subsections, only discrete phase shifts and/or discrete reflection magnitudes of IRSs can be designed practically, leading to degrading communication performance compared to the ideal case of continuous phase shift with flexible reflection amplitudes (Wu and Zhang 2019a; Dong et al. 2021; Zhang and Zhang 2020; Najafi et al. 2020). Furthermore, these constraints also complicate the passive beamforming design and make the corresponding optimization problem more difficult to solve than that with continuous phase shifts. Although some optimization techniques, such as branch and bound (BB) methods, can be adopted to obtain optimal solutions with reduced average complexity, their computational complexity is still exponential in N.

To tackle this problem, one heuristic approach is to relax such constraints and solve the problem in (P1), and then the resulting solution is quantized to the closest value in the corresponding discrete set. In general, this approach can significantly reduce computational complexity, but it may suffer from various performance penalties due to quantization/round-off errors. To explore the performance loss of IRS employing discrete phase shifts compared to that employing continuous phase shifts, a modification of (P1) is studied by ignoring the direct link between the BS and the user. Therefore, the user receive power maximization problem with discrete phase shifts can be expressed as

$$\text{(P1-DP)}: \max_{\theta} \left| \sum_{n=1}^{N} h_{r,n} g_n e^{j\theta_n} \right|^2 \tag{4.28}$$

$$\text{s.t.} \quad \theta_n \in \mathcal{F}_\theta, n = 1, \dots, N, \tag{4.29}$$

where $\mathcal{F}_\theta = \{0, \Delta\theta, \dots, (K_\theta - 1)\Delta\theta\}$ and $\Delta\theta = 2\pi/K_\theta$.

To facilitate our analysis, we solve (P1-DP) with a quantization method, first obtain the optimal continuous phase shifts, and then quantize them independently to the closest value in $\mathcal{F}_\theta$. The corresponding objective value is denoted by $P_r(b_\theta)$. In this case, $P_r(b_\theta)$ serves as a lower bound on the optimal value of (P1). Hence, when $b_\theta \to \infty$, the performance loss for IRS with b_θ phase shifters compared to the ideal case with continuous phase shifts can be characterized by a ratio, i.e., $\eta(b_\theta) \triangleq P_r(b_\theta)/P_r(\infty)$. As shown in (Wu and Zhang 2019b), this performance loss ratio is given by

$$\eta(b_\theta) = \left(\frac{2^{b_\theta}}{\pi} \sin\left(\frac{\pi}{2^{b_\theta}}\right) \right)^2. \tag{4.30}$$

Accordingly, the power ratio $\eta(b_\theta)$ depends only on the number 2^{b_θ} of discrete phase shifts, but not on the N. Therefore, even with discrete phase shifts, the same asymptotic receive power scaling order $\mathcal{O}(N^2)$ can still be achieved with a practical IRS as in the continuous phase case given in (4.10).

4.3.6 Other Related Works and Future Directions

In the last subsection, we overview other related works on IRS reflection design and point out promising directions for future work.

Besides passive beamforming design for IRSs, there are various new architectures of intelligent surfaces, such as active IRS (Khoshafa et al. 2021), omni-surface (Zhang et al. 2022b), and beyond diagonal IRS (Li et al. 2022). The corresponding beamforming design deserves further investigation. For instance, different from the passive counterpart, an active IRS is likely to be an active reflector that directly reflects the incoming signals with power amplification at the electromagnetic level. It is worth noting that active IRSs still take advantage of the IRS without the need for complex and power-hungry RF chain components (Khoshafa et al. 2021). However, the active IRS also accidentally amplifies the noise at the IRS, leading to correlated noises at the receiver, which is generally ignored for passive IRS-aided communication.

The above passive beamforming methods are mainly designed based on the assumption of perfect CSI. However, due to channel estimation errors and feedback delays, only imperfect CSI can be obtained in practical wireless systems. Motivated by this, the robust design with imperfect CSI is significantly important in IRS-aided wireless communication. One of the most popular techniques is S-procedure (Zhou et al. 2020), which is able to transform the worst-case objective function or constraints into a tractable form with linear matrix inequalities. In addition to SINR/rate maximization, another study aimed to investigate other performance metrics of IRS-assisted systems, such as outage probability and average bit error rate (BER). In Zhang et al. (2019), the outage probability of the multi-IRS-assisted system was analyzed, where it was shown that the optimal outage probability in the high SNR regime is inversely proportional to the SNR.

Although this section focuses on single-cell systems with one IRS, it is worth extending the results to more general systems with multiple BSs and/or multiple IRSs. In this case, the reflection coefficients of multiple IRSs and the transmit beamforming vectors of multiple BSs need to be jointly designed, thereby reducing power consumption and intra/inter-cell interference. In addition to discrete reflection amplitudes/phase shifts, the IRS-aided system under other hardware constraints should be further considered, such as coupled reflection amplitudes and phase shifts. For example, it is shown in Abeywickrama et al. (2020) that the asymptotic power scaling order under the ideal phase-shift model still holds for the practical case with phase-shift dependent non-uniform IRS reflection amplitude. On the other hand, it is also important to develop more advanced and computationally efficient algorithms, such as machine learning-based methods for IRS reflection design, especially for practically large IRSs (Feng et al. 2020).

4.4 IRS Deployment

In the previous sections, we have shown the effectiveness of the IRS to improve the communication performance of nearby users by optimizing signal reflection, where the IRS is assumed to be deployed at a given location. In fact, there are various IRS deployment strategies, such as near the BS/AP/user, or dividing them into smaller IRSs and deploying them in a distributed manner. On the other hand, several practical factors of deployment implementation also need to be taken into account, such as deployment/operating costs, user requirements/distribution, space constraints, and propagation environment. Note that due to the product-distance path loss of IRSs, there are several fundamental differences between the deployment strategy for IRS and that for other active communication nodes, including APs/BSs or relays. Hence, in this section, we investigate the new IRS deployment problem to draw useful insights for practical design, including link-level and network-level deployments.

4.4.1 IRS Deployment Optimization at the Link Level

In this subsection, we consider the IRS-assisted communication link between a user and a BS and discuss three key aspects of IRS deployment to enhance its performance at the link level.

4.4.1.1 Optimal Deployment of Single IRS

Firstly, we consider the simplest case where all the N IRS elements form one single IRS. For the purpose of exposition, it is assumed that both the BS and the user are equipped with a single antenna. To illustrate the effect of the double path loss between the user and the BS, a simplified 2D system setup is considered, as shown in Fig. 4.3(a), where the BS and the user are located on a straight line with a horizontal distance of D m, and the IRS can be flexibly deployed on a line above the BS and user by H m with $H \ll D$. In Fig. 4.3(a), d denotes the horizontal distance between the IRS and the user. To focus our analysis on the effect of IRS location, we consider the case where the direct user-BS link is blocked, while the other two links to and from the IRS follow a free-space LoS channel model. Accordingly, the received SNR with the optimal IRS passive beamforming can be expressed as

$$\rho_S = \frac{P\beta_0^2 N^2}{(d^2 + H^2)((D-d)^2 + H^2)\sigma^2}, \tag{4.31}$$

where β_0 denotes the path loss at reference distance of 1 meter (m), P denotes the transmit power, and σ^2 denotes the average noise power. It can be easily shown from (4.31) that ρ_S is maximized when the distance of the user-IRS or IRS-BS link is minimized, i.e., $d = 0$ or $d = D$. In this case, the maximum receive SNR can be given by $\rho_S^\star = \frac{P\beta_0^2 N^2}{H^2(D^2+H^2)\sigma^2} \approx \frac{P\beta_0^2 N^2}{H^2 D^2 \sigma^2}$ as $H \ll D$.

4.4.1.2 Single IRS versus Multiple Cooperative IRSs

Besides forming one single IRS with all elements, the IRS elements can form multiple smaller-size IRSs, as illustrated in Fig. 4.3(b). Such deployment strategy

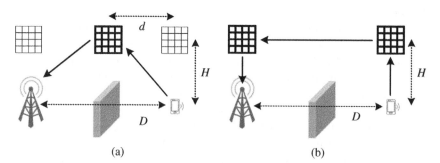

(a)   (b)

Figure 4.3 IRS deployment in a point-to-point communication system. (a) Single IRS. (b) Two cooperative IRSs.

has both pros and cons as compared to the single-IRS deployment. Specifically, an increase in the number of IRSs in the user-BS link leads to more inter-IRS reflections, resulting in higher path loss; while multiple IRSs offer new opportunities to achieve cooperative passive beamforming. In the following, we will discuss whether splitting the elements into multiple collaborative IRSs is better than combining them into a single IRS.

In Han et al. (2020), two IRS are deployed directly above the user and the BS, each with $N/2$ reflective elements, as shown in Fig. 4.3(b). All links except the user-IRS 1-IRS 2-BS link are assumed to be blocked and all channels involved follow the free-space LoS model. It is shown in Han et al. (2020) that when the distance between these two IRSs satisfies the far-field condition, the inter-IRS channel is of rank one, a passive beamforming gain of order $\mathcal{O}((N/2)^4)$ can be achieved by properly aligning the passive beamforming directions of these two IRSs. As such, the receive SNR at the BS can be expressed as

$$\rho_{\mathrm{D}} = \frac{P\beta_0^3 N^4}{16H^4 D^2 \sigma^2}. \tag{4.32}$$

Based on (4.32), under the same BS/user setup as in Fig. 4.3(b), the receive SNR with two cooperative IRSs is larger than that with a single IRS if the total number of reflecting elements is larger than a threshold. i.e.,

$$N > \frac{4H}{\sqrt{\beta_0}}. \tag{4.33}$$

This result is expected since the beamforming gain of two cooperative IRSs increases significantly as N compared to a single IRS, while the additional path loss in the two IRS cases is fixed by a given H.

4.4.1.3 LoS versus Non-LoS (NLoS)

As shown above, a major design target for IRS deployments with single-antenna BS and users is to minimize their path loss established by IRSs. In addition to the product distance, the path loss is also closely dependent on the path loss exponent, which characterizes the average channel attenuation over distance. As such, the IRS actually needs to be placed in LoS conditions with the BS/user to obtain the minimum path loss exponent, which is significantly important in mmWave/THz communications (Tan et al. 2018).

On the other hand, however, LoS channels are usually of low rank, which limits the achievable spatial multiplexing gain of IRS-assisted multiuser systems with the multi-stream transmission, resulting in their low-capacity high-SNR mechanisms (Zhang and Zhang 2020). Thus, deploying IRS at locations with comparable NLoS (random) and LoS (deterministic) channel components can be beneficial in improving the spatial power distribution of the MIMO channels, in terms of rank and condition number. As such, to maximize IRS-aided system capacity, there exists a fundamental trade-off for selecting IRS locations to achieve balanced LoS vs. NLoS propagation and passive beamforming vs. spatial multiplexing gain.

4.4.2 IRS Deployment at the Network Level: Distributed or Centralized?

In this subsection, we discuss more general IRS deployments for multiuser networks, where a BS and K users are sufficiently far apart. In this case, there are two general deployment strategies in the network: Distributed deployment, where the IRS elements form multiple distributed IRSs, each located near a user, as shown in Fig. 4.4(a); or centralized deployment, where all IRS elements form a large IRS located near the BS, as shown in Fig. 4.4(b). For the single-user case, i.e., $K = 1$, these two deployment strategies are equivalent. For the multiuser case with $K > 1$, the user-BS channels are different for these two deployment strategies. Specifically, N IRS elements can serve all users under centralized deployment; while under distributed deployment, each user is only served by its nearest IRS since the signal reflected by unassociated IRSs is too weak due to greater path loss. In the following, we compare the achievable rate of K-user multiple access channel (MAC) in uplink communication with the above two IRS deployment strategies. For ease of exposition, we consider a simple setup where both the BS and users are equipped with a single antenna, and the direct links between the BS and users are blocked.

Furthermore, for a fair comparison between these two deployment strategies, we assume that their respective user-IRS-BS channels are similar to each other (Zhang and Zhang 2021), as shown in Fig. 4.4. First, we consider time division multiple access (TDMA), where K users communicate with the BS in orthogonal time slots. For ease of analysis, we assume that N IRS elements are equally divided into K IRSs, each located near a different user in a distributed deployment, which

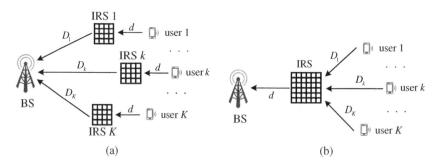

(a) (b)

Figure 4.4 IRS deployment in a multiuser communication network. (a) Distributed IRS deployment. (b) Centralized IRS deployment.

leads to $\mathcal{O}((N/K)^2)$ passive beamforming gain per user. Differently, under centralized deployment, each user can share a passive beamforming gain of $\mathcal{O}(N^2)$ with all N IRS elements. As a result, it can be seen that under the twin-channel condition in Zhang and Zhang (2021), the achievable rate region of centralized deployment includes that of distributed deployment. In particular, by assuming equal time allocation for users and free-space LoS models for all channels, it can be shown that the total rate achievable by K users under centralized/distributed deployment is given by $R_{\text{cen}} = \frac{1}{K} \sum_{k=1}^{K} \log_2\left(1 + \frac{\bar{P}\beta_0^2 N^2}{d^2 D_k^2 \bar{\sigma}^2}\right)$ and $R_{\text{dis}} = \frac{1}{K} \sum_{k=1}^{K} \log_2\left(1 + \frac{\bar{P}\beta_0^2 (N/K)^2}{d^2 D_k^2 \bar{\sigma}^2}\right)$, respectively, where $\bar{P}$ and $\bar{\sigma}^2$ denote the transmit power at each user and noise power at the BS receiver, respectively. Thus, when $N \to \infty$, the asymptotic rate difference can be expressed as

$$R_{\text{cen}} - R_{\text{dis}} \to 2\log_2 K. \tag{4.34}$$

Therefore, as K increases, the centralized deployment has a more pronounced rate gain than the distributed deployment, which is intuitive because each user is assigned a smaller number of IRS elements for distributed deployment.

Furthermore, Zhang and Zhang (2021) demonstrate that under twin-channel conditions, centralized deployment outperforms distributed deployment in terms of multiuser capacity/rate region for non-orthogonal multiple access (NOMA) and frequency division cases access (FDMA), respectively. This is because the larger number of reflective elements in a centralized IRS provides greater flexibility to trade-off users' individual achievable rates compared to a distributed IRS where each assists only one user. While centralized IRS deployments are generally more favorable than the distributed counterparts in terms of achievable rates, it is worth noting that there may be other practical factors to consider when implementing them. First, distributed deployment requires more IRSs and thus more backhaul links between BSs and IRS controllers to exchange information, resulting in more overhead for the network. Second, due to space constraints, it may not always be feasible to deploy a large centralized IRS near the BS; in contrast, it is usually more flexible to deploy multiple distributed IRSs near each user. Third, the channel statistics under the two deployment strategies may be quite different, in terms of LoS probability, NLoS fading distribution, channel correlation, etc., which may lead to different comparison results under different scenarios.

4.4.3 Other Related Work and Future Direction

In addition to the above discussion revealing some important considerations for deploying IRS in practical systems, this subsection discusses other related work, as well as some promising directions for future work.

First, it is worth extending the results in Han et al. (2020) and Zhang and Zhang (2021) discussed above to more general system setups, such as multiple cooperative IRSs, MIMO channel models, hybrid distributed and centralized IRS deployments, etc. Furthermore, it is also important to study the deployment of IRS in general multi-cell networks with inter-cell interference. For instance, Pan et al. (2020) considered a simplified setup where one IRS serves two adjacent cells and shows that the IRS should be deployed at the intersection of their edges to maximize the benefit of both cells. However, if multiple IRSs are available to serve multiple cells, the optimal IRS deployment/association strategy generally remains unknown.

Moreover, from the perspective of designing a multi-cell wireless network, the following interesting new problem arises: what is the fundamental capacity limit of IRS-aided wireless networks? Although ultra-dense network is undoubtedly the most significant driver for the astounding capacity increase in 5G networks, it is known that for wireless networks with active BSs only, increasing the spatial density of BSs beyond a certain threshold will reduce the network capacity asymptotically to zero, due to the increasingly more severe interference (Andrews et al. 2016). However, this pessimistic result is not applicable to the IRS-aided wireless network with hybrid active BSs and passive IRSs, as shown in Fig. 4.5.

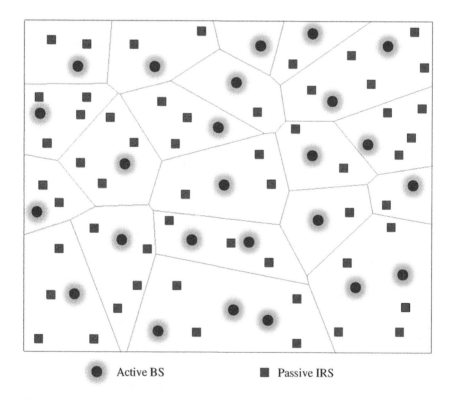

Active BS   Passive IRS

Figure 4.5 Illustration of a hybrid wireless network with active BSs and passive IRSs.

Since IRSs are passive and thus do not increase the network interference level, while they help enhance the user performance in their locally covered areas by intelligent reflection, it is expected that the network capacity will be significantly improved by adding IRSs as compared to the conventional network with active BSs only. It is also plausible that increasing the spatial density of passive IRSs may not lead to network capacity degradation due to increased interference in the conventional network. This thus opens a new avenue for investigating the optimal wireless network architecture and its new capacity limit in future work.

4.5 Conclusion

In this chapter, the main challenges of the new intelligent surface technology and the corresponding solutions are provided, including channel estimation, passive beamforming design, and deployment strategy.

Bibliography

Abeywickrama, S., Zhang, R., Wu, Q., and Yuen, C. (2020). Intelligent reflecting surface: practical phase shift model and beamforming optimization. *IEEE Transactions on Communications* 68 (9): 5849–5863.

Alexandropoulos, G.C. and Vlachos, E. (2020). A hardware architecture for reconfigurable intelligent surfaces with minimal active elements for explicit channel estimation. *ICASSP 2020 - 2020 IEEE International Conference on Acoustics, Speech and Signal Processing (ICASSP)*, 9175–9179. https://doi.org/10.1109/ICASSP40776.2020.9053976.

Andrews, J.G., Zhang, X., Durgin, G.D., and Gupta, A.K. (2016). Are we approaching the fundamental limits of wireless network densification? *IEEE Communications Magazine* 54 (10): 184–190.

Dong, L., Wang, H.-M., and Xiao, H. (2021). Secure cognitive radio communication via intelligent reflecting surface. *IEEE Transactions on Communications* 69 (7): 4678–4690.

Feng, K., Wang, Q., Li, X., and Wen, C.-K. (2020). Deep reinforcement learning based intelligent reflecting surface optimization for MISO communication systems. *IEEE Wireless Communications Letters* 9 (5): 745–749.

Gao, S., Dong, P., Pan, Z., and Li, G.Y. (2021). Deep multi-stage CSI acquisition for reconfigurable intelligent surface aided MIMO systems. *IEEE Communications Letters* 25 (6): 2024–2028. https://doi.org/10.1109/LCOMM.2021.3063464.

Han, Y., Zhang, S., Duan, L., and Zhang, R. (2020). Cooperative double-IRS aided communication: beamforming design and power scaling. *IEEE Wireless Communications Letters* 9 (8): 1206–1210.

Hu, X., Zhang, R., and Zhong, C. (2022). Semi-passive elements assisted channel estimation for intelligent reflecting surface-aided communications. *IEEE Transactions on Wireless Communications* 21 (2): 1132–1142. https://doi.org/10.1109/TWC.2021.3102446.

Jiang, T., Cheng, H.V., and Yu, W. (2021). Learning to reflect and to beamform for intelligent reflecting surface with implicit channel estimation. *IEEE Journal on Selected Areas in Communications* 39 (7): 1931–1945.

Khoshafa, M.H., Ngatched, T.M.N., Ahmed, M.H., and Ndjiongue, A.R. (2021). Active reconfigurable intelligent surfaces-aided wireless communication system. *IEEE Communications Letters* 25 (11): 3699–3703.

Li, H., Shen, S., and Clerckx, B. (2022). Beyond diagonal reconfigurable intelligent surfaces: from transmitting and reflecting modes to single-, group-, and fully-connected architectures. *IEEE Transactions on Wireless Communications* 22 (4): 2311–2324.

Liu, M., Li, X., Ning, B. et al. (2023). Deep learning-based channel estimation for double-RIS aided massive MIMO system. *IEEE Wireless Communications Letters* 12 (1): 70–74. https://doi.org/10.1109/LWC.2022.3217294.

Mishra, D. and Johansson, H. (2019). Channel estimation and low-complexity beamforming design for passive intelligent surface assisted MISO wireless energy transfer. *ICASSP 2019 - 2019 IEEE International Conference on Acoustics, Speech and Signal Processing (ICASSP)*, 4659–4663. https://doi.org/10.1109/ICASSP.2019.8683663.

Najafi, M., Jamali, V., Schober, R., and Poor, H.V. (2020). Physics-based modeling and scalable optimization of large intelligent reflecting surfaces. *IEEE Transactions on Communications* 69 (4): 2673–2691.

Pan, C., Ren, H., Wang, K. et al. (2020). Multicell MIMO communications relying on intelligent reflecting surfaces. *IEEE Transactions on Wireless Communications* 19 (8): 5218–5233.

So, A.M.-C., Zhang, J., and Ye, Y. (2007). On approximating complex quadratic optimization problems via semidefinite programming relaxations. *Mathematical Programming* 110 (1): 93–110.

Taha, A., Alrabeiah, M., and Alkhateeb, A. (2021). Enabling large intelligent surfaces with compressive sensing and deep learning. *IEEE Access* 9: 44304–44321. https://doi.org/10.1109/ACCESS.2021.3064073.

Tan, X., Sun, Z., Koutsonikolas, D., and Jornet, J.M. (2018). Enabling indoor mobile millimeter-wave networks based on smart reflect-arrays. *IEEE INFOCOM 2018-IEEE Conference on Computer Communications*, 270–278. IEEE.

Tse, D. and Viswanath, P. (2005). *Fundamentals of Wireless Communication*. Cambridge University Press.

Wang, Z., Liu, L., and Cui, S. (2020). Channel estimation for intelligent reflecting surface assisted multiuser communications: framework, algorithms, and analysis. *IEEE Transactions on Wireless Communications* 19 (10): 6607–6620. https://doi.org/10.1109/TWC.2020.3004330.

Wu, Q. and Zhang, R. (2019a). Intelligent reflecting surface enhanced wireless network via joint active and passive beamforming. *IEEE Transactions on Wireless Communications* 18 (11): 5394–5409. https://doi.org/10.1109/TWC.2019.2936025.

Wu, Q. and Zhang, R. (2019b). Beamforming optimization for wireless network aided by intelligent reflecting surface with discrete phase shifts. *IEEE Transactions on Communications* 68 (3): 1838–1851.

Wu, Q., Zhang, S., Zheng, B. et al. (2021). Intelligent reflecting surface-aided wireless communications: a tutorial. *IEEE Transactions on Communications* 69 (5): 3313–3351.

Yang, Y., Zheng, B., Zhang, S., and Zhang, R. (2020). Intelligent reflecting surface meets OFDM: protocol design and rate maximization. *IEEE Transactions on Communications* 68 (7): 4522–4535.

Yao, J., Wu, T., Zhang, Q., and Qin, J. (2020). Proactive monitoring via passive reflection using intelligent reflecting surface. *IEEE Communications Letters* 24 (9): 1909–1913.

Zhang, S. and Zhang, R. (2020). Capacity characterization for intelligent reflecting surface aided MIMO communication. *IEEE Journal on Selected Areas in Communications* 38 (8): 1823–1838.

Zhang, S. and Zhang, R. (2021). Intelligent reflecting surface aided multi-user communication: capacity region and deployment strategy. *IEEE Transactions on Communications* 69 (9): 5790–5806.

Zhang, Z., Cui, Y., Yang, F., and Ding, L. (2019). Analysis and optimization of outage probability in multi-intelligent reflecting surface-assisted systems. arXiv preprint arXiv:1909.02193.

Zhang, L., Wang, Q., Wang, H. et al. (2022a). Manifold optimization based multi-user rate maximization aided by intelligent reflecting surface. arXiv preprint arXiv:2201.09031.

Zhang, Y., Di, B., Zhang, H. et al. (2022b). Dual codebook design for intelligent OMNI-surface aided communications. *IEEE Transactions on Wireless Communications* 21 (11): 9232–9245.

Zheng, B. and Zhang, R. (2020). Intelligent reflecting surface-enhanced OFDM: channel estimation and reflection optimization. *IEEE Wireless Communications Letters* 9 (4): 518–522. https://doi.org/10.1109/LWC.2019.2961357.

Zheng, B., You, C., and Zhang, R. (2020). Intelligent reflecting surface assisted multi-user OFDMA: channel estimation and training design. *IEEE Transactions on Wireless Communications* 19 (12): 8315–8329. https://doi.org/10.1109/TWC.2020 .3021434.

Zhou, G., Pan, C., Ren, H. et al. (2020). Robust beamforming design for intelligent reflecting surface aided MISO communication systems. *IEEE Wireless Communications Letters* 9 (10): 1658–1662.

Part II

IRS for 6G Wireless Systems

5

Overview of IRS for 6G and Industry Advance

Ruiqi (Richie) Liu[1,2], Konstantinos D. Katsanos[3], Qingqing Wu[4], and George C. Alexandropoulos[3,5]

[1] State Key Laboratory of Mobile Network and Mobile Multimedia Technology, Shenzhen, China
[2] Wireless and Computing Research Institute, ZTE Corporation, Beijing, China
[3] Department of Informatics and Telecommunications, National and Kapodistrian University of Athens, Athens, Greece
[4] Department of Electronic Engineering, Shanghai Jiaotong University, Shanghai, China
[5] Technology Innovation Institute, Masdar City, Abu Dhabi, United Arab Emirates

5.1 IRS for 6G

5.1.1 Potential Use Cases

Traditionally, the main goal of the design of wireless systems is to optimize spectral efficiency, while approaching the theoretical Shannon capacity, under the assumption that wireless signal propagation is uncontrollable, in the sense that it is determined by the usually uncontrollable geometry and physical features of the wireless medium. The advent of 5G introduced several other wireless performance indicators, e.g., energy efficiency, localization, latency, and reliability, whose list is envisioned to be largely enriched with various others, as per the latest requirements for 6G networks (Samsung 2020), e.g., sensing, electromagnetic field exposure, and secrecy. The common assumption for meeting the currently demanding values of those performance objectives has been that the wireless propagation medium cannot be artificially manipulated, thus, advances in signal transmission and reception are necessary.

The recent advances in reconfigurable metamaterials (Alexandropoulos et al. 2020) have been recognized, in both academia and industry, as a means to overcome the limitation of uncontrollable over-the-air (OTA) propagation of information-bearing electromagnetic signals, paving the way to the new era of smart radio environments (SREs) (Huang et al. 2019; Di Renzo et al. 2019). Those programmable wireless environments, which are actually incentivized

Intelligent Surfaces Empowered 6G Wireless Network, First Edition.
Edited by Qingqing Wu, Trung Q. Duong, Derrick Wing Kwan Ng, Robert Schober, and Rui Zhang.

by the reconfigurable intelligent surface (RIS) technology (Jian et al. 2022), are expected to offer extra degrees-of-freedom to the designers of wireless systems and applications, enabling the joint optimization of signal transmission, reception, and OTA propagation. The increased capabilities of RISs and the numerous advantages of this technology in comparison with common techniques, such as deploying relays, are actually bringing up new directions in the evolution of the wireless network infrastructure and architecture.

Various visions of the forthcoming 6G networks point toward low-power, high-throughput, low-latency, and flexible connect-and-compute technologies to support future innovative services and the corresponding use cases (Strinati et al. 2021a,2021b). We next divide the use cases, where the RIS consideration can play a significant role, into two major categories, namely indoor and outdoor applications (Huang et al. 2020). Within each category, aiming to capture all possible ways of profiting from metasurface-based SREs in future wireless systems, a comprehensive list of RIS-enabled and RIS-boosted use cases is presented.

5.1.1.1 Indoor Use Cases

Due to massive multipath propagation in indoor environments as well as signal blocking, which usually originate from the presence of multiple scatters as well as obstacles (e.g., walls/furniture), there are remarkable issues in achieving seamless wireless communications. In addition, this behavior is more problematic when the demands for wireless resources (e.g., high throughput coverage) increase, which happens when more users try to connect to the desirable access points. However, by deploying RISs at specific points inside a room, or generally an indoor area, such as coating the walls of the room, it has been shown to offer significant enhancements in coverage (Alexandropoulos et al. 2021a), without sacrificing the quality of experience of intended users. In addition, apart from the enhanced coverage in buildings, RISs have attracted much attention due to their benefits on accurate indoor positioning and particularly their increased potential for localization and sensing (Alexandropoulos et al. 2022a). Commonly, the conventional Global Positioning System (GPS) fails not only to provide accurate positioning, but to entirely work as well, due to the extreme signal attenuation in indoor facilities. Nevertheless, accurate spatial resolution can be achieved by deploying large RISs, offering even continuous apertures for constant and increased monitoring (Gong et al. 2021). At last, RISs are also capable of enhancing various network objectives, such as energy/spectral efficiency, and many other common performance metrics intended for wireless communications, guaranteeing enhanced quality of service and providing further degrees of freedom in meeting future demands according to the wireless standards and expectations.

5.1.1.2 Outdoor Use Cases

RISs constitute a promising technology for various outdoor scenarios since they are capable of extending the coverage from outdoor base stations (BSs) to mobile user equipment (UE). In contrast to other available possible solutions, such as the relaying paradigm using various protocols (e.g., amplify-and-forward or decode-and-forward), RISs can provide targeted metrics with smaller deployment cost, hardware foorptint, and power consumption (Huang et al. 2019). Moreover, there are cases where relays and RISs offer great advantages when they are combined to act in a synergistic manner. For instance, according to studies, the achievable rates can be remarkably enhanced when RISs and decode-and-forward relays are working complementary to each other, rather than being considered as competing technologies (Yildirim et al. 2021; Abdullah et al. 2022). Therefore, it is important to highlight the various use cases that RISs can offer to outdoor environments.

To begin with, properly located RISs can extend the coverage in places where the BSs' direct links to the users are severely blocked by obstacles or do not even exist. To this end, RISs controlled mainly by the corresponding BSs can offer significant advantages at low cost of both deployment and operation, since metasurfaces, which are the main technology behind RISs, possess such properties. Furthermore, energy-efficient beamforming is a main characteristic of RISs, because their main capability is to recycle ambient electromagnetic waves and focus them toward intended users by the effective tuning of their unit elements. As a result, the transmitted signals impinged on an RIS can be forwarded to desired locations via passive beamforming that compensates for the signal attenuation from the corresponding BS. Additionally, the co-channel interference from neighboring BSs to the desired user(s) can be suppressed by dedicated passive beamforming in RIS-enabled SREs. Also, RISs have the ability to offer wireless power transfer by collecting ambient electromagnetic waves and directing them to low-power IoT devices and sensors, enabling simultaneous wireless information transmission. Other possible RIS-oriented applications lie in the field of physical-layer security, where RISs can be deployed to cancel out reflections of the BS signals to eavesdroppers' directions, providing boosted performance in terms of the achievable secrecy rates or potential advantages on secret key generation. Consequently, there are many envisioned RIS-enabled applications, in a cost-effective adaptation of the wireless environment.

All above use cases, rely mainly on the assumption that RISs are placed on immovable objects (e.g., buildings). Based on this usage model, RISs can be configured to steer the non-line-of-sight (non-LOS) links into low coverage areas, generally due to obstacles. In addition, RIS configuration can be such that it permits to redirect undesirable signals for interference mitigation, or even construct jamming signals toward potential malicious nodes. According to this stationary way of deploying RISs, several examples arise harmonizing the presented indoor and outdoor use cases. For instance, smart homes with RISs placed in the interior walls, could provide reliable coverage whenever direct links fail to serve the end-to-end users, and enhance spectral efficiency via constructive interference. Extending this paradigm, smart buildings, factories, or hospitals could benefit from the RIS technology in various aspects, leading in this way to the interconnection of different facilities to finally meet the requirements and purpose for the envisioned smart cities (Kisseleff et al. 2020). On the other hand, apart from the above stationary approach, RIS-empowered systems can be extended to mobility applications as well, because LOS connectivity is highly probable to happen in such cases instead of fixing the RISs' positions. One typical example is to assist communications by attaching RISs to ground vehicles of unmanned aerial vehicles, by improving the degrees of freedom associated with these two combined technologies. To address the challenges that originate from permanent synchronization requirements, there are quite many proposed mobility types for relevant RIS applications, such as stochastic, steerable, predictable, and hybrid mobility, that are still open problems in the current literature. However, all above use cases for both stationary and mobility models, can offer considerable improvements in future wireless communications systems, especially when combined.

Last, but not least, omnidirectional RISs (Zhang et al. 2022) can be deployed to enable various outdoor-to-indoor or indoor-to-outdoor applications, offering improved signal coverage in scenarios where it was otherwise impossible, due to the inevitable attenuation introduced by the material lying in the indoor/outdoor boundary. An indicative application is the simultaneous indoor and outdoor localization framework of He et al. (2022a), where a simultaneously transmitting (refracting) and reflecting RIS (Liu et al. 2021b) has been optimized to enable the localization of indoor and outdoor users.

5.1.2 Deployment Scenarios

In this section, various deployment scenarios for RIS-enabled SREs are presented. The scenarios are firstly related to network performance objectives, where RISs are deployed to offer improved services for the end users, by proper control of the wireless medium, in places where the coverage from BSs is either blocked by obstacles or even totally absent (Strinati et al. 2021b). Apart from the common objectives seeking notable improvement in beyond 5G wireless communications, scenarios for sustainability and secrecy are also presented to further highlight the advantages of RISs (Alexandropoulos et al. 2022a). Then, localization and sensing scenarios, for which RISs are able to offer significant advantages especially for indoor environments, are also described.

Focusing on the downlink case as an example (the uplink case can be readily extended following the same approach), an indicative taxonomy for RIS-empowered connectivity and reliability scenarios follows.

1. *Connectivity and reliability boosted by a single RIS*: In this scenario, illustrated in Fig. 5.1, it is assumed that the single- or multi-antenna UEs lie in an area where it is desired to enhance its connectivity and reliability. By placing an RIS either close to this area or the multi-antenna BS, connectivity is further boosted via a single RIS. The phase configuration/profile of the RIS can be optimized via a dedicated controller, who interacts with the BS, for desired connectivity and reliability levels. Also, it is noted that in this considered case there exist direct links between the BS and the UEs.
2. *Connectivity and reliability boosted by multiple individually controlled RISs*: In this scenario, multiple BSs aim at boosting connectivity and reliability to their respective UEs, as depicted in Fig. 5.2. The deployed RISs are assigned to BS-UE pairs, and each pair is capable to control and optimize the phase profile of its assigned individual RIS. In this case, the multiple RISs might result in multiplicative interference that needs to be carefully handled.
3. *Connection reliability enabled by multiple RISs*: In this scenario, the direct link(s) between the multi-antenna BS and the single- or multi-antenna UE(s) is (are) blocked due to obstacles, and connectivity is enabled only via the single

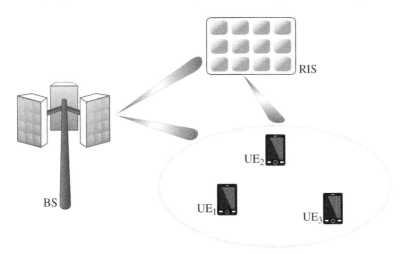

Figure 5.1 Connectivity and reliability boosted by a single RIS.

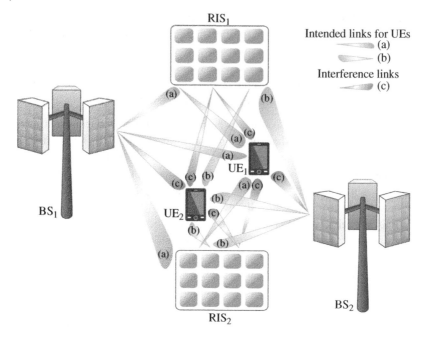

Figure 5.2 RIS-empowered downlink communication of two BS-UE pairs, where each RIS can be controlled individually by each pair.

RIS, as shown in Fig. 5.3. Similar to scenario 1, described above, the phase profile(s) of the RIS(s) can be optimized for desired reliability levels.

4. *Connectivity and reliability boosted by a single multi-tenant RIS*: For this scenario, demonstrated in Fig. 5.4, we consider pairs of BS-UE(s) and a single RIS. The RIS is now considered as a shared resource, dynamically controlled by the infrastructure and commonly accessed by the BS-UE(s) pairs. The phase profile of the RIS can be commonly optimized by the BSs to serve their own UE(s) simultaneously. Alternatively, the control of the RIS may be time-shared among the BS-UE(s) pairs. Of course, the control channel envisioned by this scenario will have to be thoroughly investigated in future activities. A special case of this scenario is the one that considers a setup where the communication is enabled by multiple cellular BSs, each one serving a distinct set of UEs. When the UE(s) move across the cell boundaries of two or more BSs, they might change their serving BS(s) frequently (i.e., yielding frequent handovers). Shared RISs among BSs can be placed in the cell boundaries in order to dynamically extend the coverage of the serving BSs (i.e., reducing the number of handovers).

The RIS technology is able to provide significant advantages in applications that are related to localization and sensing, where conventional systems (i.e., with

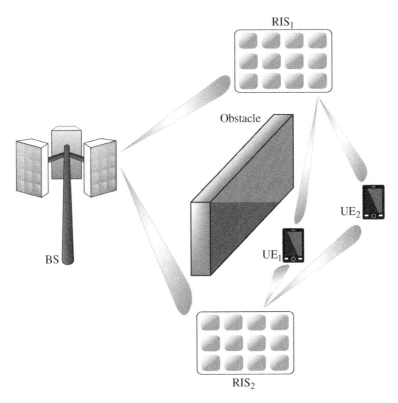

Figure 5.3 RIS-aided systems where connectivity is enabled by multiple RISs.

no RISs) need either to be enhanced or even fail (Keykhosravi et al. 2022), due to lacking the required design degrees of freedom. Specifically, conventional systems for 5G, or prior systems, for both uplink and downlink operation cases, have synchronization requirements that are based on the applicability of multi-way peer-to-peer ranging transactions to solve the timing offset in order to provide reliable results. In addition, at least three BSs are required to localize or sense one static or moving UE, and the overall process should be based on the solution of demanding nonlinear optimization problems, while the outcome depends on the quality of the measurements or the relative locations of the BSs. On the other hand, RISs can either assist or boost conventional localization solutions depending not only on their operating mode (Jian et al. 2022) (e.g., reflective, refracting, transmitting, etc.), but also on the type of localization they can offer. Accordingly, the benefits of enabling RISs for such purposes, can be also classified to *enabled*, where localization becomes feasible again, *boosted* in terms of improving the performance, and *low-profile* localization requires much lower resources; all these

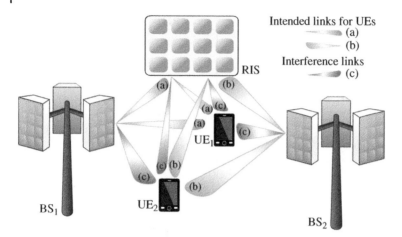

Figure 5.4 A multi-tenancy scenario with two BS-UE pairs and a shared RIS that is optimized to simultaneously boost reliable communications.

targeted benefits could be covered simultaneously. Next, detailed deployment scenarios are presented where RISs provide performance benefits for localization and sensing.

1. *Unambiguous localization under favorable problem geometry with a minimal number of active BSs*: According to this scenario, illustrated in Fig. 5.5, the conventional infrastructure is not sufficient to enable localization. However, when deploying a planar RIS operating in reflection mode, even for a single-antenna BS and considering that the BS's and the RIS's locations as known, then localization can be enabled, while a multi-RIS deployment setting can also contribute to boost performance, by a combination of time and angle measurements. To this end, the RIS(s) should be synchronized to the BS, the same holds for the location and orientation of the RIS(s), as well as the variable RIS phase-shift configurations of each RIS's elements are important requirements. Some challenges are studied in (Liu et al. 2022) with possible ways to tackle them.

2. *Non-Line-of-Sight (LOS) mitigation for better service coverage and continuity in far-field conditions*: In cases where the UE and the BS are assumed to be in the far field of the RISs, similarly to the previously described scenario, conventional systems would fail to provide an accurate location estimate for the UE, since there is no direct link between itself and the BS, as depicted in Fig. 5.6. Nevertheless, when information about the location and orientation of each RIS is available at their side combined with BS's location, then even under this setting and assuming that the two reflected paths from each RIS are properly resolved at the UE side, it is possible to estimate UE's 3D location. Therefore, whenever the minimum number of BSs (anchors) in visibility is not

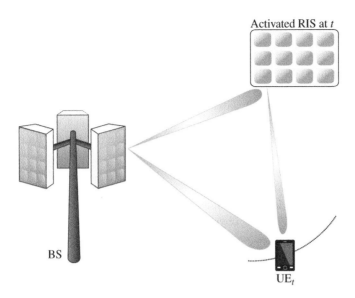

Figure 5.5 RIS-enabled localization and sensing.

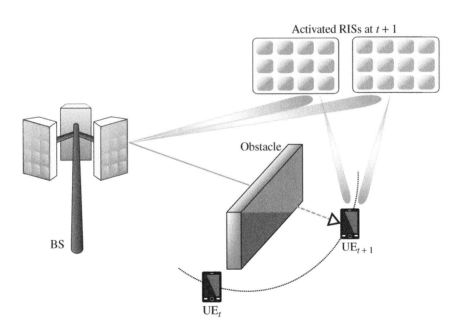

Figure 5.6 Multiple RIS-enabled localization and sensing in the vicinity of LOS links between BS and UE.

fulfilled, the UE location can be estimated via narrowband signals reflected from multiple RISs.

3. *Non-LOS mitigation for better service coverage and continuity in near-field conditions*: This scenario constitutes the extension of the previous one, under the assumption that the UE is in the near-field of one single planar RIS. Hence, the UE location can be estimated allowing to exploit signal wavefront curvature for direct positioning, unlike in the far-field case where, for instance, separate estimation is needed for the direction of departure from the RIS and the time of arrival of the RIS-reflected path.

4. *On-demand multiuser and multi-accuracy service provision*: In a multi-BS multi-user environment, the deployment of multiple RISs makes possible: (i) the (on-demand) provision of various classes of localization services to different UEs sharing the same physical environment, depending on the needs they express locally/temporarily, while (ii) spatially controlling both the localization accuracy and the geometric dilution of precision in different dimensions (i.e., both the sizes and orientations of the location uncertainty ellipses). In contrast to conventional scenarios, that demand large amounts of radio resources for allocation, or a larger number of BSs, to meet the necessary localization requirements, when RISs are combined with them provide more degrees of freedom, by offering the possibility to dedicate to specific users or even to small geographical areas.

5. *Opportunistic detection/sensing of passive objects through multi-link radio activity monitoring*: In this scenario, the opportunistic detection of either static or mobile passive objects, can be enabled by multiple RISs, by monitoring the time evolution of multipath profiles over links between the BS and several UEs, in both downlink and uplink directions. The dynamic and selective control of RISs is able to offer improvements in multipath diversity as well as richer location-dependent information to improve the performance in terms of detection, activity classification, and location estimation. Also, in highly reverberant environments, multiple RISs can create configurational diversity through wavefront shaping, even with single-antenna single-frequency measurements.

6. *RIS-assisted search-and-rescue operations in emergency scenarios*: RISs may support and overcome the shadowing effect induced by rubble in such scenarios by building ad-hoc controllable propagation conditions for the cellular signals employed in the measurement process. In addition, lightweight and low-complexity RISs mounted on drones can be used to bring connectivity capabilities to hard-to-reach locations, supporting first responders.

7. *Localization without BSs using a single or multiple RISs*: Assuming that the RISs operate in receive mode, that is, each RIS is equipped with a single or few reception radio-frequency chains, angle-of-arrival estimations can be obtained

and then combined in order to estimate the UE(s) locations, based on the multi-RIS localization architecture of Alexandropoulos et al. (2022b). A framework with co-located receiving RIS has been also devised (He et al. 2022b).

8. *RIS-aided radio environment mapping for fingerprinting localization*: RISs equipped with minimal reception circuitry for sensing enable the cartography of the electromagnetic power spatial density in a specific area of interest, accentuating the location-dependent features of the RIS-enhanced radio signatures stored in the database.

9. *RIS lens*: In this scenario, the RIS operates in transmit mode and is placed in front of a single-antenna transmitter. The user position is estimated at the UE side via the narrowband received signal, exploiting wavefront curvature (Abu-Shaban et al. 2021).

10. *Radar localization/detection of passive target(s) with hybrid RIS*: A radar is assisted by multiple hybrid RISs in order to localize/detect both static or moving target(s). Hybrid RISs simultaneously receive and reflect (Alexandropoulos et al. 2021b), giving the system the capability to localize UEs as well as the radar itself.

As the demands for connectivity and reliability are increasing for 5G-Advanced wireless communications systems, the requirements for enhanced sustainability and secrecy become more stringent and, in parallel, more urgent to be met. RIS-empowered environments are a key enabling technology, which according to recent studies, can provide remarkable improvements in various relevant performance criteria, by reducing the amount of unnecessary radiations in non-intended directions or in the directions of non-intended UEs (e.g., exposed UEs or eavesdroppers). Some of the main metrics that can be improved are D2.2 (2021): (i) the energy efficiency (EE), which can be defined as the ratio of the achievable data rate at the target UE and BS's transmit power, (ii) the electromagnetic field (EMF) utility, whose one possible definition is the attained data rate at the target UE divided by the largest EMF exposure at another UE; and (iii) the Secrecy Spectral Efficiency (SSE) that can be defined as the achievable data rate of the targeted UE subtracted by the data rate of the eavesdropper. Depending on the considered application, each one of the above metrics is able to characterize the area close to an RIS, where the latter can be optimally configured to maximize the corresponding criterion, leading to "areas with high EE," "areas with low EMF exposure," or "areas with enhanced secrecy." On the contrary, without the deployment of RISs, in other words in conventional scenarios, the above area-oriented targets cannot be met in highly signal-attenuated places, is limited. In the sequel, some representative scenarios are presented, taking into account single-BS examples in the presence of RIS(s), where downlink transmit beamforming is used to boost the received power at the intended UE and to reduce the received power at the non-intended UE. Depending on the nature of the UE, that is, if it is either an exposed UE or an eavesdropper, the obtained link is a low EMF link or a secured link, respectively.

1. *The BS-to-intended UE link is in LOS*: In this scenario, illustrated in Fig. 5.7, the RIS artificially adds propagation paths to the channel, which are coherently combined with other "natural" paths to boost the received power at the target UE (indicated by UE_1 in the figure), keeping reduced the received power at the non-intended UE (indicated by UE_2) by non-coherent combination with other "natural" paths.

2. *The BS-to-intended UE link is blocked by an obstacle and the intended UE is in the near-field of an RIS*: In this case, as illustrated in Fig. 5.8, there are obstacles blocking the BS to the intended UE (i.e., UE_1) link, and it is assumed that the latter node is placed in the near-field of the activated RIS, while the non-intended UE, that is UE_2, is closer to the BS. In such cases, the RIS adds a propagation path aiming at reducing the received power at the non-intended UE.

3. *The BS-to-intended UE link is blocked by an obstacle and the intended UE is in the far field of at least one RIS*: In this scenario, depicted in Fig. 5.9, at least two RISs are activated in order to enable sustainable communications since the intended UE (i.e., UE_1) is placed in the far field of them. Therefore, several artificially controlled RISs may become useful to reduce the energy at the non-intended UE (i.e., UE_2) which is closer to the BS.

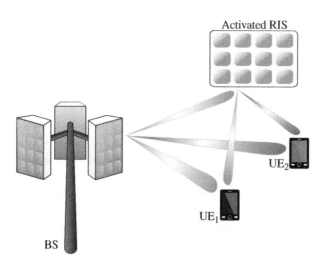

Figure 5.7 RIS-empowered sustainability and secrecy where the BS-to-intended link, that is the BS-to-UE_1 link, is in LOS.

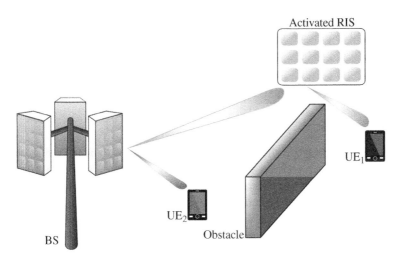

Figure 5.8 RIS-aided sustainability and secrecy where the BS-to-intended link, that is the BS-to-UE$_1$ link, is blocked and the activated RIS plays a central role in extending the connection between them.

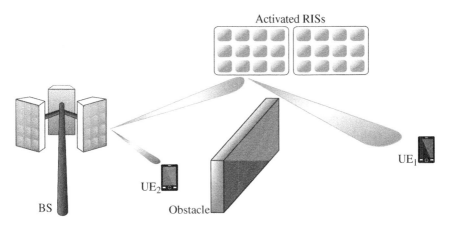

Figure 5.9 RIS-aided sustainability and secrecy where the BS-to-intended link, that is the BS-to-UE$_1$ link, is blocked, whereas UE$_1$ is located in the far-field of the activated RISs.

For all above scenarios, thanks to the RIS(s), the non-intended UE eavesdrops less (or is less exposed by) the intended UE's downlink data, because the former is on a location where propagation has been weakened artificially by each system's design. In addition, if the non-intended UE is absent, the proposed systems, which can be extended to the uplink case as well, only boost the received power at the target UE, and hence, boost the EE metric. Moreover, the aforementioned scenarios can be extended to multi-BS scenarios accordingly, where the BSs should synchronize and coordinate among themselves to perform joint beamforming.

5.2 Industrial Progresses

5.2.1 Funded Projects

As an important step toward the advanced research and pre-standardization of RISs, funding agencies worldwide are generously providing fuel to this evolving technology via supporting relevant research and innovation projects. Starting from the year 2012, RIS-dedicated research projects have contributed significantly to the theoretical study and engineering development of the RIS-empowered wireless networking concept. Some projects have already ended with notable outcomes, including theoretical studies and fabricated prototypes, while some others are still ongoing generating new ideas and fundamental research, deployment scenarios, integration schemes, and novel applications tailored for various frequency bands. The funded projects studying RISs (either entirely or by devoting some relevant research activities) for enhancing the performance of wireless networks and applications are summarized in Table 5.1.

Under the European Union (EU) funding program Horizon 2020, some projects have been supported to study physical-layer aspects, as well as devise novel software and hardware approaches to substantiate the idea of SRE through the incorporation of RISs. The VISORSURF Project (European Commission n.d.), one of the earliest future emerging technologies projects on the topic that started in January 2017, published a series of research papers focusing on designing software modules for controlling RISs, while exploring the properties of different base materials for fabricating metasurfaces. In its objective, the consortium aimed to build two experimental prototypes: a switch-based fabric array as the medium controller and a graphene-based one, making use of its exquisite properties to provide finer control. VISORSURF proposed a hardware platform, named the HyperSurface, that acts as a reconfigurable metasurface whose properties can be changed via a software interface. This project is one of the few that incorporated hardware designs with software development to deliver a compact prototype that can be tested in real-life demonstrations. The project received a total funding of €5,748,000 and ended in the December of 2020.

ARIADNE (European Commission n.d.), a three-year research and innovation action started on the 1st November 2019 and was financially supported by the Horizon 2020 program. The project, whose concluded its work in October 2022, focused on communications with high radio frequencies, and especially in the D-band ranging from 110 to 170 GHz. The project aimed to tackle the complexity

Table 5.1 Summary of research projects dedicated to RIS.

Project title	Objective and outcome	Start	End	Funding agency	Overall budget
Manipulating Terahertz Waves Using Three-Dimensional Metamaterials	Manipulation of the polarization of THz waves using passive or active metamaterial based devices	09/2012	08/2016	NSF	US$267,000
A Hardware Platform for Software-driven Functional Metasurfaces (VISORSURF)	Joint hardware and software designing for RIS with two experimental prototypes proposed	01/2017	12/2020	Horizon 2020	€5,748,000
Assessing the Feasibility of Programming the Ambient Wireless Environment	Exploring ways of changing the perceived channel along the wireless link to create more favorable conditions for communication	05/2018	04/2020	NSF	US$191,278
Liquid Metal Tuned Flexible Metasurfaces	Deployable, transportable and conformable metasurfaces that can be tuned on-demand and in real-time for radio signals	09/2019	08/2022	NSF	US$290,999
Artificial Intelligence Aided D-band Network for 5G Long Term Evolution (ARIADNE)	Merging a novel high-frequency advanced radio architecture with an artificial intelligence network processing and management approach into a new type of intelligent communications system beyond 5G	11/2019	10/2022	Horizon 2020	€5,968,393.75
Scaling WLANs to TB/sec: THz Spectrum, Architectures, and Control	A first-of-its-kind pixelated metasurface waveguide to dynamically steer a THz beam via electrical switching of the meta-elements	07/2020	06/2025	NSF	US$171,453
Enabling Seamless Coexistence between Passive and Active Networks Using Reconfigurable Reflecting Surfaces	Developing a novel framework that enables seamless co-existence among multiple passive and active wireless systems, by exploiting the RIS concept to suppress interference at the wireless receivers	10/2020	09/2023	NSF	US$320,000

(Continued)

Table 5.1 (Continued)

Project title	Objective and outcome	Start	End	Funding agency	Overall budget
Future Wireless Communications Empowered by Reconfigurable Intelligent Metamaterials (META WIRELESS)	Enabling the manipulation of wireless propagation environments by re-configurable intelligent surfaces	12/2020	11/2024	Horizon 2020	€3,995,128.44
Hybrid Reconfigurable Intelligent Surfaces: Hardware Design and Signal Processing for Beyond 5G Wireless Communications	Design hybrid RIS hardware and relevant signal processing techniques, being capable to simultaneously reflect the incident waves via low-cost beamforming, while performing local sensing	11/2021	11/2022	Royal Society, UK	£5650
Reconfigurable Intelligent Sustainable Environments for 6G Wireless Networks (RISE-6G)	Several of studies, engineering and standardization practices which can bring the technically advanced vision on RIS into industrial exploitation	01/2021	12/2023	Horizon 2020	€6,499,613.75
Harnessing Multipath Propagation in Wireless Networks: A Metasurface Transformation of Wireless Networks into Smart Reconfigurable Radio Environments (Pathfinder)	Setting the theoretical and algorithmic bases of RIS-empowered wireless 2.0 networks that will lead to further transformations of wireless networks	05/2021	04/2023	Horizon 2020	€184,707.84
Surface Waves in Smart Radio Frequency Environments (Surfer)	Pioneer the theoretic foundation and experimental validation of surface wave communications for indoor communications	02/2022	01/2024	Horizon 2020	€184,707.84
Terahertz Reconfigurable Metasurfaces for Ultra-High-Rate Wireless Communications (TERRAMETA)	Design high-performance hardware for THz RISs and holographic transmitters/receivers, as well as network analysis/optimizations techniques	01/2023	12/2025	Horizon Europe	€6,327,480

issue introduced by new frequency ranges via artificial intelligence, while RISs were considered as a powerful weapon to overcome the associated obstacles. The total budget of this project was €5,968,393.

The first research and innovation project of the Horizon 2020 program focusing completely on the RIS technology and on the resulting reconfigurable intelligent sustainable environments for 6G wireless networks is the ongoing project RISE-6G, which kick-started on the 1st of January 2021. Besides, fundamental scientific research on the topic, te project is also poised to actively participate in standardization bodies and bring its technically advanced vision into planned industrial exploitation. Optimized algorithms to enable fast and accurate controlling of RISd, as well as RIS prototyping and testing are also within the scope of this project, which receives a total award of €6,499,613. The deep involvement of network operators in this project indicates that RIS are being matured to a new stage, where testing in the current 5G network becomes possible and necessary.

Another project supported by Horizon 2020 is META WIRELESS (European Commission n.d.), led by a consortium of universities and enterprises over several European countries. The project emphasizes on the properties of metamaterials and studies how metasurfaces can manipulate radio signals to pave a path to the 6G era. A multidisciplinary effort is expected involving wireless communications, physics, electromagnetic theory, and computational learning to complete the objective of the project. This ongoing project has an overall budget of €3,995,128 and will end in November 2024.

On December 2021, UK's Royal Society started financially supporting a one-year collaborative project between the University of Hertfordshire and the National and Kapodistrian University of Athens, entitled as "Hybrid Reconfigurable Intelligent Surfaces: Hardware Design and Signal Processing for Beyond 5G Wireless Communications," which aimed at investigating novel hybrid RIS hardware designs and relevant signal processing techniques, being capable to simultaneously reflect the incident waves via low-cost beamforming, while performing local sensing, enabling channel estimation. A wideband RIS operating in the millimeter wave band was designed, including new architectures, reducing the complexity of its controlling circuits.

In the new era of the Horizon Europe, which constitutes the successor of the EU Horizon 2020 funding program, research projects are still being supported to further advance the RIS technology. According to the recent results for the latest open call, one project studying RIS for terahertz (THz) stands out from fierce competition. The project of THz reconfigurable metasurfaces for ultra-high-rate wireless communications (TERRAMETA) aims to study advanced wireless and signal processing techniques to support RIS for THz bands, which can potentially provide large bandwidth for communication and are likely to support ultra-high data rates. The TERRAMETA project will design high-performance hardware for THz networking components, including RISs and holographic transmitters/receivers, as well as advanced network analysis/optimizations techniques using the designed components. The project will be driven by 6G usage scenario requirements, and indoor, outdoor, as well as indoor-to-outdoor scenarios will be demonstrated in a real factory setting and a telecom testing field. TERRAMETA will start on the 1st of January 2023 and its total budget is €6,327,480.

In the United States (US), several projects are underway and supported by different departments of the federal government. Supported partly by the Office of Naval Research, the Air Force Office of Scientific Research and the Army Research Office, a team at Princeton University is working on optimizing beamforming using RIS for THz frequency bands (Princeton University n.d.). As a proof-of-concept, the Princeton researchers tested tile arrays measuring two-by-two with 576 programmable elements and demonstrated beam control by projecting THz holograms. Besides assisting THz communications, the project also aims to explore possibilities of using RISs to help gesture recognition, imaging, as well as industrial automation and security.

The National Science Foundation (NSF) supported a research project aiming at manipulating THz waves using three-dimensional metamaterials (SBIR STTR n.d.). This project was awarded US$267,000 and ended in August 2016 after a four-year span. The outcome of this project included studies on manipulating the polarization of THz waves and both passive and active metamaterial devices were studied.

In another five-year NSF-supported study that started in July 2020 https://www .nsf.gov/awardsearch/showAward?AWDID=1955004, the involved researchers focus on providing underlying building blocks for steerable and highly directional beams from 100 to 500 GHz. A first-of-its-kind pixelated metasurface waveguide will be developed to dynamically steer a THz beam via electrical switching of the meta-elements.

There are also projects focusing on enabling RIsS with new metamaterials as well as manipulating waves in new domains. The liquid metal-tuned flexible meta-surfaces, a three-year project funded by the NSF and started in September 2019, included manipulating both optical and radio signals within the scope of study https://www.nsf.gov/awardsearch/showAward?AWDID=1908779. The project exploited new techniques that can be applied to realize liquid-metal-enabled tunable and mechanically flexible multi-layered metasurfaces. The fabrication goal was to integrate movable liquid metals that can provide a wide tuning range with minimal loss, supporting a wide range of applications that demand control-lable and stable high-frequency response, when operating under a dynamic radio environment.

There are some other projects supported by the US government that target on developing concepts and prototypes for RISs with the intended use case being the manipulation of the propagation environment for visible lights or laser (Perry 2014) (Energy Efficiency & Renewable Energy n.d.) (NSF 2018), which might eventually result in rewarding findings in radio communications as well. Two NSF-supported projects also revealed that metamaterials can also be used to alter acoustic propagation conditions to realize dynamic control of the directionality of sound wave propagation using programmable materials https://www.nsf.gov/awardsearch/showAward?AWDID=1641084 https://www.nsf.gov/awardsearch/showAward?AWDID=1951221.

Horizon Europe HEU (n.d.), the funding program chaired by the EU and the successor of Horizon 2020, has released several rounds of project calls already. The budget of Horizon Europe is significantly increased compared to that of Horizon 2020 and is open for worldwide participation through multiple mechanisms. There are several clusters of challenging research topics, identified as the mission of Horizon Europe program, one of which is digital, industry, and space. The destinations under this cluster include digital and emerging technologies for competitiveness and fit for the green deal, with multi-sensing systems and 6G as foreseen topics. Research projects on RISs can fit in this cluster and can be supported by the ongoing Horizon Europe funding program till 2027.

5.2.2 White Papers

As in-depth researches on RIS are still ongoing, more and more white papers on 5G advanced or 6G mention RISs as a key technology to encounter challenging issues in future wireless networks. Out of 43 white papers published from 2019 till now, 17 of them include content on RIS, either by providing theoretical analysis, or depicting new applications in future wireless networks enabled by RISs, or both. 11 of them contain comprehensive analyses in chapters dedicated to RISs, which are summarized in Table 5.2. Almost all papers acknowledge RIS to be able to alter the radio environment and manipulate the amplitude, phase shift, frequency, and polarization mode of the waveform. Most of the white papers mention that RIS will be a promising solution to assist communications in cell-edge, dead zones, and under NLOS conditions, as a more cost-effective solution than traditional relays. In addition, many white papers point out that joint communications and sensing can be realized by deploying RIS into communication systems. Enhancing physical-layer security is also a promising use case stressed by some white papers. Common challenges pointed out by different researchers around the world include the maturity of manufacturing reflective panels, channel modeling, and channel estimation and feedback, especially when RIS is assumed to operate in a completely passive mode. How to control the large amount of reflective elements to achieve an optimized control in terms of fine beamforming is also raised as a major challenge. To move forward and solve these challenging points, the authors of the white papers suggest to further study channel modeling, channel estimation and feedback, machine-learning (ML)-based algorithms for RIS and approaches to manufacture metamaterials with high reliability and low cost, as future focal points.

Besides the ones with dedicated chapters on RIS summarized in Table 5.2, RISs have also been mentioned as a key candidate technology of numerous white papers on 6G, such as the ones released by the 5G Public Private Partnership (Test and KPIs Validation Working Group n.d.).

5.2.3 Prototyping and Testing

Based on the theoretical research results, some universities pioneered in building proof-of-concepts of RISs as well as conducting tests in lab or outdoor environments. In Arun and Balakrishnan (2020), the prototype RFocus was demonstrated along with its controller, algorithm, and experimental results. This fully passive prototype was tested in a typical indoor environment, achieving a median of 9.5 dB improvement in signal strength and twice the non-RIS channel capacity. An RIS prototype that can perform amplitude-and-phase-varying modulation to facilitate multiple-input multiple-output (MIMO) transmissions was presented by Tang et al. (2020). The system was evaluated OTA in real time, while taking into account the hardware constraints and their impact on the system. The authors of Dai et al. (2020) developed a low-cost and high-gain RIS panel that comprised 256 elements, where each element can be configured by 2-bit control signaling. The designed RIS could provide 21.7 dBi antenna gain at 2.3 GHz and 19.1 dBi antenna gain at 28.5 GHz. The test results on beams were acquired in a radio-frequency (RF) anechoic chamber, while the results on system performance were collected through tests indoors. Orthogonal frequency division multiplexing (OFDM) signals have been used in all the tests. In Pei et al. (2021), OTA test results were presented for RIS-assisted communication systems operating in both indoor and outdoor scenarios. The RIS comprised 1100 tunable elements and could provide a 26 dB power gain in indoor scenarios, and a 27 dB power gain in short-range outdoor tests. The RIS was able to provide a gain of 14 dB in receiving power when the distance between the transmitter and the receiver was 500 m. The signals used in the tests were also OFDM, while the transmit power was 13 dBm for 50-m transmission distance and 23 dBm for 500-m distance.

The wireless communication industry has also been building different kinds of prototypes and testing them in different environments. The trials started with non-configurable metasurfaces. In November 2018, a communication system in the 28 GHz band using a metasurface reflect-array was demonstrated (NTT DOCOMO 2018), which achieved a downlink data rate of 560 megabits per second

Table 5.2 Summary of white papers dedicated to RIS (google scholar, search: Reconfigurable Intelligent Surface Technology White Paper).

Source	Title	Publication date	Identified use cases	Future study	Challenges
6G Flagship, University of Oulu	Key Drivers and Research Challenges for 6G Ubiquitous Wireless Intelligence	09/2019	Controlling signal reflections and refractions, enabling holographic radio	Fundamental analysis to understand the performance of RIS, environmental AI whereby smart surfaces learn and autonomously reconfigure their material parameters	Focusing signals with different angles of incidence and training ML-driven smart surfaces in real time
6G Flagship, University of Oulu	Broadband Connectivity in 6G	06/2020	Replacing traditional beamforming, cognitive radio, wireless power transfer, physical-layer security, and backscattering. The most suitable use case is still an open question	Hardware implementation, channel modeling, experimental validation, and real-time control	A balance between the channels of different frequency sub-bands which complicates the joint active and passive beamforming optimization. The control interface and channel estimation aredifficult when using a passive surface that cannot send or receive pilot signals
6G Flagship, University of Oulu	Research Challenges for Trust, Security and Privacy	06/2020	Direct the beamformed signal toward the user and destructively send it toward the eavesdropper, and thus help to improve security of the network	—	To improve security, the network needs to detect and locate the eavesdropper which is not a trivial issue. If the location of the IRS is exposed, a passive attacker can also locate itself near the RIS to exploit a correlated channel for eavesdropping

(Continued)

Table 5.2 (Continued)

Source	Title	Publication date	Identified use cases	Future study	Challenges
6G Flagship, University of Oulu	6G White Paper on Localization and Sensing	06/2020	Enabling low-complexity and energy-efficient communication paradigms and further enhancing localization and sensing	Appropriate models that describe the properties of the constituent materials and the radio propagation features of the incident signal waves and novel energy-efficient methods to estimate the channel properties	Novel signal processing techniques that optimize the performance of RIS-assisted joint communications, sensing, and localization are required which is especially challenging at high frequencies
Southeast University	6G Research	08/2020	Enable smart radio environment	Channel modelling	—
Guangdong Communications and Networks Institute et al.	Hot Technologies for 6G	09/2020	Controlling signal reflections and refractions	Improving controlling approaches, building platforms to enable accurate simulations and lower the price, power consumption and weight of the RIS	Manufacturing of metamaterials, combining metamaterials to RF components and antennas, and high-dimensional radio control
FuTURE FORUM	Wireless Technology Trends toward 6G	11/2020	Passive beamforming, joint precoding, and ultra-massive MIMO when deployed near the transmitter, NLOS coverage, edge user enhancement, and high-precision positioning when deployed near the receiver	Design and modeling theory of the reflective element, channel modeling, channel estimation, and feedback mechanisms	—

Organization	Title	Date	Application	Technical focus	Challenges
FuTURE FORUM	Novel Antenna Technologies toward 6G	11/2020	Provide service to users at dead zones or cell edge, realize physical-layer security, assist device-to-device communications and wireless information and power transmission	Information and propagation theory for RIS-assisted communication systems, channel estimation and feedback	—
UNISOC	White Paper on 6G	11/2020	Alter radio environment and provide service to users in dead zones	Improve the controlling of individual reflective elements and further study the channel characteristics	How to realize channel modeling, channel estimation and feedback. How to produce metasurfaces with low cost and high reliability
Nokia Bell Labs	Communications in the 6G Era	12/2020	Improve communications	AI or ML techniques on modeling and control	Determine optimal control of the reflective elements and model signal propagation incorporating the collective effects of RISs, which in turn depend on how they are controlled
Lenovo	White Paper on 5G / 6G	02/2021	Improve communications under NLOS conditions, in indoor scenarios, and cell edge. Assist robust transmission and high-accuracy wireless positioning	Compressed sensing and AI based channel estimation	Channel estimation and feedback, networking with other network infrastructures
RIS TECH Alliance	Reconfigurable Intelligent Surface Technology White Paper	03/2023	Assisting high-frequency communication, space communication, integrated sensing and communication, internet-of-things, wireless edge computing, physical layer securityfull duplex and semantic communication	Industrial implementation, standardization and construction of the whole ecosystem	An integrated and closely related ecosystem is needed for RIS, taking all stakeholders into account

(Mbps) in lieu of 60 Mbps when no reflector is deployed. The metasurface used very small structures, compared to the free-space wavelength, that were arranged across the array to have different shapes based on their position within it. The direction of the reflected wave gain and the shape of the beam were determined by the meta-structure of the metasurface, which was not configurable once manufactured. The trial demonstrated the ability of such non-configurable metasurface to increase coverage by reflecting radio waves to users out of sight. In January 2020, the same team conducted a trial of a transparent dynamic metasurface with 5G radio signals in the 28 GHz band (ntt 2020). The metasurface appeared to be a normal transparent glass, but it was capable of reflecting radio signals in predefined directions with three modes. The designed metasurface could reflect the incident radio signals with full or partial power, or could let the signal penetrate with close-to-zero power losses. The portion of the power reflected was manipulated by the separation distance between the metasurface substrate and the movable transparent layer, which could be adjusted from 0 to more than 200 μm. The distance between the two substrates was manually controlled in the test, while plans to use piezoelectric actuators were suggested. It's still worth studying the speed of changes between the modes that such a control mechanism can support, and whether it can meet requirements for 5G-Advanced or next-generation networks. The first experiment for backscattering systems empowered by an RIS has also been recently conducted by (Fara et al. 2021). The proposed RIS was able to realize reflected beams, which belonged to a codebook, that improved the tag-to-reader bit error rate. The test was conducted in a simple lab environment where the signal source, tag, RIS, and reader were all placed on a table and close to each other. Recently, the design of a liquid-crystal-based metasurface that works on 28 GHz was revealed (KDD 2021). The prototype was designed to solve current problems in 5G networks, such as coverage holes in high frequencies. The start-up company (Popov et al. 2021) presented the design of an RIS prototype with 1600 elements operating at around 29.5 GHz. The prototype was tested indoors with horn antennas used as the transmitter and receiver. The transmitter was placed in an office room, while the receiver was outside in the corridor, such that there is no LOS path between them. The results showed that the RIS can effectively enhance the constellation diagram of quadrature phase shift keying (QPSK) for the above scenario. In Liu et al. (2022), the research team at ZTE Corporation presented a variety of RIS prototypes including 1-bit single-polarized, 1-bit dual-polarized, 1.6-bit single-polarized, 2-bit dual-polarized, and 4-bit dual-polarized panels with 20×20, 32×32, 64×64 or 128×128 elements, working at 2.6, 4.9, or 26 GHz. The paper reported world's first trials of RISs in 5G networks as the prototypes were tested in 5G networks with off-the-shelf 5G user equipment. Some of the tests were done in a sophisticated environment, where pedestrians and vehicles were passing by during the test. Simulations results were also provided to validate that the tests achieve similar results with theoretical analysis. As demonstrated by the trial results, RISs can increase the Reference Signal Receive Power (RSRP) in different reception scenarios around 15–35 dB, depending on the detail test setup and the implementation of the RIS. In all scenarios, RISs successfully demonstrated the capability of supporting a more robust wireless communication in a more cost- and energy-efficient way, compared to their counterparts, such as relays.

5.2.4 Standardization Progress

With extensive research efforts and activities dedicated to the RIS technology, standardization has been kicked off both regionally and globally. Notably, a first proposal to establish a study item on RIS has been submitted to 3GPP, the most successful and influential global standard development organization (SDO) in the wireless industry, in its meeting in March 2021 (ZTE Corporation, Sanechips 2021). There are also other ongoing or planned standard activities in other SDOs which are introduced with details in Liu et al. (2021a) and Liu et al. (2022).

In China, RIS-related standardization work has been kicked off in two different standard organizations, namely, the China Communications Standards Association (CCSA) and the FuTURE Mobile Communication Forum. CCSA is the official standard organization established by enterprises and institutes in China for carrying out standardization activities in the field of Information and Communications Technology (ICT) across the country. During the 55th meeting of the technical committee 5 of the working group 6 (TC5 WG6), the proposal to establish a study item on RIS has been approved (CCSA 2020). The study item started immediately after approval and ended in June 2022 with a published technical report. This study item aimed to investigate numerous aspects that would help to realize RIS-assisted wireless communication systems, including channel modeling, channel estimation and feedback, and beamforming with RIS, RIS, and AI, as well as networking protocols for RIS-assisted networks. In the FuTURE forum, a working group to study how to integrate RISs into next-generation wireless networks has been established in December 2020 (FuTURE Forum n.d.). This working group is studying the potential use cases and key technologies to support RISs to be standardized and commercialized, and will publish a white paper to summarize the technical trends and further promote RIS-assisted wireless systems.

In June 2021, a new industry specification group (ISG) on RIS was approved by the European Telecommunications Standards Institute (ETSI). During the second plenary meeting of that ISG, three new work items were approved, which focus on channel and system modeling, use cases and deployment scenarios, and the impact on current standards. All work item (WI) descriptions and related contributions can be found on the ETSI portal (ETSI n.d.). The details of the three

Table 5.3 Summary of completed work items in ETSI ISG RIS.

WI code	RIS-001	RIS-002	RIS-003
WI title	Use Cases, Deployment Scenarios and Requirements	Technological Challenges, Architecture and Impact on Standardization	Communication Models, Channel Models, and Evaluation Methodology
Release date	31 December 2022	01 April 2023	31 December 2022

ongoing work items are provided in Table 5.3. The ISG is focusing on studies and the generation of informative reports till the year of 2023, after which normative specifications may be considered.

The regional efforts to standardize RISs now have echoes on global platforms. During the ITU-R WP 5D meeting in October 2020, countries submitted their proposed candidates for the next-generation wireless technologies, which will be considered to be included into the IMT Future Technology Trends report. During the meeting, China submitted a proposal on the preliminary draft new report, which included RIS as a key technology for future wireless networks (ITU n.d.). In 3GPP, the most successful and influential global standard development organization (SDO) in the wireless industry, a first proposal to establish a study item on RIS has been submitted in its meeting in March 2021 (ZTE Corporation, Sanechips 2021). As 3GPP is starting the second phase of 5G with Release 18, which is officially named 5G-Advanced, many companies are actively proposing to accept RIS as a formal work item. Due to the concerns of the maturity of RIS fabrications, and whether RIS panels can perform stably and uniformly under all conditions (temperatures, weather, humidity, etc.), RIS has not been explicitly accepted as an independent work item to Release 18. However, there is one approved study item on network-controlled repeaters (ZTE Corporation 2021), approved in December 2021, which can be seen as a first step towards standardization of the RIS technology. Network-controlled repeaters focus on extension of network coverage. The study scope covers identification of side control information for network-controlled repeaters, taking into account practical factors such as maximum transmission power with beamforming information, timing information to align transmission/reception boundaries of network-controlled repeaters, information on TDD configurations, ON-OFF information for efficient interference management and improved energy efficiency, as well as power control information for efficient interference management. Detailed signaling to carry the side control information as well as management of such repeaters are also studied. The study item generates a technical report with all the results obtained during the study phase (ZTE Corporation 2022). Following this study item, a formal work item has been approved to start from September 2022 (ZTE, Sanechips 2022), to define specifications needed to support all the functionalities of network-controlled repeaters. It is anticipated that the study and work on network-controlled repeaters can be used as a baseline for 3GPP to consider standardization of RISs.

Bibliography

Abdullah, Z., Alexandropoulos, G.C., Kisseleff, S. et al. (2022). Combining relaying and reflective surfaces: power consumption and energy efficiency analysis. *Proceedings of the IEEE GLOBECOM*, 1–6. Rio de Janeiro, Brazil, December 2022.

Abu-Shaban, Z., Keykhosravi, K., Keskin, M.F. et al. (2021). Near-field localization with a reconfigurable intelligent surface acting as lens. *Proceedings of the IEEE ICC*, 1–6. Montreal, Canada, June 2021.

Alexandropoulos, G.C., Lerosey, G., Debbah, M., and Fink, M. (2020). Reconfigurable intelligent surfaces and metamaterials: the potential of wave propagation control for 6G wireless communications. *IEEE ComSoc TCCN Newsletter* 6 (1): 25–37.

Alexandropoulos, G.C., Shlezinger, N., and del Hougne, P. (2021a). Reconfigurable intelligent surfaces for rich scattering wireless communications: recent experiments, challenges, and opportunities. *IEEE Communications Magazine* 59 (6): 28–34.

Alexandropoulos, G.C., Shlezinger, N., Alamzadeh, I. et al. (2021b). Hybrid reconfigurable intelligent metasurfaces: enabling simultaneous tunable reflections and sensing for 6G wireless communications. https://arxiv.org/abs/2104.04690.

Alexandropoulos, G.C., Crozzoli, M., Phan-Huy, D.-T. et al. (2022a). Smart wireless environments enabled by RISs: deployment scenarios and two key challenges. *2022 Joint European Conference on Networks and Communications & 6G Summit (EuCNC/6G Summit)*, 1–6, June 2022a.

Alexandropoulos, G.C., Vinieratou, I., and Wymeersch, H. (2022b). Localization via multiple reconfigurable intelligent surfaces equipped with single receive RF chains. *IEEE Wireless Communications Letters* 11 (5): 1072–1076. https://doi.org/10.1109/LWC.2022.3156427.

Arun, V. and Balakrishnan, H. (2020). RFocus: beamforming using thousands of passive antennas. *17th Symposium on Networked Systems Design and Implementation (NSDI 20)*, 1047–1061.

CCSA (2020). Meeting Detail of TC5 WG6 Meeting 55, September 2020. URL http://www.ccsa.org.cn/meetingDetail/91?title=TC5WG6%E7%AC%AC55%E6%AC%A1%E5%B7%A5%E4%BD%9C%E7%BB%84%E4%BC%9A%E8%AE%AE.

Dai, L., Wang, B., Wang, M. et al. (2020). Reconfigurable intelligent surface-based wireless communications: antenna design, prototyping, and experimental results. *IEEE Access* 8: 45913–45923. https://doi.org/10.1109/ACCESS.2020.2977772.

Di Renzo, M., Debbah, M., Phan-Huy, D.-T. et al. (2019). Smart radio environments empowered by reconfigurable AI meta-surfaces: an idea whose time has come. *EURASIP Journal on Wireless Communications and Networking* 2019 (1): 129.

Energy Efficiency & Renewable Energy (n.d.). Project Profile: University of California San Diego (CSP: COLLECTS). https://www.energy.gov/eere/solar/project-profile-university-california-san-diego-csp-collects (accessed 16 August 2023).

ETSI (n.d.). Work item descriptions of new WIs. https://portal.etsi.org (accessed 16 August 2023).

European Commission (n.d.). Future wireless communications empowered by reconfigurable intelligent meta-materials (meta wireless). https://cordis.europa.eu/project/id/956256 (accessed 16 August 2023).

European Commission (n.d.). Artificial intelligence aided D-band network for 5G long term evolution. https://cordis.europa.eu/project/id/871464.

European Commission (n.d.). VisorSurf - a hardware platform for software-driven functional metasurfaces. https://cordis.europa.eu/project/id/736876.

Fara, R., Phan-Huy, D.-T., Ratajczak, P. et al. (2021). Reconfigurable intelligent surface -assisted ambient backscatter communications –experimental assessment. *2021 IEEE International Conference on Communications Workshops (ICC Workshops)*.

FuTURE Forum (n.d.). http://www.future-forum.org/en/index.asp; https://www.nsf.gov/awardsearch/showAward?AWDID=1232081.

Gong, T., Vinieratou, I., Ji, R. et al. (2021). Holographic MIMO communications: theoretical foundations, enabling technologies, and future directions. arXiv preprint arXiv:2212.01257.

He, J., Fakhreddine, A., and Alexandropoulos, G.C. (2022a). Simultaneous indoor and outdoor 3D localization with STAR-RIS-assisted millimeter wave systems. *Proceedings of the IEEE VTC-Fall*, 1–6. London/Beijing, UK/China, September 2022a.

He, J., Fakhreddine, A., Vanwynsberghe, C. et al. (2022b). 3D localization with a single partially-connected receiving RIS: positioning error analysis and algorithmic design. arXiv preprint arXiv:2212.02088.

Horizon Europe (n.d.). The commission's proposal for Horizon Europe, strategic planning, implementation, news, related links. https://ec.europa.eu/info/horizon-europeen. https://research-and-innovation.ec.europa.eu/funding/funding-opportunities/fundingprogrammes-and-open-calls/horizon-europe_en. (assessed on ...)

Huang, C., Zappone, A., Alexandropoulos, G.C. et al. (2019). Reconfigurable intelligent surfaces for energy efficiency in wireless communication. *IEEE Transactions on Wireless Communications* 18 (8): 4157–4170.

Huang, C., Hu, S., Alexandropoulos, G.C. et al. (2020). Holographic MIMO surfaces for 6G wireless networks: opportunities, challenges, and trends. *IEEE Wireless Communications* 27 (5): 118–125.

ITU WP 5D (n.d.). https://www.itu.int/md/R19-WP5D.AR-C/en.

Jian, M., Alexandropoulos, G.C., Basar, E. et al. (2022). Reconfigurable intelligent surfaces for wireless communications: overview of hardware designs, channel models, and estimation techniques. *Intelligent and Converged Networks* 3 (1): 1–32. https://doi.org/10.23919/ICN.2022.0005.

KDDI Research (2021). Development of the world's first direction-variable liquid crystal meta-surface reflector, October 2021. https://www.kddi-research.jp/english/newsrelease/2021/100702.html (accessed 16 August 2023).

Keykhosravi, K., Denis, B., Alexandropoulos, G.C. et al. (2022). Leveraging RIS-enabled smart signal propagation for solving infeasible localization problems. arXiv preprint arXiv:2204.11538.

Kisseleff, S., Martins, W.A., Al-Hraishawi, H. et al. (2020). Reconfigurable intelligent surfaces for smart cities: research challenges and opportunities. *IEEE Open Journal of the Communications Society* 1: 1781–1797.

Liu, R., Wu, Q., Di Renzo, M., and Yuan, Y. (2021a). A path to smart radio environments: an industrial viewpoint on reconfigurable intelligent surfaces. CoRR, abs/2104.14985. https://arxiv.org/abs/2104.14985.

Liu, Y., Mu, X., Xu, J. et al. (2021b). STAR: simultaneous transmission and reflection for 360deg coverage by intelligent surfaces. *IEEE Wireless Communications* 28 (6): 102–109.

Liu, R., Dou, J., Li, P. et al. (2022). Simulation and field trial results of reconfigurable intelligent surfaces in 5G networks. *IEEE Access* 10: 122786–122795. https://doi.org/10.1109/ACCESS.2022.3223447.

Liu, R., Jian, M., and Zhang, W. (2022). A TDoA based Positioning Method for Wireless Networks assisted by Passive RIS. *IEEE Globecom Workshops (GC Wkshps)*, Rio de Janeiro, Brazil, 2022, pp. 1531–1536. doi: 10.1109/GCWkshps56602.2022.10008550.

Liu, R., Alexandropoulos, G.C., Wu, Q., Jian, M. et al. (2022). How Can Reconfigurable Intelligent Surfaces Drive 5G-Advanced Wireless Networks: A Standardization Perspective. *IEEE/CIC International Conference on Communications in China (ICCC Workshops)*, Sanshui, Foshan, China, 2022, pp. 221–226. doi: 10.1109/ICCCWorkshops55477.2022.9896658.

Measurement Test and 5GPPP KPIs Validation Working Group (n.d.). Beyond 5G/6G KPIs and target values. https://5g-ppp.eu/wp-content/uploads/2022/06/whitepaperb5g-6g-kpis-camera-ready.pdf (accessed 17 August 2023).

NSF (2018). Fabrication of high performance metasurfaces by nanoimprinting of refractive index, September 2018. https://grantome.com/grant/NSF/CMMI-1825787 (accessed 16 August 2023).

NTT (2020). Press releases: DOCOMO conducts world's first successful trial of transparent dynamic metasurface, January 2020. https://www.nttdocomo.co.jp/english/info/mediacenter/pr/2020/011700.html (accessed 16 August 2023).

NTT DOCOMO (2018). https://www.businesswire.com/news/home/20181204005253/en/NTT-DOCOMO-and-Metawave-Announce-Successful-Demonstration-of-28GHz-Band-5G-Using-Worlds-First-Meta-Structure-Technology.

Pei, X., Yin, H., Tan, L. et al. (2021). RIS-aided wireless communications: prototyping, adaptive beamforming, and indoor/outdoor field trials. *IEEE Transactions on Communications* 69 (12): 8627–8640.

Perry, C. (2014). Collaborative "metasurfaces" grant to merge classical and quantum physics, May 2014. https://www.seas.harvard.edu/news/2014/05/collaborative-metasurfaces-grant-merge-classical-and-quantum-physics (accessed 17 August 2023).

Popov, V., Odit, M., Gros, J.-B. et al. (2021). Experimental demonstration of a mmWave passive access point extender based on a binary reconfigurable intelligent surface. *Frontiers in Communications and Networks* 2: https://doi.org/10.3389/frcmn.2021.733891.

Princeton University (n.d.). Super surfaces use terahertz waves to help bounce wireless communication into the next generation — Office of the Dean for Research. https://research.princeton.edu/news/super-surfaces-use-terahertz-waves-help-bounce-wireless-communication-next-generation.

RISE-6G Deliverable D2.2 (2021). Metrics and KPIs for RISE wireless systems analysis: first results, June 2021.

Samsung (2020). The next hyper-connected experience for all. White Paper, June 2020.

SBIR STTR (n.d.). Metasurface photonics. https://www.sbir.gov/node/1413395 (accessed 16 August 2023).

Strinati, E.C., Alexandropoulos, G.C., Sciancalepore, V. et al. (2021a). Wireless environment as a service enabled by reconfigurable intelligent surfaces: the RISE-6G perspective. *2021 Joint European Conference on Networks and Communications & 6G Summit (EuCNC/6G Summit)*, 562–567, June 2021a.

Strinati, E.C., Alexandropoulos, G.C., Wymeersch, H. et al. (2021b). Reconfigurable, intelligent, and sustainable wireless environments for 6G smart connectivity. *IEEE Communications Magazine* 59 (10): 99–105.

Tang, W., Dai, J.Y., Chen, M.Z. et al. (2020). MIMO transmission through reconfigurable intelligent surface: system design, analysis, and implementation. *IEEE Journal on Selected Areas in Communications* 38 (11): 2683–2699. https://doi.org/10.1109/JSAC.2020.3007055.

Yildirim, I., Kilinc, F., Basar, E., and Alexandropoulos, G.C. (2021). Hybrid RIS-empowered reflection and decode-and-forward relaying for coverage extension. *IEEE Communications Letters* 25 (5): 1692–1696.

Zhang, S., Zhang, H., Di, B. et al. (2022). Intelligent omni-surfaces: ubiquitous wireless transmission by reflective-refractive metasurfaces. *IEEE Transactions on Wireless Communications* 21 (1): 219–233.

ZTE Corporation (2021). New SI: Study on NR Network-controlled Repeaters, December 2021. https://www.3gpp.org/ftp/TSG_RAN/TSG_RAN/TSGR_94e/Docs/RP-213700.zip.

ZTE Corporation (2022). Study on NR Network-controlled Repeaters, May 2022. https://www.3gpp.org/dynareport/38867.htm (accessed 17 August 2023).

ZTE Corporation, Sanechips (2021). Support of Reconfigurable Intelligent Surface for 5G Advanced, March 2021. https://www.3gpp.org/ftp/tsg_ran/TSG_RAN/TSGR_91e/Docs/RP-210618.zip.

ZTE, Sanechips (2022). New WID on NR network-controlled repeaters, September 2022. https://www.3gpp.org/ftp/tsg_ran/TSG_RAN/TSGR_97e/Docs/RP-222673.zip.

6

RIS-Aided Massive MIMO Antennas[*]

Stefano Buzzi[1,2,3], Carmen D'Andrea[1,2], and Giovanni Interdonato[1,2]

[1]*Department of Electric and Information Engineering (DIEI), University of Cassino and Southern Latium, Cassino (FR), Italy*
[2]*Consorzio Nazionale Interuniversitario per le Telecomunicazioni (CNIT), Parma, Italy*
[3]*Dipartimento di Elettronica, Informazione e Bioingegneria, Politecnico di Milano, Milano, Italy*

6.1 Introduction

One of the key physical layer technologies used to boost capacity in currently under deployment 5G wireless networks is massive multiple-input multiple-output (MIMO) (Marzetta et al. 2016). Indeed, the use of the joint coherent transmission/reception from several antenna elements, and the fully digital beamforming at the base station (BS), permits an aggressive spatial multiplexing of a large number of user equipments (UEs) in the same time-frequency slot. Large numbers of antennas at the BSs bring pleasant phenomena deriving from the *law of the large numbers*, which simplifies signal processing and resource allocation and thus helps in reducing the hardware complexity and the circuit power consumption. However, uncontrollably increasing the number of active antenna elements (hence of RF chains) to improve the data rates can be, in general, an expensive and not energy-efficient solution (Ng et al. 2012). For this reason, engineers and researchers worldwide started looking for an alternative that couples the benefit of the massive MIMO technology with reduced complexity. To this purpose, reconfigurable intelligent surfaces (RISs) have emerged as one of the most striking innovations for the evolution of 5G systems into 6G systems. Indeed,

[*]This work has been supported by Italian Ministry of Education University and Research (MIUR) though the program "Dipartimenti di Eccellenza 2018-2022" and through the Progetti di Ricerca di Interesse Nazionale (PRIN) 2017 "LiquidEDGE" project. The work of C. D'Andrea has been also partially supported by the "Starting Grant 2020" (PRASG) Research Project funded by the University of Cassino and Southern Latium, Italy.

they constitute an emerging affordable solution to aid other technologies in implementing energy-efficient communications systems, and better coping with harsh propagation environments (Basar et al. 2019; Huang et al. 2019; Wu and Zhang 2019). An RIS is a meta-surface consisting of low-cost, typically passive, tiny reflecting elements that can be properly configured in real time to either focus the energy towards areas where coverage or additional capacity is needed or to null the interfering energy in specific spatial points. Basically, an RIS has the ability to shape the propagation channel and create a smart radio environment (Di Renzo et al. 2020) so as to improve the overall performance of the communication system. Technically, each RIS element introduces a configurable phase shift to an impinging electromagnetic wave, so that the resulting reflected beam is steered toward the desired direction. Beamforming is carried out by the environment in addition to the active radiative system where the impinging wave is generated. While initially available RIS technology enabled the tuning of the phase shift only of the reflected impinging waves, recently new types of RISs, nicknamed active RIS, have been introduced, with the capability of controlling both the amplitude and the phase of the reflected waves[1] (Long et al. 2021; Zhang et al. 2023). Several studies have recently appeared showing the benefits that RISs can bring in different applications such as, mobile edge computing networks (Bai et al. 2020), physical layer security systems (Hong et al. 2020), cognitive radio networks (Zhang et al. 2020) and radar systems (Buzzi et al. 2021b). In this chapter, we discuss the case in which an RIS is placed at short distance and in front of an active antenna array (see Fig. 6.1), with a non-large number of elements. The UEs to be served are placed in the backside of the active antenna array. The structure is to be intended as a low-complexity approximation of a massive MIMO system. A similar structure, but in a different setting and under different hypotheses, was also considered in Jamali et al. (2020). In the considered RIS-aided massive MIMO architecture, each active antenna is equipped with a dedicated RF chain and illuminates the RIS. Each passive element at the RIS receives a superposition of the signals transmitted by the active antennas and adds a desired phase shift to the overall signal. Leveraging the additional system optimization parameters introduced by the tunable reflecting elements of the RIS, a simpler transceiver architecture (i.e., less antennas and less RF chains) can be used than that usually employed in conventional massive MIMO systems with large-scale antenna array. The chapter is organized as follows. We firstly provide a signal and channel model suitable for the analysis of such a system, then we propose a channel estimation procedure which, exploiting only the active antennas and using different configurations of the RIS phase shifts, is capable

---

1 Notice that the amplification is realized through simple reflection-type amplifiers, and no RF chains are needed (see (Zhang et al. 2023) for further details).

to estimate the channel from a large number of RIS elements to all the UEs. We derive a closed-form expression for an achievable downlink spectral efficiency (SE) per user, by using the popular hardening lower-bound, and formulate a generic optimization problem that can be used for two purposes: (i) the minimization of the cross-correlation among channel vectors to reduce the multiuser interference, and (ii) striking a balance between the conflicting requirements that the RIS phases should not lower the norms of the composite channels, while they should minimize the cross-correlation among channel vectors. In the numerical results, we show the effectiveness of the considered architecture assuming both omnidirectional and directional active antennas, which allow us to overcome the traditional system with only active antennas.

Notation

We use non-bold letters for scalars, a and A, lowercase boldface letters, $\mathbf{a}$, for vectors, and uppercase lowercase letters, $\mathbf{A}$, for matrices. The transpose, the inverse, and the conjugate transpose of a matrix $\mathbf{A}$ are denoted by $\mathbf{A}^\mathsf{T}$, $\mathbf{A}^{-1}$, and $\mathbf{A}^\mathsf{H}$, respectively. The imaginary unit is denoted as i, the trace and the main diagonal of the matrix $\mathbf{A}$ are denoted as tr$(\mathbf{A})$ and diag$(\mathbf{A})$, respectively. The diagonal matrix obtained by the scalars $a_1, \ldots, a_N$ is denoted by diag $(a_1, \ldots, a_N)$. The N-dimensional identity matrix is denoted as $\mathbf{I}_N$, the $(N \times M)$-dimensional matrix with all zero entries is denoted as $\mathbf{0}_{N \times M}$ and $\mathbf{1}_{N \times M}$ denotes a $(N \times M)$-dimensional matrix with unit entries. The vectorization operator is denoted by vec$(\cdot)$ and the Kronecker product is denoted by $\otimes$. Given the matrices $\mathbf{A}$ and $\mathbf{B}$, with proper dimensions, the horizontal concatenation is denoted by $[\mathbf{A}, \mathbf{B}]$. The (m, ℓ)-th entry and the ℓ-th column of the matrix $\mathbf{A}$ are denoted as $[\mathbf{A}]_{(m,\ell)}$ and $[\mathbf{A}]_{(:,\ell)}$, respectively. The block-diagonal matrix obtained from matrices $\mathbf{A}_1, \ldots, \mathbf{A}_N$ is denoted by blkdiag $(\mathbf{A}_1, \ldots, \mathbf{A}_N)$. The statistical expectation and variance operators are denoted as E$\{\cdot\}$ and var$\{\cdot\}$ respectively; $\mathcal{CN}\left(\mu, \sigma^2\right)$ denotes a complex circularly symmetric Gaussian random variable with mean μ and variance σ^2.

6.2 System Model

Consider a single-cell system where a base station is equipped with an RIS-aided antenna and serves K single-antenna UEs. We denote by N_A the elements of the active antenna array, and by $N_R > N_A$ the number of configurable reflective elements of the RIS. The uplink channel between the k-th UE and the RIS is represented by the N_R-dimensional vector $\mathbf{h}_k$, while the matrix-valued channel from the RIS to the receive antenna array is represented through the $(N_A \times N_R)$-dimensional matrix $\mathbf{H}$. The tunable phase shifts introduced by the reflective elements of the RIS are represented through a diagonal $(N_R \times N_R)$-dimensional matrix $\boldsymbol{\Phi}$ whose ℓ-th coefficient is $e^{j\phi_\ell}$, with ϕ_ℓ the phase shift associated to the ℓ-th element of the RIS. Accordingly, the composite uplink channel from the k-th UE to the active array is the N_A-dimensional vector $\overline{\mathbf{h}}_k = \mathbf{H}\boldsymbol{\Phi}\mathbf{h}_k$, representing the composite channel from the active antenna array to the users through the RIS with phase-shift configuration represented by $\boldsymbol{\Phi}$. Note that in the following we omit the dependence of $\overline{\mathbf{h}}_k$ from $\boldsymbol{\Phi}$ for ease of notation. Under the rich-scattering assumption, usually verified at sub-6 GHz frequencies, the channel vector $\mathbf{h}_k$ is modeled as a complex Gaussian distributed vector with zero mean and covariance matrix $\beta_k \mathbf{I}_{N_R}$, with β_k representing the large-scale fading coefficient for the k-th UE.

6.2.1 Channel Model

We model the RIS-to-active array $(N_A \times N_R)$-dimensional matrix $\mathbf{H}$. To this aim, and for the sake of simplicity, we consider the scheme represented in Fig. 6.1, where we have an active array that illuminates a large RIS placed at a short distance, that we denote by D; the UEs to be served are placed in the backside of the active antenna array. We make the following assumptions:

1) We neglect the blockage that the active array can introduce on the electromagnetic radiation reflected by the RIS. In practical 3D scenarios, the active array can be placed laterally with respect to the RIS so as to simply avoid the blockage problem (see Fig. 6.1(a)).
2) Since the network UEs are placed at some distance from the back of the active antenna array, the direct signal from the active antenna array to the mobile UEs will be considerably weaker than the RIS-reflected signal, and will be thus neglected.
3) Given the short distance between the RIS and the active antenna array, the channel $\mathbf{H}$ is modeled as a deterministic quantity, since we can neglect contributions from nearby scatterers.
4) We neglect the mutual coupling between the elements of the active array and between the elements of the RIS. This is an assumption that is usually verified provided that the element spacing at the active array and at the RIS, i.e., d_A and d_R, respectively, does not fall below the typical $\lambda/2$ value (Qian and Di Renzo 2021), with λ the wavelength corresponding to the radiated frequency.
5) We assume that each element of the RIS is in the far field of each element of the active antenna array. However, the whole RIS is not required to be in the far field of the whole active antenna array. This implies that the radiation coming from the i-th active antenna and impinging on the j-th RIS element has a

power that falls proportionally to $1/d_{i,j}^2$, with $d_{i,j}$ the distance between the i-th active antenna and the j-th RIS element; however, the plane-wave assumption is locally fulfilled when considering the single RIS element, but, when considering the entire RIS, spherical wave propagation is to be accounted for. The received power density is thus not constant across the RIS elements.

With regard to assumption 5) we note that the far-field condition among the generic i-th transmit antenna and j-th RIS element is $d_{i,j} > 2\max\{\Delta_A^2, \Delta_R^2\}/\lambda$ and $d_{i,j} \gg \max\{\Delta_A, \Delta_R, \lambda\}$, with Δ_A and Δ_R the size of the active antenna and RIS element, respectively (Stutzman and Thiele 2012). The far-field condition among the whole RIS and the whole active array would have instead required that the fulfillment of the previous conditions with $d_{i,j}$, Δ_A, and Δ_R replaced by D, $N_A d_A$, and $N_R d_R$. We also note that, as will be clarified in the sequel of the chapter, the absence of the far-field assumption between the whole active array and the RIS, although making the matrix $\mathbf{H}$ modeling somewhat more involved, is crucial in order to have a full-rank matrix channel that will permit communicating along many signal space directions; indeed, when D grows large, the line-of-sight condition leads in the absence of strong scatterers to a rank-one channel matrix that provides no spatial diversity. Based on the above assumptions, and according to standard electromagnetic radiation theory, the (i, j)-th entry of $\mathbf{H}$, say $\mathbf{H}(i, j)$, representing the uplink channel coefficient from the j-th element of the RIS to the i-th active antenna, can be written as:

$$\mathbf{H}(i, j) = \sqrt{\rho G_A(\theta_{i,j}) G_R(\theta_{i,j})} \frac{\lambda}{4\pi d_{i,j}} e^{-\mathrm{i}2\pi d_{i,j}/\lambda}. \tag{6.1}$$

In the above equation, $G_A(\theta_{i,j})$ and $G_R(\theta_{i,j})$ represent the i-th active antenna element and the j-th passive RIS element gains corresponding to the look angle $\theta_{i,j}$ (see Fig. 6.1(b)), while $\rho < 1$ is a real-valued term modeling the RIS efficiency in reflecting the impinging waves. Fig. 6.2 shows the eigenvalues of the matrix $\mathbf{H}$ as a function of the ratio D/λ for the case of $N_A = 8$, $N_R = 64$ and $d_A = d_R = \lambda/2$. The value $\rho = 1$ has been taken and we have assumed $G_A(\theta) = G_R(\theta) = 2$, for any $\theta \in [-\pi/2, \pi/2]$. The figure shows that, for $D \leq 10\lambda$, the eigenvalues are very close, which denotes the fact that the matrix $\mathbf{H}$ is full rank and all its eigendirections have approximately the same strength, whereas for $D > 10\lambda$ the eigenvalues get more spread and tend to take smaller and smaller values. This behavior indicates that the matrix is converging to the rank-one configuration.

6.2.2 Active Antenna Configuration

In order to see the impact of the active antenna configuration, we consider two kinds of active antennas: *omnidirectional* and *directional*. In the first case, the active antennas are omnidirectional and radiate power in all the angular

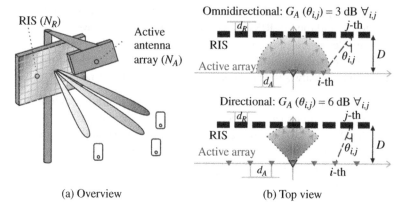

(a) Overview   (b) Top view

Figure 6.1 (a) A possible configuration of the considered RIS-aided antenna structure in 3D space. A small planar array with N_A active antennas is mounted at close distance from an RIS with N_R elements. The relative positioning of the active array and of the RIS is such that no obstruction is created between the RIS and the UEs on the ground. (b) Proposed arrangement of active array and RIS for a 2D layout, in case of (top) omnidirectional and (bottom) directional antennas radiating in the angular sector $[-\pi/2, \pi/2]$ and $[-\pi/4, \pi/4]$, respectively.

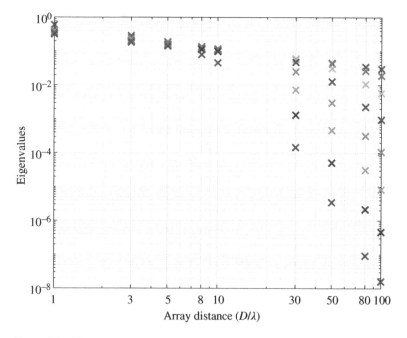

Figure 6.2 Eigenvalues of **H** as a function of D/λ, for the case $N_A = 8$, $N_R = 64$ and $d_A = d_R = \lambda/2$.

directions, i.e., $G_A(\theta) = 3$ dB $\forall \theta \in [-\pi/2, \pi/2]$. One other possibility is to assume that the antennas of the active array radiate in a narrower angular sector (see lower part of Fig. 6.1(b)); denoting by α the width of such angular sector, the gain $G_A(\theta)$ is taken equal to

$$G_A(\theta) = \begin{cases} g_D(\alpha) = 10 \log_{10}\left(\dfrac{2\pi}{\alpha}\right), & \text{if } \theta \in [-\alpha/2, \alpha/2], \\ 0, & \text{otherwise} \end{cases} \tag{6.2}$$

Note that, given the RIS and active array relative position depicted in Fig. 6.1(b), if $N_R \gg N_A$ some RIS element are external to the angular sector in which each active antenna radiates power. This situation strongly degrades the performance of the system because, actually, the "effective" elements of the RIS are only the ones that receive power from active antennas. In this latter situation, elements of the active antenna array are to be spaced at a distance $d_A > \lambda/2$ in order to ensure that each RIS element is illuminated by the main beam of at least one element of the active antenna array. Based on simple geometrical considerations, a suitable antenna spacing in this case is

$$d_A = \frac{(N_R - 1)d_R - 2D\tan(\alpha/2)}{N_A - 1}. \tag{6.3}$$

6.3 Uplink/Downlink Signal Processing

6.3.1 Uplink Channel Estimation

During the uplink training phase, each UE transmits a known pilot sequence; we denote by $\boldsymbol{\varphi}_k$ the pilot sequence, of length τ_p and squared norm τ_p, assigned to the k-th UE. Since, in this chapter, we assume $N_R > N_A$, in order to allow the active array to estimate the channels $\{\mathbf{h}_1, \ldots, \mathbf{h}_K\}$, we use a similar approach as in Buzzi et al. (2021a). In particular, we assume that the pilot sequences are transmitted by the UEs Q times, each one with a different RIS configuration. Without loss of generality, we assume $Q = N_R/N_A$ in order to guarantee generation of a number of observables that is not smaller than the number of unknown coefficients. Pilots are drawn from a set of orthogonal sequences, which, for the sake of simplicity, are assumed to be real; clearly, the cardinality of such set cannot exceed τ_p, and, for $K > \tau_p$, the same pilot sequence will be assigned to more than one UE, causing the so-called "pilot contamination" effect, which degrades the accuracy of channel estimation. Denoting by $\eta_k^{(u,t)}$ the uplink transmit power from the k-th UE during the training phase, the data observed at the active antenna array when the RIS assumes the q-th configuration can be arranged in the following

$(N_A \times \tau_p)$-dimensional matrix:

$$\mathbf{Y}^{(q)} = \sum_{k=1}^{K} \sqrt{\eta_k^{(u,t)}} \mathbf{H}\mathbf{\Phi}_T^{(q)} \mathbf{h}_k \boldsymbol{\varphi}_k^{\mathsf{T}} + \mathbf{N}^{(q)}, \tag{6.4}$$

where $\mathbf{\Phi}_T^{(q)}$ is the q-th configuration of the RIS assumed

$$\mathbf{\Phi}_T^{(q)} = \mathrm{diag}\left(\boldsymbol{\phi}^{(q)}\right), \tag{6.5}$$

and $\boldsymbol{\phi}^{(q)}$ is a N_R-dimensional vector whose ℓ-th entry is $e^{i\tilde{\phi}_{\ell,q}}$ with $\tilde{\phi}_{\ell,q}$ a random angle chosen from a quantized set of phase shifts available at the RIS. In order to estimate the k-th UE channel $\mathbf{h}_k$, the data $\mathbf{Y}^{(q)}$ is projected along the k-th UE pilot sequence, leading to

$$\mathbf{y}_{k,q} = \sqrt{\eta_k^{(u,t)}} \tau_p \mathbf{H}\mathbf{\Phi}_T^{(q)} \mathbf{h}_k + \sum_{\substack{j=1 \\ j \neq k}}^{K} \sqrt{\eta_j^{(u,t)}} \mathbf{H}\mathbf{\Phi}_T^{(q)} \mathbf{h}_j \rho_{j,k} + \mathbf{n}_{k,q}, \tag{6.6}$$

with $\rho_{j,k} = \boldsymbol{\varphi}_j^{\mathsf{T}} \boldsymbol{\varphi}_k$, and $\mathbf{n}_k = \mathbf{N}\boldsymbol{\varphi}_k \sim \mathcal{CN}(0, \sigma^2 \tau_p \mathbf{I}_{N_A})$. These observables are collected for all $q = 1, \ldots, Q$ as the following $N_R \times 1$-dimensional vector

$$\tilde{\mathbf{y}}_k = \left[\mathbf{y}_{k,1}^{\mathsf{T}}, \, \ldots, \, \mathbf{y}_{k,Q}^{\mathsf{T}} \right]^{\mathsf{T}}. \tag{6.7}$$

After some manipulations, Eq. (6.7) can be written as

$$\tilde{\mathbf{y}}_k = \sqrt{\eta_k^{(u,t)}} \tau_p \tilde{\mathbf{H}}_T \mathbf{h}_k + \sum_{\substack{j=1 \\ j \neq k}}^{K} \sqrt{\eta_j^{(u,t)}} \tilde{\mathbf{H}}_T \mathbf{h}_j \rho_{j,k} + \tilde{\mathbf{n}}_k, \tag{6.8}$$

where

$$\tilde{\mathbf{H}}_T = \left[\left(\mathbf{H}\mathbf{\Phi}_T^{(1)} \right)^{\mathsf{T}}, \, \ldots, \, \left(\mathbf{H}\mathbf{\Phi}_T^{(Q)} \right)^{\mathsf{T}} \right]^{\mathsf{T}}, \tag{6.9}$$

$$\tilde{\mathbf{n}}_k = \left[\mathbf{n}_{k,1}^{\mathsf{T}}, \, \ldots, \, \mathbf{n}_{k,Q}^{\mathsf{T}} \right]^{\mathsf{T}}. \tag{6.10}$$

In order to estimate the *essential directions* of the channels $\{\mathbf{h}_1, \ldots, \mathbf{h}_K\}$, we use the *economy size* singular valued decomposition (SVD) of the matrix $\tilde{\mathbf{H}}_T$:

$$\tilde{\mathbf{H}}_T = \mathbf{U}\mathbf{\Lambda}\mathbf{V}^{\mathsf{H}} \approx \tilde{\mathbf{U}}\tilde{\mathbf{\Lambda}}\tilde{\mathbf{V}}^{\mathsf{H}}, \tag{6.11}$$

where $\mathbf{U}, \mathbf{\Lambda}, \mathbf{V}$ are the N_R-dimensional matrices corresponding to the regular SVD and, say q the number of significant eigenvalues of the matrix $\tilde{\mathbf{H}}_T$, $\tilde{\mathbf{U}}$ and $\tilde{\mathbf{V}}$ contains the columns of $\mathbf{U}$ and $\mathbf{V}$ corresponding to the q highest eigenvalues and $\tilde{\mathbf{\Lambda}}$ is the diagonal matrix containing these eigenvalues. Note that the approximation in Eq. (6.11) holds if the matrix $\tilde{\mathbf{H}}_T$ has q eigenvalues significantly higher with respect to the remaining ones. In our scenario, it holds because of the structure of matrix $\mathbf{H}$. Practically q can be choosen as the number of the main eigenvalues which "contains" the 99% of the whole energy of the eigenvalues of $\tilde{\mathbf{H}}_T$. Using the above relationships, we can write the q-dimensional vector

$$\bar{\mathbf{y}}_k = \tilde{\mathbf{U}}^H \tilde{\mathbf{y}}_k = \sqrt{\eta_k^{(u,t)}} \tau_p \tilde{\mathbf{\Lambda}} \tilde{\mathbf{V}}^H \mathbf{h}_k + \sum_{\substack{j=1 \\ j \neq k}}^{K} \sqrt{\eta_j^{(u,t)}} \tilde{\mathbf{\Lambda}} \tilde{\mathbf{V}}^H \mathbf{h}_j \rho_{j,k} + \tilde{\mathbf{U}}^H \tilde{\mathbf{n}}_k. \tag{6.12}$$

Given the properties of the SVD, we have

$$\mathbf{h}_k = \mathbf{V} \mathbf{V}^H \mathbf{h}_k = \mathbf{V} \mathbf{v}_k, \tag{6.13}$$

where $\mathbf{v}_k = \mathbf{V}^H \mathbf{h}_k = \left[\tilde{\mathbf{v}}_k^T, \bar{\mathbf{v}}_k^T \right]^T$ and given the approximation in Eq. (6.11) the following holds

$$\mathbf{h}_k \approx \tilde{\mathbf{V}} \tilde{\mathbf{v}}_k. \tag{6.14}$$

Thus, we perform a channel estimation technique aimed at determining the strongest directions of the channels, corresponding to the highest eigenvalues, that we can estimate with highest power and consequently recover the channel exploiting Eq. (6.14). Based on Eq. (6.14), we can rewrite (6.12) as

$$\bar{\mathbf{y}}_k \approx \sqrt{\eta_k^{(u,t)}} \tau_p \tilde{\mathbf{\Lambda}} \tilde{\mathbf{v}}_k + \sum_{\substack{j=1 \\ j \neq k}}^{K} \sqrt{\eta_j^{(u,t)}} \tilde{\mathbf{\Lambda}} \tilde{\mathbf{v}}_j \rho_{j,k} + \tilde{\mathbf{U}}^H \tilde{\mathbf{n}}_k. \tag{6.15}$$

Thus, the linear minimum mean square error (MMSE) estimate of $\tilde{\mathbf{v}}_k$ can be expressed as

$$\hat{\tilde{\mathbf{v}}}_k = \mathbf{R}_{\tilde{\mathbf{v}}_k \bar{\mathbf{y}}_k} \mathbf{R}_{\bar{\mathbf{y}}_k \bar{\mathbf{y}}_k}^{-1} \bar{\mathbf{y}}_k, \tag{6.16}$$

where

$$\mathbf{R}_{\tilde{\mathbf{v}}_k \bar{\mathbf{y}}_k} \triangleq E\left\{ \tilde{\mathbf{v}}_k \bar{\mathbf{y}}_k^H \right\} = \sqrt{\eta_k^{(u,t)}} \tau_p \beta_k \tilde{\mathbf{\Lambda}}^H, \tag{6.17}$$

and

$$\mathbf{R}_{\bar{\mathbf{y}}_k \bar{\mathbf{y}}_k} \triangleq E\left\{ \bar{\mathbf{y}}_k \bar{\mathbf{y}}_k^H \right\} = \left[\sum_{j=1}^{K} \eta_j^{(u,t)} \beta_j \rho_{j,k}^2 \right] \tilde{\mathbf{\Lambda}} \tilde{\mathbf{\Lambda}}^H + \sigma^2 \tau_p \mathbf{I}_q. \tag{6.18}$$

The channel estimate $\hat{\mathbf{h}}_k$, exploiting Eqs. (6.14) and (6.16) can be thus written as

$$\hat{\mathbf{h}}_k \approx \tilde{\mathbf{V}} \mathbf{R}_{\tilde{\mathbf{v}}_k \bar{\mathbf{y}}_k} \mathbf{R}_{\bar{\mathbf{y}}_k \bar{\mathbf{y}}_k}^{-1} \bar{\mathbf{y}}_k. \tag{6.19}$$

It is easy to realize that the LMMSE estimation of the vector containing the strongest directions of the channel $\hat{\mathbf{h}}_k$ is Gaussian distributed with zero mean and covariance matrix

$$\mathbf{R}_{\hat{\mathbf{h}}_k \hat{\mathbf{h}}_k} = \tilde{\mathbf{V}} \mathbf{R}_{\tilde{\mathbf{v}}_k \bar{\mathbf{y}}_k} \mathbf{R}_{\bar{\mathbf{y}}_k \bar{\mathbf{y}}_k}^{-1} \mathbf{R}_{\tilde{\mathbf{v}}_k}^H \tilde{\mathbf{V}}^H. \tag{6.20}$$

Moreover, the estimation error $\tilde{\mathbf{h}}_k = \mathbf{h}_k - \hat{\mathbf{h}}_k$ is statistically independent from the channel estimate $\hat{\mathbf{h}}_k$ and is itself a zero-mean complex Gaussian vector with covariance matrix $\mathbf{R}_{\tilde{\mathbf{h}}_k \tilde{\mathbf{h}}_k} = \beta_k \mathbf{I}_{N_R} - \mathbf{R}_{\hat{\mathbf{h}}_k \hat{\mathbf{h}}_k}$. Notice also that if two UEs, k and j, are assigned the same pilot sequence during the uplink training phase, i.e., $\boldsymbol{\varphi}_k = \boldsymbol{\varphi}_j$, then we will have $\mathbf{y}_k = \mathbf{y}_j$, and the channel estimates $\hat{\mathbf{h}}_k$ and $\hat{\mathbf{h}}_j$ will be proportional since the following two relations hold:

$$
R_{\tilde{v}_k \tilde{y}_k} = \frac{\beta_k \sqrt{\eta_k^{(u,t)}}}{\beta_j \sqrt{\eta_j^{(u,t)}}} R_{\tilde{v}_j \tilde{y}_j} \implies \hat{\mathbf{h}}_k = \frac{\beta_k \sqrt{\eta_k^{(u,t)}}}{\beta_j \sqrt{\eta_j^{(u,t)}}} \hat{\mathbf{h}}_j \tag{6.21}
$$

6.3.2 Downlink Data Transmission

During the downlink data transmission phase, the complex envelope of the signal received by the k-th UE in the n-th symbol interval can be shown to be expressed as

$$
r_k(n) = \sqrt{\eta_k^{(d)}} \mathbf{h}_k^\mathsf{T} \boldsymbol{\Phi}^\mathsf{T} \mathbf{H}^\mathsf{T} \mathbf{w}_k x_k(n) + \sum_{\substack{j=1 \\ j \neq k}}^{K} \sqrt{\eta_j^{(d)}} \mathbf{h}_k^\mathsf{T} \boldsymbol{\Phi}^\mathsf{T} \mathbf{H}^\mathsf{T} \mathbf{w}_j x_j(n) + z_k(n),
$$

$$\tag{6.22}$$

where $\eta_k^{(d)}$ and $\mathbf{w}_k$ represent the downlink transmit power and the downlink beamformer for UE k, respectively, while $x_k(n)$ is the data symbol, drawn from a constellation with unit average energy, intended for UE k in the n-th signaling interval. The term $z_k(n)$ is the AWGN term, modeled as $\mathcal{CN}(0, \sigma_k^2)$.

6.4 Performance Measures

6.4.1 SINR and Spectral Efficiency under Perfect Channel State Information (CSI)

For benchmarking purposes, we start considering the ideal case in which perfect CSI is available both at the active array and at the UEs. Given (6.22), it can be easily shown that the k-th UE downlink Signal-to-Interference plus Noise Ratio (SINR), for perfect CSI, is written as:

$$
\gamma_k^{(d)} = \frac{\eta_k^{(d)} \left| \mathbf{h}_k^\mathsf{T} \boldsymbol{\Phi}^\mathsf{T} \mathbf{H}^\mathsf{T} \mathbf{w}_k \right|^2}{\displaystyle\sum_{\substack{j=1 \\ j \neq k}}^{K} \eta_j^{(d)} \left| \mathbf{h}_k^\mathsf{T} \boldsymbol{\Phi}^\mathsf{T} \mathbf{H}^\mathsf{T} \mathbf{w}_j \right|^2 + \sigma_k^2.} \tag{6.23}
$$

Consequently, assuming Gaussian-distributed codewords, the SE for the k-th UE can be obtained by using the classical Shannon expression, i.e.,:

$$R_k^{(d)} = \log_2(1 + \gamma_k^{(d)}), \quad [\text{bit/s}]. \tag{6.24}$$

6.4.2 SINR and Spectral Efficiency under Imperfect Channel State Information (CSI)

In general, when the channel coefficients are not perfectly known, it is not clear what is "signal" and what is "interference" in Eq. (6.22). In particular, for imperfect CSI, the intuitive notion of the SINR in Eq. (6.23) is in general not rigorously related to a corresponding notion of information theoretic achievable rate (Caire 2018).

6.4.2.1 The Upper-Bound (UB) to the System Performance

An upper bound to the downlink SE, that can be easily evaluated numerically and can give us an intuitive idea of the system performance can be written as (Caire 2018)

$$R_{k,\text{UB}}^{(d)} = \bar{\xi}\mathbb{E}\left[\log_2(1 + \gamma_k^{(d)})\right], \quad [\text{bit/s}], \tag{6.25}$$

where the pre-log factor $\bar{\xi} < 1$ accounts for the fraction of the channel coherence interval used for downlink data transmission. In Eq. (6.25) the expectation is computed over the fast fading channel realizations and the beamforming vectors $\mathbf{w}_1, \ldots, \mathbf{w}_K$ in Eq. (6.23) are evaluated exploiting the channel estimates.

6.4.2.2 The Hardening Lower-Bound (LB) to System Performance

What can be done is to derive also lower bound (LB) to the SE, as detailed in the following (Marzetta 2016). Under the assumption of Rayleigh fading, and using a conjugate beamformer, i.e., letting $\mathbf{w}_k = \mathbf{H}^*\mathbf{\Phi}^*\hat{\mathbf{h}}_k^*$, a lower bound to the achievable SE for UE k can be found by resorting to the popular hardening lower bound, by treating all the interference and noise contributions as uncorrelated effective noise. In particular, we can rewrite (6.22) as follows:

$$
r_k(n) = \underbrace{\sqrt{\eta_k^{(d)}}E\left\{\mathbf{h}_k^\mathsf{T}\mathbf{\Phi}^\mathsf{T}\mathbf{H}^\mathsf{T}\mathbf{w}_k\right\}x_k(n)}_{\text{DS}_k}
$$

$$
+ \underbrace{\sqrt{\eta_k^{(d)}}\left[\mathbf{h}_k^\mathsf{T}\mathbf{\Phi}^\mathsf{T}\mathbf{H}^\mathsf{T}\mathbf{w}_k - E\left\{\mathbf{h}_k^\mathsf{T}\mathbf{\Phi}^\mathsf{T}\mathbf{H}^\mathsf{T}\mathbf{w}_k\right\}\right]x_k(n)}_{\text{BU}_k}
$$

$$
+ \underbrace{\sum_{\substack{j=1\\j\neq k}}^{K}\sqrt{\eta_j^{(d)}}\mathbf{h}_k^\mathsf{T}\mathbf{\Phi}^\mathsf{T}\mathbf{H}^\mathsf{T}\mathbf{w}_j x_j(n)}_{\text{UI}_{k,j}} + z_k(n),
\tag{6.26}
$$

The hardening lower bound is thus written as

$$R_{k,\text{LB}}^{(d)} = \overline{\xi} \log_2 \left(1 + \gamma_{k,\text{LB}}^{(d)}\right),$$ (6.27)

with $\gamma_{k,\text{LB}}^{(d)}$ is

$$\gamma_{k,\text{LB}}^{(d)} = \frac{\left|\text{DS}_k\right|^2}{E\left\{\left|\text{BU}_k\right|^2\right\} + \sum_{\substack{j=1 \\ j \neq k}}^{K} E\left\{\left|\text{UI}_{k,j}\right|^2\right\} + \sigma_k^2}.$$ (6.28)

The expressions in Eqs. (6.27) and (6.28) hold for any choice of beamforming vectors and RIS configuration. A closed-form expression of $\gamma_{k,\text{LB}}^{(d)}$ can be obtained with conjugate beamforming and no-channel dependent $\mathbf{\Phi}$, and it is reported in Eq. (6.29) at the top of this page and the complete derivation is provided in Appendix. Finally, we can note that the analysis of the considered structure is, from a mathematical point of view, different from that required to analyze conventional massive MIMO systems with one active antenna array with a large number of elements. As it is well known, in a conventional massive MIMO system, when the number an antenna elements grow without bound the channel vectors *hardens* (i.e., their norm tends to be deterministic) and channel vectors belonging to different UEs become almost orthogonal. Although the RIS may be helpful in making the composite channel vectors $\left\{\overline{\mathbf{h}}_1, \dots, \overline{\mathbf{h}}_K\right\}$ almost orthogonal, letting N_R grow without bounds and keeping N_A fixed does not provide the above described effect. Otherwise stated, it can be shown that the channel hardening property holds *only* if the number of active antennas increases without bounds, we omit this proof here for due to the lack of space.

$$\gamma_{k,\text{LB}}^{(d)} = \frac{\eta_k^{(d)} \text{tr}^2 \left[\mathbf{A}_\Phi \mathbf{R}_{\hat{\mathbf{h}}_k \hat{\mathbf{h}}_k}^*\right]}{\left\{ \begin{array}{l} \eta_k^{(d)} \text{tr}\left(\beta_k \mathbf{A}_\Phi \mathbf{R}_{\hat{\mathbf{h}}_k \hat{\mathbf{h}}_k}^* \mathbf{A}_\Phi\right) + \sum_{j \in \mathcal{P}_k} \eta_j^{(d)} \left(\dfrac{\beta_j \sqrt{\eta_j^{(u,t)}}}{\beta_k \sqrt{\eta_k^{(u,t)}}}\right)^2 \left[\text{tr}^2(\mathbf{A}_\Phi \mathbf{R}_{\hat{\mathbf{h}}_k \hat{\mathbf{h}}_k}^*)\right] \\ + \beta_k \text{tr}(\mathbf{A}_\Phi \mathbf{R}_{\hat{\mathbf{h}}_k \hat{\mathbf{h}}_k}^* \mathbf{A}_\Phi)\right] + \sum_{j \notin \mathcal{P}_k} \eta_j^{(d)} \beta_k \text{tr}\left[\mathbf{A}_\Phi \mathbf{R}_{\hat{\mathbf{h}}_j \hat{\mathbf{h}}_j}^* \mathbf{A}_\Phi\right] + \sigma_k^2 \end{array} \right\}}$$ (6.29)

6.5 Optimization of the RIS Phase Shifts

In order to optimize the system performance for finite N_A, we define on the matrix

$$\mathbf{Q}(\mathbf{\Phi}) = \mathbf{S}(\mathbf{\Phi})\mathbf{S}(\mathbf{\Phi})^\mathsf{H},$$ (6.30)

with

$$S(\boldsymbol{\Phi}) = \mathbf{H}\boldsymbol{\Phi}\overline{\mathbf{H}}, \tag{6.31}$$

and $\overline{\mathbf{H}} = [\mathbf{h}_1, \dots, \mathbf{h}_K]$. The entries of matrix $\mathbf{Q}(\boldsymbol{\Phi})$ give us an idea on the potential interference that the UEs suffer for a given RIS configuration $\boldsymbol{\Phi}$. We thus focus on the matrix $\mathbf{Q}(\boldsymbol{\Phi})$ in order to formulate two optimization problems aimed at the allocation of the RIS phase offsets. Firstly, we define one objective function as

$$f_1(\boldsymbol{\Phi}) = \sum_{k=1}^{K} \sum_{k'=k+1}^{K} |\overline{\mathbf{h}}_k^H \overline{\mathbf{h}}_{k'}|. \tag{6.32}$$

Otherwise stated, $f_1(\boldsymbol{\Phi})$ is the sum of the off-diagonal entries of the matrix $\mathbf{Q}(\boldsymbol{\Phi})$. Then, we define the second objective function as

$$f_2(\boldsymbol{\Phi}) = \frac{\displaystyle\sum_{k=1}^{K} \sum_{k'=k+1}^{K} \left|\overline{\mathbf{h}}_k^H \overline{\mathbf{h}}_{k'}\right|}{\displaystyle\sum_{k=1}^{K} \|\overline{\mathbf{h}}_k\|^2}, \tag{6.33}$$

otherwise stated, $f_2(\boldsymbol{\Phi})$ is the ratio between the sum of the off-diagonal entries and the sum of the diagonal entries of the matrix $\mathbf{Q}(\boldsymbol{\Phi})$. The minimization problem with objective function $f_1(\boldsymbol{\Phi})$ is aimed at the minimization of the cross-correlation among channel vectors to reduce interference. On the other hand, the one with objective function $f_2(\boldsymbol{\Phi})$ strikes a balance between the conflicting requirements that the RIS phases should not lower the norms of the composite channels, while at the same time should minimize cross-correlation among channel vectors. The generic optimization problem can be written as

$$\min_{\boldsymbol{\Phi}} \quad f_{(\cdot)}(\boldsymbol{\Phi}) \tag{6.34a}$$

$$\text{s.t. } |\phi_i| = 1, \quad i = 1, \dots, N_R \tag{6.34b}$$

where $(\cdot) \in \{1, 2\}$. The optimization problem (6.34) is not convex, in the following we propose a solution based on the alternating maximization theory (Bertsekas 1997). We consider the N_R optimization problems, $i = 1, \dots, N_R$,

$$\min_{\phi_i} f_{(\cdot)}(\phi_i) \tag{6.35a}$$

$$\text{s.t. } |\phi_i| = 1, \tag{6.35b}$$

with ϕ_ℓ, $\ell = 1, \dots, i-1, i+1, \dots, N_R$ remain fixed. This optimization problem can be solved using an exhaustive search. It is proved in (Bertsekas 1997, Proposition 2.7.1) that iteratively solving (6.35) monotonically decreases the value of the objective of (6.34), and converges to a first-order optimal point if the solution of (6.35) is unique for any i. The optimization algorithms with the objective functions defined in Eqs. (6.32) and (6.32) can be summarized as in Algorithm 6.1.

Algorithm 6.1 Optimization of the RIS Phase Shifts

1: Choose a starting random RIS configuration of the phase shifts $\mathbf{\Phi}$;
2: **repeat**
3: **for** $i = 1 \to N_R$ **do**
4:   Solve Problem (6.35);
5:   Set the (i, i)-th entry of the matrix $\mathbf{\Phi}$ as the solution of Problem (6.35), ϕ_i^* say.
6: **end for**
7: **until** convergence of the objective function

6.6 Numerical Results

In our simulation setup, we consider a communication bandwidth of $W = 20$ MHz centered over the carrier frequency $f_0 = 1.9$ GHz. The additive thermal noise is assumed to have a power spectral density of -174 dBm/Hz, while the front-end receiver at the active array and at the UEs is assumed to have a noise figure of 5 dB. The heights of active array and RIS are 10 m while the height of the UEs is 1.5 m. The $K = 8$ UEs are uniformly distributed in an angular region of $120°$ in front of the RIS at distances in 10, 400 m. We model the large-scale fading coefficients β_k from the UEs to the RIS according to 3GPP (2019) and the matrix $\mathbf{H}$ according to Eq. (6.1) with $D = 5\lambda$, $\rho = 1$, $d_A = \lambda/2$; in the case of omnidirectional active antennas $G_R(\theta) = 3$ dB $\forall\theta$ and $d_R = \lambda/2$; in the case of directional active antennas d_R follows Eq. (6.3) with $g_D(\alpha)$ and α following Eq. (6.2) specified in the figures' legends and captions. With regard to the channel estimation procedure, q is chosen as the number of eigenvalues which collects the 98% of the total sum of eigenvalues, orthogonal sequences with length $\tau_p = 16$ are assumed at the UEs and the power transmitted during the uplink training is $\eta_k = 800$ mW. The channel coherence time is assumed (in samples) to be $\tau_c = 200$ and $\bar{\xi} = (\tau_c - \tau_p)/\tau_c$. The maximum power transmitted by the active array is $P_{\max}^A = 7$ dBW and it is equally distributed between the UEs, i.e., $\eta_k^{(d)} = P_{\max}^A/K\|\mathbf{w}_k\|^2$. We start considering the normalized mean squares error (NMSE) defines as

$$\text{NMSE} = \frac{\sum_{k=1}^{K} \mathbb{E}\left[\left\|\mathbf{h}_k - \hat{\mathbf{h}}_k\right\|^2\right]}{\sum_{k=1}^{K} \mathbb{E}\left[\|\mathbf{h}_k\|^2\right]}. \tag{6.36}$$

In Fig. 6.3, we show the NMSE versus the number or RIS elements for different values of number of antennas at the active array for omnidirectional and bidirectional cases. For the latter case, we consider two different values of $g_D(\alpha)$ and α for

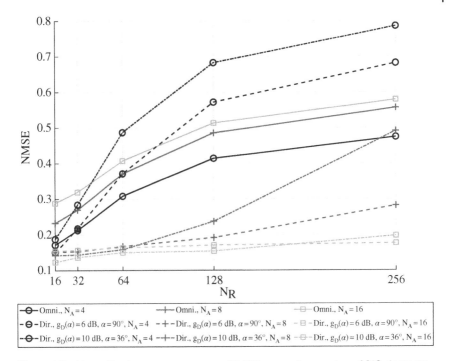

Figure 6.3 Normalized mean squares error (NMSE) versus the number of RIS elements, N_R, for different values of N_A in the case of omnidirectional and directional active antennas.

the directional active antennas, specified in the legend. We can see that, when N_R increases, the performance of the channel estimation decreases for low values of active antennas, especially in the cases of omnidirectional active antennas. This behavior is a consequence of the fact that the number of parameters to estimate increases and the number of active antennas is too low to obtain a good estimation of them. When the number of active antennas is 16, the channel estimates obtained with our procedure are more accurate and the NMSE remains quite constant when the number of RIS elements increases.

In Fig. 6.4, we report the cumulative distribution function (CDF) of the SE per user using the expressions discussed in Section 6.4 for perfect CSI (PCSI) and imperfect CSI (ICSI). In this figure, we assume a random configuration of the RIS phase shifts. We can see that in the case of directional active antennas the performance in terms of SE increases with respect to the omnidirectional case due to the higher values in the matrix **H**, which models the communication between the RIS and the active array. With the considered system, we can note that the hardening LB is not as tight as in the case of classical massive MIMO, and this is due to

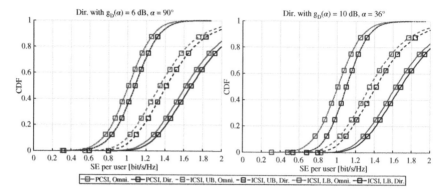

Figure 6.4 CDF of the Spectral efficiency (SE) per user in the case of omnidirectional (Omni.) and directional (Dir.) active antennas with $N_R = 64$ and $N_A = 16$. Comparison between the case of perfect CSI (PCSI), and imperfect CSI (ICSI) with upper-bound (UB) and lower-bound (LB) of the performance.

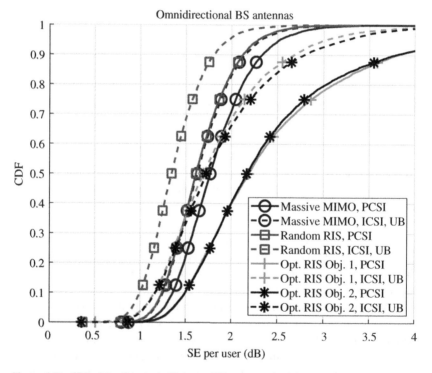

Figure 6.5 CDF of the Spectral efficiency (SE) per user in the case of omnidirectional active antennas with $N_R = 64$ and $N_A = 16$. Comparison between the case of perfect CSI (PCSI), and imperfect CSI (ICSI) with upper-bound (UB) of the performance with RIS phase-shifts optimization with objective functions $f_1(\cdot)$ and $f_2(\cdot)$.

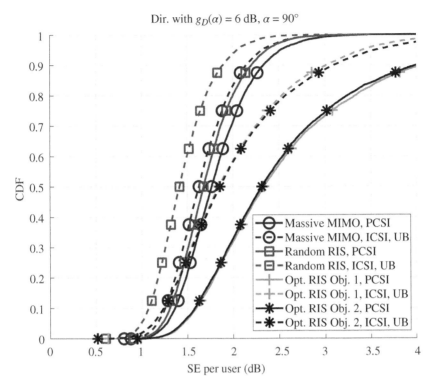

Figure 6.6 CDF of the Spectral efficiency (SE) per user in the case of directional active antennas with $g_D(\alpha) = 6$ dB, $\alpha = 90°$, $N_R = 64$ and $N_A = 16$. Comparison between the case of perfect CSI (PCSI), and imperfect CSI (ICSI) with upper-bound (UB) of the performance with RIS phase-shifts optimization with objective functions $f_1(\cdot)$ and $f_2(\cdot)$.

the fact that the channel hardening assumption does not hold increasing *only* the number of RIS elements N_R.

In Figs. 6.5–6.7, we report the CDF of the SE per user in the case of PCSI and ICSI using the UB of the performance. We report the cases of RIS with random configuration of phase shifts (Random RIS) and in the cases of phase shift optimization (Opt. RIS) detailed in Section 6.5 where we denote by Obj. 1 the Problem (6.34) with objective function $f_1(\cdot)$ and Obj. 2 the one with objective function $f_2(\cdot)$. We compare the performance with the traditional Massive MIMO with N_A antennas. We can see that the optimized RIS with directional active antennas strongly increases the performance of the system compared with the traditional massive MIMO, especially in the PCSI case. Additionally, we can also see that the proposed channel estimation procedure is effective and that the phase-shift optimization works well also in this case, especially when directional active antennas are assumed.

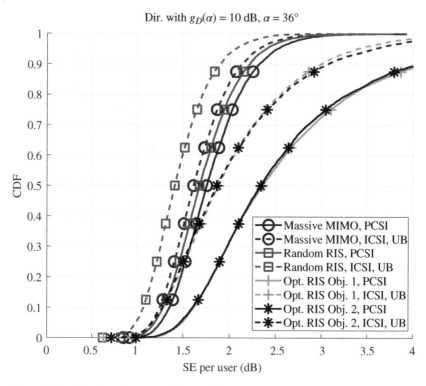

Figure 6.7 CDF of the Spectral efficiency (SE) per user in the case of directional active antennas with $g_D(\alpha) = 10$ dB, $\alpha = 36°$, $N_R = 64$ user in the case of directional, and $N_A = 16$. Comparison between the case of perfect CSI (PCSI), and imperfect CSI (ICSI) with upper-bound (UB) of the performance with RIS phase-shifts optimization with objective functions $f_1(\cdot)$ and $f_2(\cdot)$.

6.7 Conclusions

In this chapter, we considered an RIS-aided massive MIMO system in which an RIS is placed at short distance and in front of an active antenna array, with a not-large number of elements. The UEs to be served are placed in the backside of the active antenna array. We provided a signal and channel model suitable for the analysis of such a system and proposed a channel estimation procedure that, using different configurations of the RIS phase shifts, is capable to estimate the channel from the large number of RIS elements and all the UEs exploiting only the active antennas. We derived a closed-form expression for the hardening lower bound of the downlink performance and formulated a generic optimization problem which can be used for two purposes: reducing interference and maximizing the composite channel norms. In the numerical results, we showed the effectiveness of the considered architecture, assuming both omnidirectional and directional active antennas, and of the phase shifts allocation algorithm.

6.A Appendix

To obtain Eq. (6.29), we start by computing $|\mathrm{DS}_k|$, and in order to make the following formulas more compact, we let $\mathbf{A}_\Phi = \boldsymbol{\Phi}^\mathsf{T}\mathbf{H}^\mathsf{T}\mathbf{H}^*\boldsymbol{\Phi}^*$; we have:

$$
\begin{aligned}
\mathrm{DS}_k &= \sqrt{\eta_k^{(d)}} E\left\{\mathbf{h}_k^\mathsf{T}\mathbf{A}_\Phi\hat{\mathbf{h}}_k^*\right\} \\
&= \sqrt{\eta_k^{(d)}}\,\mathrm{tr}\left[E\left\{\mathbf{A}_\Phi\hat{\mathbf{h}}_k^*(\hat{\mathbf{h}}_k^\mathsf{T} + \tilde{\mathbf{h}}_k^\mathsf{T})\right\}\right] \\
&= \sqrt{\eta_k^{(d)}}\,\mathrm{tr}\left[E\left\{\mathbf{A}_\Phi\hat{\mathbf{h}}_k^*\hat{\mathbf{h}}_k^\mathsf{T}\right\}\right] \\
&= \sqrt{\eta_k^{(d)}}\,\mathrm{tr}\left[\mathbf{A}_\Phi\mathbf{R}_{\hat{\mathbf{h}}_k\hat{\mathbf{h}}_k}^*\right].
\end{aligned}
\tag{6.A.37}
$$

In (6.A.37), we have exploited the fact that $E\left\{\hat{\mathbf{h}}_k^*\tilde{\mathbf{h}}_k^\mathsf{T}\right\} = \mathbf{0}$, and that

$$
E\left\{\hat{\mathbf{h}}_k^*\hat{\mathbf{h}}_k^\mathsf{T}\right\} = \mathbf{R}_{\hat{\mathbf{h}}_k\hat{\mathbf{h}}_k}^* = E\left\{\mathbf{h}_k^*\mathbf{y}_k^\mathsf{T}\right\}E\left\{\mathbf{y}_k\mathbf{y}_k^\mathsf{H}\right\}^{-1}E\left\{\mathbf{y}_k^*\mathbf{h}_k^\mathsf{T}\right\}.
$$

In order to compute BU_k, it is useful to start computing the following statistical expectation:

$$
\begin{aligned}
E\left\{\left|\mathbf{h}_k^\mathsf{T}\mathbf{A}_\Phi\hat{\mathbf{h}}_k^*\right|^2\right\} &= E\left\{\left|(\hat{\mathbf{h}}_k^\mathsf{T} + \tilde{\mathbf{h}}_k^\mathsf{T})\mathbf{A}_\Phi\hat{\mathbf{h}}_k^*\right|^2\right\} \\
&= \underbrace{E\left\{\left|\hat{\mathbf{h}}_k^\mathsf{T}\mathbf{A}_\Phi\hat{\mathbf{h}}_k^*\right|^2\right\}}_{A} + \underbrace{E\left\{\left|\tilde{\mathbf{h}}_k^\mathsf{T}\mathbf{A}_\Phi\hat{\mathbf{h}}_k^*\right|^2\right\}}_{B} \\
&\quad + \underbrace{2\Re\left[E\left\{\hat{\mathbf{h}}_k^\mathsf{T}\mathbf{A}_\Phi\hat{\mathbf{h}}_k^*\hat{\mathbf{h}}_k^\mathsf{T}\mathbf{A}_\Phi^\mathsf{T}\tilde{\mathbf{h}}_k^*\right\}\right]}_{C}
\end{aligned}
\tag{6.A.38}
$$

It is easy to realize that the third summand, i.e., C, is zero since $\tilde{\mathbf{h}}_k$ has zero mean and is independent of $\hat{\mathbf{h}}_k$. Regarding the term B, we have:

$$
\begin{aligned}
E\left\{\left|\tilde{\mathbf{h}}_k^\mathsf{T}\mathbf{A}_\Phi\hat{\mathbf{h}}_k^*\right|^2\right\} &= E\left\{\mathrm{tr}\left[\mathbf{A}_\Phi\hat{\mathbf{h}}_k^*\hat{\mathbf{h}}_k^\mathsf{T}\mathbf{A}_\Phi^\mathsf{H}\tilde{\mathbf{h}}_k^*\tilde{\mathbf{h}}_k^\mathsf{T}\right]\right\} \\
&= \mathrm{tr}\left[\mathbf{A}_\Phi\mathbf{R}_{\hat{\mathbf{h}}_k\hat{\mathbf{h}}_k}^*\mathbf{A}_\Phi\left(\beta_k\mathbf{I}_{N_R} - \mathbf{R}_{\hat{\mathbf{h}}_k\hat{\mathbf{h}}_k}^*\right)\right].
\end{aligned}
\tag{6.A.39}
$$

In order to compute the first summand in (6.A.38), i.e., A, we first note that, letting $\hat{\mathbf{t}}_k \sim \mathcal{CN}(0, \mathbf{I}_{N_R})$, and considering the following eigenvalue decomposition $\mathbf{R}_{\hat{\mathbf{h}}_k\hat{\mathbf{h}}_k} =$

$\mathbf{V}_k\mathbf{\Sigma}_k\mathbf{V}_k^{\mathsf{H}}$, we can write $\widehat{\mathbf{h}}_k = \mathbf{V}_k\mathbf{\Sigma}_k^{1/2}\widehat{\mathbf{t}}_k$. Accordingly, we have:

$$
E\left\{\left|\widehat{\mathbf{h}}_k^{\mathsf{T}}\mathbf{A}_\Phi\widehat{\mathbf{h}}_k^*\right|^2\right\} = E\left\{\left|\widehat{\mathbf{t}}_k^{\mathsf{T}}\underbrace{\mathbf{\Sigma}_k^{1/2}\mathbf{V}_k^{\mathsf{T}}\mathbf{A}_\Phi\mathbf{V}_k^*\mathbf{\Sigma}_k^{1/2}}_{\triangleq \mathbf{Q}_k=\mathbf{U}_k\mathbf{\Lambda}_k\mathbf{U}_k^{\mathsf{H}}}\widehat{\mathbf{t}}_k^*\right|^2\right\}
$$

$$
= E\left\{\left|\widehat{\mathbf{t}}_k^{\mathsf{T}}\mathbf{\Lambda}_k\widehat{\mathbf{t}}_k^*\right|^2\right\} = E\left\{\left(\sum_{i=1}^{N_R}\mathbf{\Lambda}_k(i,i)|\widehat{\mathbf{t}}_k(i)|^2\right)^2\right\}
$$

$$
= \left(\sum_{i=1}^{N_R}\mathbf{\Lambda}_k(i,i)\right)^2 + \sum_{i=1}^{N_R}\left(\mathbf{\Lambda}_k(i,i)\right)^2 = \mathrm{tr}^2(\mathbf{Q}_k) + \mathrm{tr}(\mathbf{Q}_k\mathbf{Q}_k^{\mathsf{H}}).
$$

$$(6.A.40)$$

In the above equation, $\mathbf{U}_k$ and $\mathbf{\Lambda}_k$ are the eigenvectors and eigenvalues matrices obtained from the eigenvalue decomposition of the matrix $\mathbf{Q}_k$; moreover, we have exploited the fact that the vector $\mathbf{U}_k^{\mathsf{H}}\widehat{\mathbf{t}}_k^*$ is statistically undistinguishable from $\widetilde{\mathbf{t}}_k^*$, and that the quantities $\left\{|\widehat{\mathbf{t}}_k(i)|^2, \ i = 1,\ldots,N_R\right\}$ are a sequence of i.i.d. exponentially-distributed random variates with unit parameter. Using (6.A.40), (6.A.39), and (6.A.38), we can now finally obtain, upon simple algebra, $E\{|\mathrm{BU}_k|^2\}$ as follows:

$$
E\{|\mathrm{BU}_k|^2\} = \eta_k^{(d)}\left[\underbrace{E\left\{\left|\mathbf{h}_k^{\mathsf{T}}\mathbf{A}_\Phi\widehat{\mathbf{h}}_k^*\right|^2\right\}}_{(6.38)} - \underbrace{E\left\{\mathbf{h}_k^{\mathsf{T}}\mathbf{A}_\Phi\widehat{\mathbf{h}}_k^*\right\}^2}_{\propto(6.37)}\right]
$$

$$
= \eta_k^{(d)}\mathrm{tr}\left(\beta_k\mathbf{A}_\Phi\mathbf{R}_{\widehat{\mathbf{h}}_k\widehat{\mathbf{h}}_k}^*\mathbf{A}_\Phi\right). \tag{6.A.41}
$$

Finally, we have to compute the term $E\left\{|\mathrm{UI}_{k,j}|^2\right\}$. We start by considering the case in which UEs k and j share the same pilot sequence, so that (6.21) holds. In this case, we have

$$
E\left\{|\mathrm{UI}_{k,j}|^2\right\} = \eta_j^{(d)}\left(\frac{\beta_j\sqrt{\eta_j^{(u,t)}}}{\beta_k\sqrt{\eta_k^{(u,t)}}}\right)^2 E\left\{|\mathbf{h}_k^{\mathsf{T}}\mathbf{A}_\Phi\widehat{\mathbf{h}}_k^*|^2\right\}
$$

$$
= \eta_j^{(d)}\left(\frac{\beta_j\sqrt{\eta_j^{(u,t)}}}{\beta_k\sqrt{\eta_k^{(u,t)}}}\right)^2 \left[\mathrm{tr}^2(\mathbf{A}_\Phi\mathbf{R}_{\widehat{\mathbf{h}}_k\widehat{\mathbf{h}}_k}^*) + \beta_k\,\mathrm{tr}(\mathbf{A}_\Phi\mathbf{R}_{\widehat{\mathbf{h}}_k\widehat{\mathbf{h}}_k}^*\mathbf{A}_\Phi)\right],
$$

$$(6.A.42)$$

where (6.A.38) has been exploited. If, instead, UE k and UE j are assigned orthogonal pilots, then we have

$$
\begin{aligned}
E\left\{|\mathrm{UI}_{k,j}|^2\right\} &= \eta_j^{(d)} E\left\{|\mathbf{h}_k^\mathsf{T}\mathbf{A}_\Phi\hat{\mathbf{h}}_j^*|^2\right\} \\
&= \eta_j^{(d)} E\left\{\mathbf{h}_k^\mathsf{T}\mathbf{A}_\Phi\hat{\mathbf{h}}_j^*\hat{\mathbf{h}}_j^\mathsf{T}\mathbf{A}_\Phi\mathbf{h}_k^*\right\} = \eta_j^{(d)}\beta_k \operatorname{tr}\left[\mathbf{A}_\Phi\mathbf{R}_{\hat{\mathbf{h}}_j\hat{\mathbf{h}}_j}^*\mathbf{A}_\Phi\right],
\end{aligned}
$$

$$(6.A.43)$$

where the independence between $\mathbf{h}_k$ and $\hat{\mathbf{h}}_j$ has been exploited. Denoting by $\mathcal{P}_k$ the set of UEs using the same pilot as UE k, we can now express the k-th UE downlink SINR lower bound as (6.29).

Bibliography

3GPP (2019). Study on Channel Model for Frequencies from 0.5 to 100 GHz (Release 16). Technical report, *3GPP TR 38.901*, Dec. 2019. https://www.researchgate.net/profile/Ming-Zhu-39/publication/348790466_3GPP_TR_38901_Channel_Model/links/633781d376e39959d6898826/3GPP-TR-38901-Channel-Model.pdf.

Bai, T., Pan, C., Deng, Y. et al. (2020). Latency minimization for intelligent reflecting surface aided mobile edge computing. *IEEE Journal on Selected Areas in Communications* 38 (11): 2666–2682. Jul. 2020. https://doi.org/10.1109/JSAC.2020.3007035.

Basar, E., Di Renzo, M., De Rosny, J. et al. (2019). Wireless communications through reconfigurable intelligent surfaces. *IEEE Access* 7: 116753–116773. Aug. 2019. https://ieeexplore.ieee.org/abstract/document/8796365.

Bertsekas, D.P. (1997). *Nonlinear Programming*. Athena Scientific. Jan. 1997. https://www.tandfonline.com/doi/abs/10.1057/palgrave.jors.2600425.

Buzzi, S., D'Andrea, C., Zappone, A. et al. (2021a). RIS configuration, beamformer design, and power control in single-cell and multi-cell wireless networks. *IEEE Transactions on Cognitive Communications and Networking* 7 (2): 398–411. Mar. 2021. https://doi.org/10.1109/TCCN.2021.3068414.

Buzzi, S., Grossi, E., Lops, M., and Venturino, L. (2021b). Radar target detection aided by reconfigurable intelligent surfaces. *IEEE Signal Processing Letters* 28: 1315–1319. Jun. 2021. https://doi.org/10.1109/LSP.2021.3089085.

Caire, G. (2018). On the ergodic rate lower bounds with applications to massive MIMO. *IEEE Transactions on Wireless Communications* 17 (5): 3258–3268. Feb. 2018. https://doi.org/10.1109/TWC.2018.2808522.

Di Renzo, M., Zappone, A., Debbah, M. et al. (2020). Smart radio environments empowered by reconfigurable intelligent surfaces: how it works, state of research, and the road ahead. *IEEE Journal on Selected Areas in Communications* 38 (11): 2450–2525. Jul. 2020. https://doi.org/10.1109/JSAC.2020.3007211.

Hong, S., Pan, C., Ren, H. et al. (2020). Artificial-noise-aided secure MIMO wireless communications via intelligent reflecting surface. *IEEE Transactions on Communications* 68 (12): 7851–7866. Sep. 2020. https://doi.org/10.1109/TCOMM .2020.3024621.

Huang, C., Zappone, A., Alexandropoulos, G.C. et al. (2019). Reconfigurable intelligent surfaces for energy efficiency in wireless communication. *IEEE Transactions on Wireless Communications* 18 (8): 4157–4170. Jun. 2019. https://doi .org/10.1109/TWC.2019.2922609.

Jamali, V., Tulino, A.M., Fischer, G. et al. (2020). Intelligent surface-aided transmitter architectures for millimeter-wave ultra massive MIMO systems. *IEEE Open Journal of the Communications Society* 2: 144–167. Dec. 2020. https://ieeexplore.ieee.org/ abstract/document/9310290.

Long, R., Liang, Y.-C., Pei, Y., and Larsson, E.G. (2021). Active reconfigurable intelligent surface-aided wireless communications. *IEEE Transactions on Wireless Communications* 20 (8): 4962–4975. Mar. 2021. https://doi.org/10.1109/TWC.2021 .3064024.

Marzetta, T.L., Larsson, E.G., Yang, H., and Ngo, H.Q. (2016). *Fundamentals of Massive MIMO*. Cambridge University Press. Nov. 2016. https://doi.org/10.1017/ CBO9781316799895.

Ng, D.W.K., Lo, E.S., and Schober, R. (2012). Energy-efficient resource allocation in OFDMA systems with large numbers of base station antennas. *IEEE Transactions on Wireless Communications* 11 (9): 3292–3304. Jul. 2012. https://doi.org/10.1109/ TWC.2012.072512.111850.

Qian, X. and Di Renzo, M. (2021). Mutual coupling and unit cell aware optimization for reconfigurable intelligent surfaces. *IEEE Wireless Communications Letters* 10 (6): 1183–1187. Feb. 2021. https://doi.org/10.1109/LWC.2021.3061449.

Stutzman, W.L. and Thiele, G.A. (2012). *Antenna Theory and Design*. Wiley. 2012. https://scholar.google.com/scholar_lookup?title=Antenna theory and design&publication_year=2012&author=W.L. Stutzman&author=G.A. Thiele.

Wu, Q. and Zhang, R. (2019). Intelligent reflecting surface enhanced wireless network via joint active and passive beamforming. *IEEE Transactions on Wireless Communications* 18 (11): 5394–5409. Aug. 2019. https://doi.org/10.1109/TWC.2019 .2936025.

Zhang, L., Wang, Y., Tao, W. et al. (2020). Intelligent reflecting surface aided MIMO cognitive radio systems. *IEEE Transactions on Vehicular Technology* 69 (10): 11445–11457. Jul. 2020. https://doi.org/10.1109/TVT.2020.3011308.

Zhang, Z., Dai, L., Chen, X. et al. (2023). Active RIS vs. passive RIS: which will prevail in 6G?. IEEE Transactions on Communications 71(3): 1707–1725. Mar. 2023. https://ieeexplore.ieee.org/abstract/document/9998527?casa_token= bnJ9IUwCpDwAAAAA:QsAB3iAsAfvyKtmFhVJC9MwC4CYqtNVvF1yYEaz6Jy Jw5XTgKGwn09_wJmrgmrWFFBikYA.

7

Localization, Sensing, and Their Integration with RISs

George C. Alexandropoulos[1,2], Hyowon Kim[4], Jiguang He[2], and Henk Wymeersch[3]

[1] Department of Informatics and Telecommunications, National and Kapodistrian University of Athens, Athens, Greece
[2] Technology Innovation Institute, Abu Dhabi, United Arab Emirates
[3] Department of Electrical Engineering, Chalmers University of Technology, Gothenburg, Sweden
[4] Department of Electronics Engineering, Chungnam National University, Daejeon, South Korea

7.1 Introduction

The ability to localize connected users has been part of cellular communication systems since early generations, with progressive improvements (del Peral-Rosado et al. 2017), which are soon expected to match global navigation satellite system (GNSS) performance (Egea-Roca et al. 2022). Localization relies on the ability to infer geometric information from the wireless channel between the user equipment (UE) and several base stations (BSs). Such information can be in the form of a complete channel response (as used in fingerprinting methods (Vo and De 2016)) or, more commonly, in the form of delays, directional angles, and Doppler shifts. The latter type of information is also used for detecting and tracking passive objects in radar-type sensing in modern radars (Geng et al. 2021). While radar and communication systems have developed separately, there is now a renewed interest in their convergence in the form of integrated sensing and communication (ISAC) (Wang et al. 2022), where reconfigurable intelligent surfaces (RISs) are expected to play an important role (Chepuri et al. 2022).

7.1.1 Localization in 5G

In 5G, since 3GPP Release 16, there have been several advances in support of accurate localization (the so-called positioning in the 3GPP nomenclature) (Dwivedi et al. 2021). Localization requires several ingredients (del Peral-Rosado

Intelligent Surfaces Empowered 6G Wireless Network, First Edition.
Edited by Qingqing Wu, Trung Q. Duong, Derrick Wing Kwan Ng, Robert Schober, and Rui Zhang.

et al. 2017): first of all, the *pilot signal* sent in uplink time (UL) or downlink (DL) between a UE and several BSs. These pilots should be designed to make optimal use of the frequency (bandwidth), temporal (time), and spatial (antenna) resources. From these pilots, the receiver estimates the line-of-sight (LoS) *geometric parameters* (e.g., time-of-arrival (ToA), downlink angle-of-departure (DL-AoD), or uplink angle-of-arrival (UL-AoA)). Additional, time-based measurements can be converted to distances when devices are synchronized, which in practice leads to time-difference-of-arrival (TDoA) measurements, when BSs are synchronized and round-trip-time (RTT) measurements if they are not. Similarly, angle measurements require knowledge of the BS orientations in a global reference frame, which requires a priori calibration. Estimation of the LoS parameters implies *filtering out of multipath signals*, which can be supported by super-resolution methods. Multipath is an important disturbance for positioning and can lead to large biases (Witrisal et al. 2016). Finally, from the LoS geometric parameters, *the UE position is recovered*, generally by solving a non-linear and non-convex maximum likelihood (ML) or least squares (LS) optimization problem. Since localization involves a multidimensional unknown, it involves several BSs. For example, using TDoA measurements (Ma et al. 2021), at least 4 BSs are needed to solve for the 3D user position and its 1D clock bias. Using RTT measurements, 3 BSs suffices, while using UL-AoA measurements, positioning is feasible with 2 BSs.

Localization performance is captured by several metrics (Behravan et al. 2023). Among those, accuracy, latency, coverage (availability), and scalability are among the most important. In particular, accuracy and coverage are connected, where accuracy measures the random error norm between the true and estimated position, while coverage captures the fraction of time, or space, a certain accuracy can be achieved (e.g., 95% coverage for 1 m accuracy). The latency is the time between the positioning request and the computed positioning being available; the pilot transmission is one part of the latency. Finally, the scalability determines how many devices can be localized in a certain volume. Requirements on each of these metrics have implications on the pilot signals (e.g., duration or broadcast versus unicast), the estimation methods (e.g., their latency), the positioning method (e.g., at the UE for low latency), the BS deployment, as well as higher layer protocols and resource allocation.

Starting with 5G, requirements for commercial use cases have been introduced in support of a variety of challenging use cases (Dwivedi et al. 2021). For instance, in Release 16, a 3 m horizontal accuracy with 80% coverage and a latency of less than one second is specified. In the vision toward 6G, even more strict requirements are foreseen, for applications such as augmented and extended reality, for human-robot interaction, and for digital twins. In addition, since radar-like sensing is expected to be part of 6G, also sensing requirements are starting to be introduced, such as range, angle, and Doppler resolution (Behravan et al. 2023).

7.1.2 RIS Key Advantages

As we will see, RISs have the potential to improve localization and sensing in two ways: to enhance/boost or to enable (Wymeersch et al. 2020). Our focus in this section will be on reflective metasurfaces, while more details on other types of RISs will be provided in Section 7.2.

7.1.2.1 Localization

An RIS provides an additional location (for time-based measurements) and orientation reference (for angle-based measurements), in addition to the BSs, provided that the RIS location and orientation are carefully calibrated (Keykhosravi et al. 2022a). An RIS also provides additional propagation path between the BSs and the UE. To extract the geometric parameters of this path, it must be detected by the channel parameter estimation routine. To facilitate this, the reconfigurability of the RIS can be utilized: the RIS phase configurations can be used to shape the multipath in space (by generating directional beams), in time (by applying different configurations over time), and in frequency (if controlled frequency-selective configurations are possible). These allow a remote receiver to separate the RIS path, allowing the metasurface to optimize its configurations in frequency, time, and space, similar to a conventional BS. Since an RIS is almost passive (Alexandropoulos et al. 2020a, 2021b), it introduces no delays, so it is naturally synchronized to any transmitting BS. In summary, an RIS behaves similar to a perfectly synchronized BS with a large analog array, providing angle and delay measurements. This allows the RIS to replace one or more BSs, overcome LoS blockages, and provide high-quality angle measurements, due to its usually large aperture.

7.1.2.2 Sensing

While arguable, localization is a special case of sensing, where RISs can also provide opportunities for improving and enabling radar-like monostatic sensing. In monostatic sensing, the transmitter and receiver are co-located, and thus, perfectly phase-synchronized. From the backscattered signal, the receiver can map passive objects by measuring their range (directly from the ToA), angle-of-arrival (AoA), and Doppler shifts. An RIS can be used to amplify the backscattered signal, and thus extend the range or provide visibility around corners. However, due to the large path loss in sensing, the use of RIS is more limited than in localization.

7.2   RIS Types and Channel Modeling

In this section, we discuss the key hardware and operation features of the RIS types that are deployed throughout this chapter for localization and sensing. In addition, we present channel models for RIS-parameterized wireless links, which are mainly used in the RIS-assisted localization and sensing literature.

7.2.1   RIS Hardware Architectures

The concept of the RIS-empowered smart wireless environments (Di Renzo et al. 2019; Huang et al. 2019) initially considered *reflective RISs* with almost zero power consumption unit elements (Alexandropoulos et al. 2020a), whose reflection characteristics are reconfigurable, enabling programmable manipulation of the incoming electromagnetic (EM) waves. Of course, an RIS needs to be orchestrated by a dedicated control unit being responsible to manage the metasurface's overall phase profile and interact with the rest of the network (Strinati et al. 2021). A fine-grained control over the reflected EM field for quasi-free space beam manipulation contributes in realizing accurate reflective beamforming. Unit elements of sub-wavelength size are a favorable choice for this, although inevitable strong mutual coupling, and well-defined gray-scale-tunable EM properties, exist. Conversely, in rich scattering environments (Alexandropoulos et al. 2021b), the wave energy is statistically equally spread throughout the wireless medium. The ensuing ray chaos implies that rays impact the RIS from all possible, rather than one well-defined, directions. The goal thus becomes the manipulation of as many ray paths as possible, which is different from the common goal of creating a directive beam. This manipulation has two kinds of aims, including tailoring those rays to create constructive interference at a target location and stirring the field efficiently. These manipulations can be efficiently realized with RISs equipped with half-wavelength-sized meta-atoms, enabling the control of more rays with a fixed amount of electronic components.

To deal with the large overhead of channel estimation in RIS-empowered wireless communications systems (Jian et al. 2022), a *receiving RIS* hardware architecture was introduced in Alexandropoulos and Vlachos (2020), which can perform explicit or implicit channel estimation at the RIS side (or its controller) with minimal number of receive radio-frequency (RF) chains. This RIS type can be deployed for signal reception and estimation (via, e.g., compressed sensing tools (Vlachos et al. 2019)), as well as reconfigurable reflections similar to reflective RISs, in a time-orthogonal manner. According to this architecture, the outputs of the RIS elements are fed to a single reception RF chain, which typically includes a low noise amplifier, a mixer downconverting the signal from

RF to baseband, and an analog-to-digital converter. In Alamzadeh et al. (2021), a hybrid metamaterial, which can simultaneously reflect a portion of the signal impinging on it in a programmable way, while feeding the other portion of that signal to a sensing unit, was designed. This design led to the introduction of the *hybrid RIS* hardware architecture, which enables simultaneous reconfigurable reflection and sensing of the impinging EM waveform. In those metasurfaces, a waveguide is deployed to couple to each RIS element and can be connected to a reception RF chain. This makes possible for the RIS to locally process a portion of the received signals in the digital domain. However, the coupling of the elements to waveguides makes it impossible for the incident wave to be perfectly reflected. Actually, the coupling level determines the ratio of the reflected energy to the absorbed energy. Its footprint can be reduced and the coupling to the sampling waveguide can be mitigated by keeping this waveguide near cutoff. Such hybrid RISs were recently considered by Alexandropoulos et al. (2021a) and Zhang et al. (2021) for facilitating explicit and implicit channel estimation as well as flexible network management with reduced overhead.

The concept of the *simultaneously transmitting and reflecting RIS (STAR-RIS)* (Xu et al. 2021), which is also referred to as the intelligent omni-surface (Zhang et al. 2022), enables wireless signals incident on the surface to be simultaneously reflected and refracted with tunable manners. Such metasurfaces can assist in achieving a full-space, i.e., with a 360-degree coverage reconfigurable wireless environment (Liu et al. 2021b). The key to achieving tunability for both the reflected and transmitted (refracted) signals is that each metamaterial element supports both electric and magnetic currents. For the realization of the functionalities, two separate series of phase shifters are usually leveraged for realizing them, i.e., one for controlling refraction and the other for controlling reflection.

In Table 7.1, the available RIS types, in terms of operational features and key hardware components, are summarized. For completeness, transmissive (Di Palma et al. 2017) and amplifying (Hu et al. 2018) RISs, as well as RIS-based antennas (Shlezinger et al. 2019), which are lately receiving increased attention as candidates for holographic-type extremely massive multiple-input-multiple-output (MIMO) systems (Gong et al. 2023), are also listed.

7.2.2 RIS-Parameterized Channel Models

The successful deployment of RISs for wireless applications largely depends on the development of efficient signal-processing tools. To this end, it is of paramount importance to ensure that the underlying end-to-end channel models faithfully capture the wave physics involved in programmable wireless environments parameterized by RISs. The introduction of physics-compliant channel models is

Table 7.1 Taxonomy of the available RIS types in terms of operational features and key hardware components.

RIS type	Mode of operation	Key hardware components
Reflective RIS (e.g., Alexandropoulos et al. (2020a))	Tunable reflection	Reflection elements, control unit
Receiving RIS Alexandropoulos and Vlachos (2020)	Time-orthogonal tunable reflection and reception	Reflection elements, reception RF chain(s), waveguides, control unit
Hybrid RIS Alexandropoulos et al. (2021a)	Simultaneous tunable reflection and reception	Reflecting elements, reception RF chain(s), waveguides, control unit
STAR-RIS (e.g., Xu et al. (2021) and Zhang et al. (2022))	Simultaneous tunable reflection and refraction	Reflection elements, refraction elements, control unit
Amplifying RIS Tasci et al. (2022)	Amplified tunable reflection	Reflection elements, power amplifier(s), control unit
Transmissive RIS Di Palma et al. (2017)	Tunable refraction, acting as lens	Refraction elements, control unit
RIS-based Antennas Shlezinger et al. (2019)	Conventional transmission and/or reception	Radiating elements, transmission/reception RF chain(s), baseband unit

an ongoing active area of research (Di Renzo et al. 2022; Faqiri et al. 2023), including models that provide different levels of flexibility for devising localization and sensing algorithms. For this chapter's targeted applications relying on statistical signal processing approaches, models describing the geometrical features of wireless links are more desirable, whereas, for unstructured and stochastic channel models, machine learning (e.g., Huang et al. (2018), Alexandropoulos et al. (2020b), and Stylianopoulos et al. (2021)) and fingerprinting approaches are usually deployed (e.g., Alexandropoulos et al. (2021b)).

Considering a triangular system with one N_t-antenna transmitter (TX), one N_r-antenna receiver (RX), and a single N_{ris}-element RIS, and following the cascade model for RIS-parameterized channels (Huang et al. 2019), the end-to-end channel matrix for each nth subcarrier (with $n = 1, 2, \dots, N$) can be expressed as follows:

$$\mathbf{H}_{e2e}[n] = \mathbf{H}_{RX\text{-}RIS}[n]\mathbf{\Omega}\mathbf{H}_{RIS\text{-}TX}[n] + \mathbf{H}_{RX\text{-}TX}[n], \qquad (7.1)$$

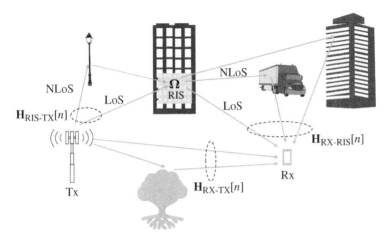

Figure 7.1 The individual channel matrices in RIS-assisted wireless links.

where $\mathbf{H}_{\text{RX-RIS}}[n] \in \mathbb{C}^{N_r \times N_{\text{ris}}}$, $\mathbf{H}_{\text{RIS-TX}}[n] \in \mathbb{C}^{N_{\text{ris}} \times N_t}$, and $\mathbf{H}_{\text{RX-TX}}[n] \in \mathbb{C}^{N_r \times N_t}$ represent the RX-RIS, RIS-TX, and the direct RX-TX channels, respectively, as depicted in Fig. 7.1, and $\mathbf{\Omega}$ is the N_{ris}-element diagonal matrix, including the tunable RIS elements. The non-zero elements of $\mathbf{\Omega}$, also known as the RIS phase profiles or phase configurations, usually follow a constant-modulus constraint for the cases of metasurfaces with non-amplifying reflective elements. The literature includes various forms of phase profiles: (i) random profiles (Dardari et al. 2022); (ii) directional profiles (Rahal et al. 2022a); (iii) adaptive profiles, where each element's phase is configured to direct the induced reflection in a slightly different direction, uniformly illuminating a variable-width area (Jamali et al. 2022; Kim et al. 2022a); and (iv) time-domain profiles (Keykhosravi et al. 2022b) used to separate controlled signal propagation paths from uncontrolled ones. The Khatri–Rao product of $\mathbf{H}_{\text{RIS-TX}}^{\mathsf{T}}[n]$ and $\mathbf{H}_{\text{RX-RIS}}[n]$, i.e., $\mathbf{H}_{\text{RIS-TX}}^{\mathsf{T}}[n] \diamond \mathbf{H}_{\text{RX-RIS}}[n]$, is the so-called cascaded channel, whose knowledge is necessary for designing the rate-optimal joint active (i.e., TX precoder) and passive (i.e., RIS phase profile) beamforming (Ardah et al. 2021; Pan et al. 2022).

In the latter cascade end-to-end channel representation, the involved matrices are usually modeled using: (i) geometric or (ii) stochastic models. The former is widely applied in millimeter-wave (mmWave) and THz wireless communications, resulting in poor scattering propagation environments, while the latter are commonly used for sub-6 GHz communications, where rich scattering propagation conditions are usually present. In the sequel, we detail these two channel model categories for generating the individual channels in $\mathbf{H}_{\text{e2e}}[n]$.

7.2.2.1 Geometric Channel Model

Geometric channels are constructed by following the geometric relationships among each pair of network nodes and the objects of the wireless propagation environment. The corresponding channel coefficients are represented as functions of the channel parameters (e.g., angles-of-departure (AoDs), AoAs, phase rotations, and propagation delays (a.k.a. ToA)), which are derived from geometric relationships. An example channel matrix comprising L signal propagation paths is given for each nth subcarrier (for $n = -(N-1)/2, \ldots, (N-1)/2$) as Heath et al. (2016):

$$\mathbf{H}_{\text{RX-TX}}[n] \triangleq \sum_{l=1}^{L} \frac{e^{-j(2\pi\tau_l \frac{nB}{N} + \varphi_l)}}{\sqrt{\rho_l}} \boldsymbol{\alpha}_r(\phi_l)\boldsymbol{\alpha}_t^{\mathsf{H}}(\theta_l), \tag{7.2}$$

including the L AoDs $\{\theta_l\}_{l=1}^{L}$, L AoAs $\{\phi_l\}_{l=1}^{L}$, L unknown phase rotations $\{\varphi_l\}_{l=1}^{L}$ (sum of phase biases at the TX and RX), and L propagation delays $\{\tau_l\}_{l=1}^{L}$ (it holds for the distances covered by each lth path that $d_l = c\tau_l$, where c is the speed of light). In this expression, B denotes the overall bandwidth spanned by all the subcarriers. Without loss of generality, the index $l = 1$ is associated with the LoS path and indices $l = 2, 3, \ldots, L$ are associated with non-LoS (NLoS) channel components. The path loss of the LoS component in (7.2) is modeled as $\rho_1 = d_1^2 f_c^2 / 10^{8.755}$, where f_c (in KHz) is the carrier frequency, defined as $f_c = \frac{c}{\lambda}$ with λ being the wavelength of the carrier. Alternatively, the standard 3GPP urban micro (UMi) path loss model can be considered, according to which $\rho_1 = 10^{2.27} d_1^{3.67} f_c^{2.6}$, where f_c needs to be included in GHz (Akdeniz et al. 2014). As shown from these two expressions, the path loss can be written as a function of the propagation delay. In addition, it can be seen that the path loss of each of the NLoS paths depends on its length as well as on the reflection/refraction/diffraction loss that path may encounter, which makes it cumbersome to accurately formulate.

The array response vectors $\boldsymbol{\alpha}_r(\phi_l) \in \mathbb{C}^{N_r}$ and $\boldsymbol{\alpha}_t(\theta_l) \in \mathbb{C}^{N_t}$ in (7.2) are functions of ϕ_l and θ_l, respectively. Under the assumption of a uniform linear array (ULA) structure with half-wavelength inter-element spacing, it holds that $\boldsymbol{\alpha}_r(\phi_l) = [1, e^{-j\pi \sin(\phi_l)}, \ldots, e^{-j\pi(N_r-1)\sin(\phi_l)}]$ and $\boldsymbol{\alpha}_t(\theta_l) = [1, e^{-j\pi \sin(\theta_l)}, \ldots, e^{-j\pi(N_t-1)\sin(\theta_l)}]$. Extension to uniform planar arrays (UPAs) are provided in He et al. (2022a), after taking into account both the azimuth (horizontal) and elevation (vertical) angles in the array response vectors. The channel representation in (7.2) can be used for modeling both the RIS-TX, RX-RIS, and the direct TX-RX channels in (7.1).

It is noted that the channel in (7.2) implies planar signal propagation in the far field, i.e., it holds $d_l \geq \frac{2D^2}{\lambda}$ $\forall l$ with D being the diameter of the antenna aperture. When focusing on near-field scenarios, and considering the LoS path as an example, each (i,j)th coefficient (with $i = 1, 2, \ldots, N_r$ and with $j = 1, 2, \ldots, N_t$) of $\mathbf{H}_{\text{RX-TX}}[n]$ can be modeled as Dardari et al. (2022):

$$\left[\mathbf{H}_{\text{RX-TX}}[n]\big|_{l=1}\right]_{i,j} = \frac{1}{\sqrt{\rho_1}} \exp\left(-j\left(\frac{2\pi f_c}{c}\|\mathbf{p}_{\text{TX},j} - \mathbf{p}_{\text{RX},i}\| + \varphi_1\right)\right), \quad (7.3)$$

where $\mathbf{p}_{\text{RX},i}$ and $\mathbf{p}_{\text{TX},j}$ are the coordinates of the ith receive and the jth transmit antennas, respectively. In fact, this near-field geometric channel model includes the far-field one as a special case. For Δ denoting the distance between the first TX antenna and the first RX antenna, the following approximation holds: $\|\mathbf{p}_{\text{TX},j} - \mathbf{p}_{\text{RX},i}\| \approx \lambda(i-1)\sin(\phi_1)/2 - \lambda(j-1)\sin(\theta_1)/2 + \Delta$. Similarly can be done for the remaining NLoS paths in $\mathbf{H}_{\text{RX-TX}}[n]$, as well as for the other channel matrices in (7.1).

7.2.2.2 Stochastic Channel Modeling

Capitalizing on the composition of the end-to-end channel matrix in (7.1) from the individual RIS-TX, RX-RIS, and TX-RX channels, the rich literature of statistical distributions, even including arbitrary spatial correlation (Björnson and Sanguinetti 2021; Alexandropoulos et al. 2007, 2010; Alexandropoulos and Mathiopoulos 2010), can be used to stochastically model diverse fading conditions for each of the channel matrix coefficients. This approach can be straightforwardly extended to scenarios with many TXs, RXs, and multiple RISs, resulting in more individual channel matrices.

For example, in a rich scattering environment, there usually exists a sufficiently large number of paths (i.e., $L \to \infty$ in (7.2)) between the TX and RX. Following the central limit theorem for the case of LoS absence (Kermoal et al. 2002), each channel coefficient becomes complex Gaussian distributed with zero mean, rendering a Rayleigh fading environment. When a LoS channel component is present, the Rician distribution (Kang and Alouini 2006) arises, which includes the K factor defined as the ratio of the power contributions of the LoS path over the NLoS ones. Besides Rayleigh and Rician fading, the versatile Nakagami-m distribution models have more diverse fading conditions. By controlling its shape parameter m, one can model signal fading conditions ranging from severe to moderate, or even no fading cases (Beaulieu and Cheng 2005). It is noted that when $m = 1$, Nakagami-m fading boils down to Rayleigh fading.

7.3 Localization with RISs

In this section, we first overview fundamentals on localization, and then, present several localization case studies enabled or boosted with different types of RISs.

7.3.1 Fundamentals on Localization

Localization refers to the process of estimating the 2D or 3D location of a UE, based on UL or DL pilot signals to or from, respectively, several BSs (del Peral-Rosado et al. 2017). Table 7.2 presents the different types of measurements resulting from the reception of pilot signals, which are used in various geometric-based localization techniques. For UE localization, any combination of angle and delay measurements suffices, but mutual synchronization among BSs and a certain number of BSs are required. Due to this synchronization issue, UE localization with pure ToA measurements is impractical in real scenarios (e.g., a 10 ns clock error corresponds to 3 m error). Examples of delay and angle measurements for localization are depicted in Fig. 7.2.

Table 7.2 Requirements for 3D localization with respect to the type of measurements.

Measurement	UL or DL	BS	Comments
ToA	Either	3	BSs should be synchronized with the UE
TDoA	Either	4	BSs should be mutually synchronized
RTT	Both	3	No synchronization needed
AoA at the BS	UL	2	Requires planar arrays at each BS
AoD from the BS	DL	2	Requires planar arrays at each BS
TDoA + UL-AoA	UL	2	BSs should be mutually synchronized
RTT + UL-AoA	Both	1	No synchronization needed

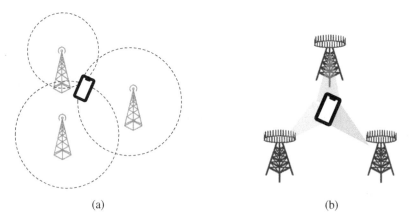

(a)   (b)

Figure 7.2 Geometric-based localization techniques via (a) ToA and (b) AoD measurements. Source: Adapted from (Wymeersch et al. 2022).

The localization pilots can be designed in the time, frequency, and space domains. In the time-frequency domain, those signals can occupy different bandwidths and are usually orthogonal across BSs; in this case, they are known as *comb signals*. In the space domain, the BS capability for directional beams allows time-frequency signals to be propagated in desired directions, providing angle measurements, such as AoA in the UL or AoD in the DL. The measurement accuracy for the ToA and AoA/AoD depends on factors, such as the bandwidth, transmission power, number of antennas, and hardware limitations (Wen et al. 2018).

7.3.2 Localization with Reflective RISs

There have been already proposed several methods combining BSs with reflective RISs for localization, aiming to deliver accurate position estimation with reduced numbers of BSs, thus offering low-cost positioning alternatives. Some representative scenarios are listed in Table 7.3 and visualized in Fig. 7.3. The first row of the table considers a baseline ToA-based localization approach with four synchronized BSs in a LoS propagation environment. The rows that follow showcase how the consideration of a reflective RIS can relax the baseline's requirements, e.g., reducing the number of BSs and removing the need for LoS and synchronization. In more details:

- **Localization with one wideband BS (Keykhosravi et al. 2021)** is enabled by an RIS through the combination of a ToA-based and an angle of departure (AoD) measurements. The cost for this estimation includes: (i) the RIS needs to sweep through several phase profiles, which leads to increased latency; and (ii) the path via the RIS is weaker than the LoS one, leading to higher error variances.

Table 7.3 Localization scenarios with reflective RISs.

	TX	RX	RIS	LoS	Comments
ToA Loc.	4 BSs	1 UE	0	Yes	4 ToAs
Keykhosravi et al. (2021)	1 BS	1 UE	1 RIS	No	1 ToA and AoD at RIS
Keykhosravi et al. (2022a)	1 BS	1 UE	2 RISs	No	2 AODs at RISs
Keykhosravi et al. (2022b)	1 UE	1 UE	1 RIS	Yes	1 ToA and AoD at RIS
Rahal et al. (2022b)	1 BS	1 UE	1 RIS	No	PEB, RIS profile design
Ghazalian et al. (2022)	1 BS	1 BS	1 RIS	No	1 ToA and AoA/AoD at RIS

PEB stands for the position error bound and the underlined node denotes the one to be localized.

Figure 7.3 Overview of selected localization scenarios with reflective RISs.

- **Localization with one narrowband BS (Keykhosravi et al. 2022a)** is enabled by at least two RISs, each of which providing an AoD measurement. To separate the signals from each RIS, orthogonal spatial codes can be used.
- **Localization without BSs (Keykhosravi et al. 2022b)** is enabled by a single RIS when the UE can be equipped with full-duplex radios (Alexandropoulos et al. 2022a). In this case, the UE can estimate the range and AoD using several phase profiles.
- **Localization without LoS to the BS (Rahal et al. 2022b)** is enabled by a single large-sized RIS, so that the UE to be localized is placed on the RIS's geometric near-field region. This allows the UE to harness the wavefront curvature.
- **RIS localization (Ghazalian et al. 2022)** is needed for all aforementioned scenarios. Hence, efficient methods to determine the RIS location are needed.

7.3.3 Localization with a Single STAR-RIS

The STAR-RIS has been mainly proposed for offering spherical communication coverage (Liu et al. 2021a,2021b), which renders it particularly relevant for application requiring the simultaneous connection of indoor and outdoor UEs, e.g., when the metasurface coats the glass facade of a building. In such cases, UEs lying in front and back of the STAR-RIS can be simultaneously served by the same cellular BS. In this section, we focus on STAR-RIS-empowered simultaneous indoor and outdoor 3D localization, and in particular in the setup illustrated in Fig. 7.4 for mmWave frequencies (He et al. 2022a), where two UEs send concurrently over the UL sounding reference signals toward the BS. The received signals across multiple time slots can be used by the BS to estimate the Cartesian coordinates of both users, enabling their 3D localization.

According to the theoretical performance analysis of (He et al. 2022a), Fig. 7.5 includes the Cramér–Rao bound (CRB) of simultaneous indoor and outdoor 3D localization for the system in Fig. 7.4. In particular, the root mean square error (RMSE) in meters with different setups for the parameters $\epsilon_1 \in [0, 1]$ (power splitting ratio for refraction) and $\eta_1 \in [0, 1]$ (power allocation ratio for the outdoor UE) is demonstrated versus the operating SNR. As shown, cm-level localization

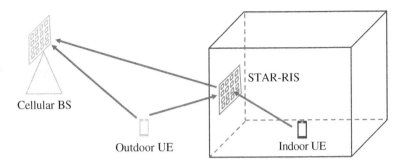

Figure 7.4 STAR-RIS-enabled simultaneous indoor and outdoor 3D localization in the UL direction.

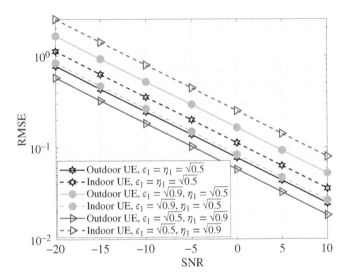

Figure 7.5 CRB on the estimation of the 3D Cartesian coordinates of the two UEs for different pairs of the power ratios ϵ_1 and η_1.

accuracy can be achieved for both UEs when the SNR value exceeds 1 dB for the balanced setup, i.e., $\epsilon_1 = \eta_1 = \sqrt{0.5}$, where the performance gap between the two UEs' position estimation is the smallest among the three setups. However, better performance is achieved for the outdoor UE, which is justified by the additional contribution brought from the existence of the for the direct channel between that user and the BS. All in all, it can be concluded that well-selected ϵ_1 and η_1 values can satisfy the quality of services and user fairness for both UEs simultaneously, playing a significant role for both communications and localization.

7.3.4 Localization with Multiple Receiving RISs

With at least three spatially distributed receiving RISs, one is capable of performing 3D user localization only via narrowband measurements. To this end, the receiving signals at each metasurface can be used for extracting AoA estimations, and then, those estimations can be fused (e.g., at a location server or at one of the RIS controllers playing the role of the master controller among those of all RISs) to obtain the UE location. In Alexandropoulos et al. (2022b) where this an approach was first presented (see (Fig. 7.6)), considering receiving RISs each equipped with a single reception RF chain, both theoretical and algorithmic results were presented. The proposed UE localization algorithm is summarized in the following two steps: (i) AOA estimations at all receiving RISs using the orthogonal matching pursuit (OMP) technique (other alternative on-the-grid and off-the-grid compressed sensing methods can also be leveraged), (ii) mapping of the AoAs to the 3D user coordinates via a maximum likelihood (ML) approach initialized by line intersection.

The RMSE of the localization error in meters versus the source transmit power in dBm considering 3 and 4 RISs, each with 8×8 elements, is illustrated in Fig. 7.7. Both far- and near-field channel models were considered in the performance evaluation (Alexandropoulos et al. 2022b), and theoretical PEB curves under the far-field model are included. It can be seen in the figure that the proposed localization scheme attains the theoretical PEBs in the low transmit power regime. The results also show an error floor due to the model mismatch in the case of the near-field, and due to multipath for the far-field case. However, it is clear that the proposed scheme is still able to provide up to cm-level localization accuracy with a small transmit power.

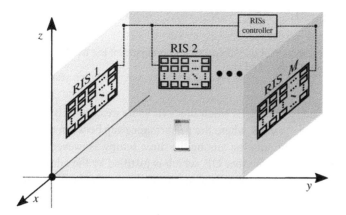

Figure 7.6 3D user localization system with multiple spatially distributed receiving RISs.

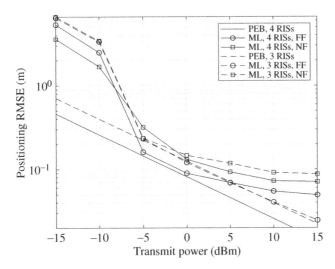

Figure 7.7 The RMSE of the localization error in meters versus the source transmit power in dBm considering 3 and 4 RISs, each with 8 × 8 elements. The proposed two-stage localization algorithm is evaluated along with the theoretical PEB, considering both a far-field (FF) and a near-field (NF) channel models.

Although the proposed localization system in Alexandropoulos et al. (2022b) implies that the true AoAs at the receiving RISs are sufficiently different with each other, due to the absence, or reduced, spatial correlation among the spatially distributed metasurfaces, the deployment cost and implementation complexity are increased compared to cases where the receiving RISs can be co-located. This potential was the subject of study in He et al. (2022b). In this work, the authors presented a partially-connected receiving RIS comprising multiple receiving sub-RISs, each equipped with a reception RF chain, as demonstrated in Fig. 7.8. This system does not require backhaul links to connect the sub-RISs, as this

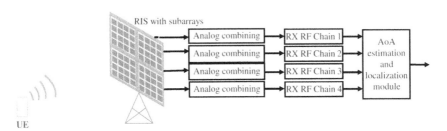

Figure 7.8 3D user localization with a partially-connected receiving RIS comprising several sub-RISs, each deploying a reception RF chain enabling baseband signal processing.

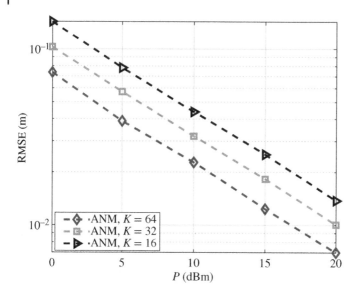

Figure 7.9 Localization performance of the ANM-based technique versus the transmit power P with a partially-connected receiving RIS with four 16-element sub-RISs for different training overhead values K.

was the case with the spatially distributed receiving RISs necessitating a central processing unit. The 3D localization performance of the partially-connected receiving RIS for different values K of the number of training time slots is illustrated in Fig. 7.9, considering four sub-RISs each with 16 metamaterial elements. Unlike the OMP-based AoA estimation technique adopted in Alexandropoulos et al. (2022b), the off-grid compressed sensing technique of the atomic norm minimization (ANM) was implemented. It is evident from the figure that, when the transmit power at the UE increases to 20 dBm, 1-cm UE localization accuracy can be achieved even when the training overhead is $K = 32$ time slots.

7.4 Sensing with RISs

In this section, the potential of reflective RISs for sensing as well as simultaneous localization and radio mapping simultaneous localization and mapping (SLAM) is discussed.

7.4.1 Link Budget Analysis

In general, sensing refers to the action of detecting and estimating the characteristics of changes or events in the environment. When considered in wireless systems, it often entails the detection and tracking of passive objects. In contrast to localization, any support for sensing has not been released in 3GPP up to date. The sensing versions are conventionally classified as follows (Wymeersch et al. 2021):

- *Monostatic sensing*: This refers to the case where the TX and RX are co-located. The former emits a waveform and the latter processes the backscattered waveform to detect the presence of targets, including their distance, bearing, and velocity. Due to the round-trip propagation time, transmission and reception happen in different time intervals, i.e., in a time-orthogonal manner. Full duplex radios can be used for simultaneous transmission and reception (Islam et al. 2022).
- *Bistatic sensing*: The TX and RX are placed in different locations and their clock difference can be synchronized via their direct LoS path. By processing the backscattered waveform at the RX side, sensing information can be obtained.
- *Multistatic sensing*:: This refers to the case where multiple RXs are used for sensing, each receiving the backscattered waveform. The sensing information obtained at the different RXs is fused to deliver the synthesized sensing of the objects in the RXs' vicinity.

Similar to localization, RISs can improve sensing by providing additional signal propagation paths, as shown in Fig. 7.10. They can be deployed for enabling sensing in challenging scenarios, thus, contributing in overcoming LoS blockage conditions (Aubry et al. 2021), as well as for boosting sensing performance (Wymeersch et al. 2020). To understand the potential of RIS-aided sensing (Buzzi et al. 2021), the link budget for single- and double-bounce signals of monostatic and bistatic radar sensing (del Peral-Rosado et al. 2017) are next studied, considering a single-antenna BS and UE. In particular, the RIS gain is defined as $G_{\mathrm{RIS}} \triangleq \mathbb{E}[|\alpha_{\mathrm{RIS}}^{\mathsf{T}}(\phi)\Omega\alpha_{\mathrm{RIS}}(\phi)|^2]$ for both investigated signal categories, where $\alpha(\phi)$ represents the array vector at the N_{ris}-element reflective RIS. When using directional RIS phase profiles, it holds $G_{\mathrm{RIS}} = N_{\mathrm{ris}}^2$, whereas $G_{\mathrm{RIS}} = N_{\mathrm{ris}}$ when random phase profiles are used. The area of each RIS element has been

Figure 7.10 A signal path enabled by an RIS, which is deployed to overcome a signal blockage. This RIS-enabled path is often termed as virtual LoS channel path.

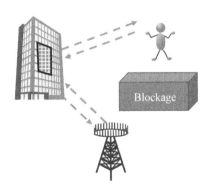

chosen as $A_{RIS} = (\lambda/4)^2$. The distances between the UE and RIS, UE and scattering point (SP), RIS and a SP, BS and RIS, and BS and SP are denoted by d_{UR}, d_{US}, d_{RS}, d_{BR}, and d_{BS}, respectively.

7.4.1.1 Monostatic Radar Sensing

A full-duplex UE transmits orthogonal frequency division multiplexing (OFDM) signals, which are reflected by the RIS or scattered by the SPs. Then, the UE receives the following single- and double-bounce channel paths: (i) UE-RIS-UE; (ii) UE-RIS-SPs-UE; (iii) UE-SPs-RIS-UE; and (iv) UE-SPs-UE. The link budgets for paths (i), (ii), and (iv) can be respectively modeled as follows:

$$\frac{P_{Rx,1}^{(M)}}{P_{Tx}} = \frac{\lambda^2 \sigma_{RCS}}{(4\pi)^3 d_{US}^4}, \quad \frac{P_{RX,2}^{(M)}}{P_{Tx}} = \frac{\lambda^2 \sigma_{RCS} A_{RIS} G_{RIS}}{(4\pi)^4 d_{US}^4 d_{UR}^2 d_{RS}^2}, \quad \frac{P_{RX,3}^{(M)}}{P_{Tx}} = \frac{\lambda^2 A_{RIS} G_{RIS}}{(4\pi)^3 d_{UR}^4}, \quad (7.4)$$

where P_{Tx} denotes the UE transmit power, $\{P_{Rx,i}^{(M)}\}_{i=1}^3$ are the UE's received powers from the signal propagation paths (i), (ii), and (iv), respectively, and σ_{RCS} represents the radar cross-section. It is noted that the link budget for path (iii) is the same with that for path (ii).

7.4.1.2 Bistatic Radar Sensing

It is assumed that a BS transmits OFDM signals, which are then reflected by the RIS or scattered by the SPs and received by a UE. The following single- and double-bounce channel paths are generated: (i) BS-RIS-UE; (ii) BS-RIS-SPs-UE; and (iii) BS-SPs-RIS-UE. It is assumed that there exists a large obstacle between the BS and UE, hence, the LoS and UE-SPs-UE channel paths are absent. The link budgets for all available paths are respectively modeled as:

$$\frac{P_{RX,1}^{(B)}}{P_{Tx}} = \frac{\lambda^2 A_{RIS} G_{RIS}}{(4\pi)^3 d_{BR}^2 d_{RS}^2}, \quad \frac{P_{RX,2}^{(B)}}{P_{Tx}} = \frac{\lambda^2 \sigma_{RCS} A_{RIS} G_{RIS}}{(4\pi)^4 d_{US}^2 d_{BR}^2 d_{RS}^2}, \quad \frac{P_{RX,3}^{(B)}}{P_{Tx}} = \frac{\lambda^2 \sigma_{RCS} A_{RIS} G_{RIS}}{(4\pi)^4 d_{BS}^2 d_{UR}^2 d_{RS}^2}.$$

$$(7.5)$$

Figure 7.11 demonstrates the fundamental limits of RIS-aided sensing. In particular, the path losses of the different signal propagation paths, defined as the ratio of the received power over the transmit one, i.e., P_{RX}/P_{TX} (P_{RX} takes any of the $P_{Rx,i}^{(M)}$ and $P_{Rx,i}^{(B)}$ values), are illustrated. For both sensing cases, the single RIS was located at the point $[0, 0]^T$ having $N_{ris} = 2500$ elements; a single antenna was mounted on the top of the UE; and the SP was placed at the point $[x_{SP}, -20]^T$ with $x_{SP} \in (-50, 50)$. In the monostatic sensing case, the UE was located at the point $[-30, -30]^T$, whereas, in the bistatic sensing case, the BS and UE were respectively placed at the points $[-30, -30]^T$ and $[30, -30]^T$. Both directional and random RIS phase profiles have been considered. As shown in Fig. 7.11(a) for the former sensing case, the UE-SP-UE path is always stronger than the UE-RIS-SP-UE one, even

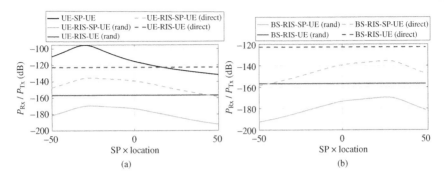

Figure 7.11 Link budgets in dB of the single- and double-bounce signals in RIS-aided (a) monostatic and (b) bistatic radar sensing systems. Both directional (direct) and random (rand) RIS phase profiles have been considered.

though the link budget of double-bounce signals with directional phase configurations is boosted by about 34 dB, as compared to when random configurations are used. It can be also seen that, in both monostatic and bistatic sensing, the signals from single-bounce paths via RIS reflection are stronger than those from the double-bounce ones. In fact, the latter paths act as interference in passive SP sensing, which can be, however, mitigated with appropriate precoding and combining in the case of MIMO UEs (e.g., a full-duplex MIMO one (Alexandropoulos et al. 2022a)). Finally, it is confirmed that, due to the severe path loss experienced at the RIS-reflected channel path, the double-bounce signals carry limited information as compared to the single-bounce ones that are not reflected by the RIS. However, in bistatic sensing scenarios where the BS-SP-UEs path is blocked due to, e.g., the existence of a large obstacle between the BS and UEs, the double-bounce signals can be rather informative for passive objects' sensing.

7.4.2 Joint Sensing and Localization with a Single RIS

Consider the scenario illustrated in Fig. 7.12 where a full duplex UE transmits OFDM signals, which are reflected by the RIS, specularly reflected by a reflection point (RP) on a reflective surface, and possibly scattered by a SP. As previously discussed, multiple-bounce signals are usually hardly detectable, hence, only the single-bounce ones are considered. The RP and SPs are regarded as landmarks and modeled by a random finite set (Mahler 2007). To this end, the UE receives signals via the following back-propagated single-bounce paths: (i) reflection from the RIS; (ii) specular reflection by the large surface (Kim et al. 2020); and (iii) scattering by the SP. The resolvable signal path is indexed by l and the path (i) can be controlled by the RIS phase profiles (called as a controlled path and denoted

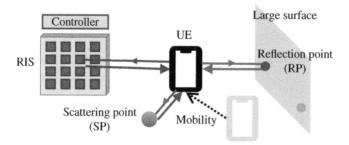

Figure 7.12 An RIS-enabled SLAM scenario with one-bounce signal propagation resulting from the (i) UE-RIS-UE, (ii) UE-RP-UE, and (iii) UE-SP-UE channel paths.

by $l = 0$), while the paths (ii) and (iii) cannot (called as uncontrolled paths and indexed using $l \neq 0$).

The UE's parameters are assumed to be dynamically changing and let the time index denoted by k. The channel parameter are represented by $\boldsymbol{\eta}_{0,k} \triangleq [\boldsymbol{\phi}_{k,0}^\mathsf{T}, \tau_{0,k}, f_{0,k}^D, \boldsymbol{\theta}_{0,k}^\mathsf{T}]^\mathsf{T}$ for $l = 0$, and $\boldsymbol{\eta}_{k,l} \triangleq [\tau_{l,k}, f_{l,k}^D, \boldsymbol{\theta}_{l,k}^\mathsf{T}]^\mathsf{T}$ for $l \neq 0$, where $\boldsymbol{\phi}_k \triangleq [\phi_k^{\text{az}}, \phi_k^{\text{el}}]^\mathsf{T}$ denotes the AoAs[1] in both azimuth and elevation at the RIS, $\boldsymbol{\theta}_{l,k} \triangleq [\theta_{l,k}^{\text{az}}, \theta_{l,k}^{\text{el}}]^\mathsf{T}$ indicates the AoD at the UE, τ_k is the time delay, and f_k^D denotes the Doppler shift. It is further assumed that the channel parameters are obtained using an adequate estimation technique (e.g., Jiang et al. (2021)), and that the estimation referring to each l-path is represented by $\mathbf{z}_{k,l} \triangleq \boldsymbol{\eta}_{k,l} + \mathbf{r}_{k,l}$ with $\mathbf{r}_{k,l}$ being the measurement noise.

UE and Landmark Estimates

Let the UE state be denoted by $\mathbf{s}_k$ and the landmarks by $\mathcal{X}$. These parameters can be estimated using the marginalized Poisson multi-Bernoulli SLAM filter (Kim et al. 2022b), consisting of prediction and update steps. Due to the considered fixed landmarks, there is no prediction. After the update step, a posterior probability density function $f(\mathbf{s}_k, \mathcal{X}|\mathcal{Z}_k)$ is obtained, by computing $f(\mathbf{s}_k|\mathcal{Z}_k)$, $\lambda(\mathbf{x}|\mathcal{Z}_k)$ for $\mathbf{x} \in \mathcal{X}^P$ and $f^i(\mathcal{X}^i|\mathcal{Z}_k)$ for $i \in \mathcal{I}$. In the latter densities, $\mathcal{X}^P$ is the set of landmarks that have never been detected and follows the Poisson process, while $\mathcal{X}^i$ is the set of landmarks that have been detected and follows the Bernoulli process. In addition, $\mathcal{I}$ is the set of indices for the detected landmarks; $f(\mathbf{s}_k|\mathcal{Z}_k)$ is the UE state density; $\lambda(\mathbf{x}|\mathcal{Z}_k)$ is the intensity function for the undetected landmarks; and $f^i(\mathcal{X}^i|\mathcal{Z}_k)$ is the Bernoulli density for the detected landmarks. The above densities and intensity are analytically presented in (Kim et al., 2022b, Appendix D).

In Fig. 7.13, the SLAM performance of the scheme in Kim et al. (2022a) is evaluated via the mean absolute error (MAE) of the UE position and the SP generalized

---

1 In monostatic sensing, the AoA is equivalent to the AoD.

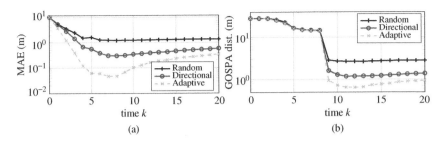

Figure 7.13 SLAM performance with respect to the time index k for different RIS phase profiles: (a) MAE of UE position and (b) GOSPA distance of SP.

optimal subpattern assignment (GOSPA) distance. Four SPs were randomly distributed near the UE and the RIS location was set to the point $\mathbf{x}_{RIS} = [40, 0, 20]^\top$. The UE state was set to $\mathbf{s}_0 = [50, -30, 0, \pi/2, 11.11]^\top$ with units in meters for the first three parameters and in rad and meters/second for the latter two parameters. The initial UE uncertainty was 10 m and the UE was considered to have dynamics following the constant turn model (Li and Jilkov 2003). The following parameters were also used: the RIS array size was $N_{ris} = 2500$ (50×50); the UE had 16 (4×4) antenna elements; the number of transmissions was set to 20; the carrier frequency was 30 GHz; the bandwidth was set to 200 MHz; the subcarrier spacing was 120 kHz; and the transmission power was 20 dBm. As clearly shown in Fig. 7.13(a) for the UE localization, the performance with the adaptive phase is always better than that with random and directional phases. In Fig. 7.13(b) with the SP GOSPA distance results, it is confirmed that the landmarks with the adaptive RIS phase can be well estimated. This behavior indicates that the SLAM filter with the adaptive RIS phase is robust to missed detections and false alarms.

7.5 Conclusion and Open Challenges

Various works have been included in this chapter regarding 3D localization and sensing, as well as their simultaneous realization, empowered by the incorporation of RISs, either operating independently or in synergy with the existing cellular systems. This fast evolving technology, in terms of hardware architectures, configuration optimization algorithms, network integration schemes, and over-the-air computing capabilities, provides various opportunities for low-cost and low-power consumption localization and sensing in beyond 5G and 6G wireless systems. However, there still exist certain challenges that need to be better understood and finally overcome.

While there has been a significant effort in modeling end-to-end RIS-parameterized channels for communications, those do not necessarily capture the physical geometry of wireless channels, nor the coherence needed in space, time, and frequency bands. Also, the frequency response of RISs, as well as a slew of hardware impairments (e.g., mutual coupling) must be understood in more detail. To this end, accurate models are necessary for the validation of theoretical methods. In addition, the large path loss of RIS-enabled propagation paths, in conjunction with the large dependence of localization and sensing on geometry, mandates that RISs should be placed in suitable locations to provide adequate localization and sensing coverage. Those placements may be different from those that are suitable for communication, so appropriate trade-offs must be considered. The RIS location and orientation must be also accurately calibrated for most localization and sensing applications. Hence, there is a need to develop such calibration methods to properly place each RIS in the global coordinate system used by the infrastructure.

Estimating the angles or spatial frequencies at the RISs demands that their phase profiles change quickly over time. The RIS hardware must be designed with this ability in mind, since a slow update rate limits the application scenarios that can be covered, due to UE and object mobility. Moreover, each RIS can comprise hundreds or thousands of metamaterials. It is likely that over the lifetime of the metasurface, these elements age or fail, get locked into certain states or do not provide the desired state. Hence, fault monitoring and mitigation methods must be developed, along with RIS failure models that accurately capture space and time-correlated errors. Most likely, localization and sensing will be more sensitive to even minor faults than communication. Last but not least, in environments equipped with a plurality of RISs, some form of coordination or interference control is necessary. This can be achieved by temporal encoding or by BS beamforming optimization. The problem is especially challenging in multi-functional, multioperator, and multifrequency settings.

Bibliography

Akdeniz, M.R., Liu, Y., Samimi, M.K. et al. (2014). Millimeter wave channel modeling and cellular capacity evaluation. 32 (6): 1164–1179. https://doi.org/10.1109/JSAC.2014.2328154.

Alamzadeh, I., Alexandropoulos, G.C., Shlezinger, N., and Imani, M.F. (2021). A reconfigurable intelligent surface with integrated sensing capability. *Scientific Reports* 11 (20737): 1–10.

Alexandropoulos, G.C. and Mathiopoulos, P.T. (2010). Performance evaluation of selection diversity receivers over arbitrarily correlated generalized Gamma fading channels. *IET Communications* 4 (10): 1253–1265.

Alexandropoulos, G.C. and Vlachos, E. (2020). A hardware architecture for reconfigurable intelligent surfaces with minimal active elements for explicit channel estimation. *Proceedings of the IEEE ICASSP*, Barcelona, Spain, May 2020.

Alexandropoulos, G.C., Sagias, N.C., and Berberidis, K. (2007). On the multivariate Weibull fading model with arbitrary correlation matrix. *IEEE Antennas and Wireless Propagation Letters* 6: 93–95.

Alexandropoulos, G.C., Mathiopoulos, P.T., and Sagias, N.C. (2010). Switch-and-examine diversity over arbitrary correlated Nakagami-*m* fading channels. *IEEE Transactions on Vehicular Technology* 59 (4): 2080–2087.

Alexandropoulos, G.C., Lerosey, G., Debbah, M., and Fink, M. (2020a). Reconfigurable intelligent surfaces and metamaterials: the potential of wave propagation control for 6G wireless communications. *IEEE ComSoc TCCN Newsletter* 6 (1): 25–37.

Alexandropoulos, G.C., Samarakoon, S., Bennis, M., and Debbah, M. (2020b). Phase configuration learning in wireless networks with multiple reconfigurable intelligent surfaces. *Proceedings of the IEEE GLOBECOM*, Taipei, Taiwan, December 2020b.

Alexandropoulos, G.C., Shlezinger, N., Alamzadeh, I. et al. (2021a). Hybrid reconfigurable intelligent metasurfaces: enabling simultaneous tunable reflections and sensing for 6G wireless communications. https://arxiv.org/pdf/2104.04690.

Alexandropoulos, G.C., Shlezinger, N., and del Hougne, P. (2021b). Reconfigurable intelligent surfaces for rich scattering wireless communications: recent experiments, challenges, and opportunities. *IEEE Communications Magazine* 59 (6): 28–34.

Alexandropoulos, G.C., Islam, M.A., and Smida, B. (2022a). Full-duplex massive multiple-input, multiple-output architectures: recent advances, applications, and future directions. *IEEE Vehicular Technology Magazine* 17 (4): 83–91. https://doi.org/10.1109/MVT.2022.3211689.

Alexandropoulos, G.C., Vinieratou, I., and Wymeersch, H. (2022b). Localization via multiple reconfigurable intelligent surfaces equipped with single receive RF chains. *IEEE Wireless Communications Letters* 11 (5): 1072–1076.

Ardah, K., Gherekhloo, S., de Almeida, A.L.F., and Haardt, M. (2021). TRICE: a channel estimation framework for RIS-aided millimeter-wave MIMO systems. *IEEE Signal Processing Letters* 28: 513–517. https://doi.org/10.1109/LSP.2021.3059363.

Aubry, A., De Maio, A., and Rosamilia, M. (2021). Reconfigurable intelligent surfaces for N-LOS radar surveillance. *IEEE Transactions on Vehicular Technology* 70 (10): 10735–10749.

Beaulieu, N. and Cheng, C. (2005). Efficient Nakagami-m fading channel simulation. *IEEE Transactions on Vehicular Technology* 54 (2): 413–424. https://doi.org/10 .1109/TVT.2004.841555.

Behravan, A., Yajnanarayana, V., Keskin, M.F. et al. (2023). Positioning and sensing in 6G: gaps, challenges, and opportunities. *IEEE Vehicular Technology Magazine* 18 (1): 40–48.

Björnson, E. and Sanguinetti, L. (2021). Rayleigh fading modeling and channel hardening for reconfigurable intelligent surfaces. *IEEE Wireless Communications Letters* 10 (4): 830–834.

Buzzi, S., Grossi, E., Lops, M., and Venturino, L. (2021). Radar target detection aided by reconfigurable intelligent surfaces. *IEEE Signal Processing Letters* 28: 1315–1319.

Chepuri, S.P., Shlezinger, N., Liu, F. et al. (2022). Integrated sensing and communications with reconfigurable intelligent surfaces. *accepted for publication in IEEE Signal Processing Magazine.* Available: arXiv:2211.01003.

Dardari, D., Decarli, N., Guerra, A., and Guidi, F. (2022). LOS/NLOS near-field localization with a large reconfigurable intelligent surface. *IEEE Transactions on Wireless Communications* 21 (6): 4282–4294. https://doi.org/10.1109/TWC.2021 .3128415.

Di Palma, L., Clemente, A., Dussopt, L. et al. (2017). Circularly-polarized reconfigurable transmitarray in Ka-band with beam scanning and polarization switching capabilities. *IEEE Transactions on Antennas and Propagation* 65 (2): 529–540.

Di Renzo, M., Debbah, M., Phan-Huy, D.-T. et al. (2019). Smart radio environments empowered by reconfigurable AI meta-surfaces: an idea whose time has come. *EURASIP Journal on Wireless Communications and Network* 2019 (1): 1–20.

Di Renzo, M., Danufane, F.H., and Tretyakov, S. (2022). Communication models for reconfigurable intelligent surfaces: from surface electromagnetics to wireless networks optimization. *Proceedings of the IEEE* 110 (9): 1164–1209.

Dwivedi, S., Shreevastav, R., Munier, F. et al. (2021). Positioning in 5G networks. *IEEE Communications Magazine* 59 (11): 38–44.

Egea-Roca, D., Arizabaleta-Diez, M., Pany, T. et al. (2022). GNSS user technology: state-of-the-art and future trends. *IEEE Access* 10: 39939–39968.

Faqiri, R., Saigre-Tardif, C., Alexandropoulos, G.C. et al. (2023). PhysFad: physics-based end-to-end channel modeling of RIS-parametrized environments with adjustable fading. *IEEE Transactions on Wireless Communications* 22 (1): 580–595.

Geng, Z., Yan, H., Zhang, J., and Zhu, D. (2021). Deep-learning for radar: a survey. *IEEE Access* 9: 141800–141818.

Ghazalian, R., Keikhosravi, K., Chen, H. et al. (2022). Bi-static sensing for near-field RIS localization. arXiv preprint arXiv:2206.13915.

Gong, T., Vinieratou, I., Ji, R. et al. (2023). Holographic MIMO communications: theoretical foundations, enabling technologies, and future directions. *accepted for publication in IEEE Communication Surveys & Tutorials*. Available: arXiv:2212.01257.

He, J., Fakhreddine, A., and Alexandropoulos, G.C. (2022a). Simultaneous indoor and outdoor 3D localization with STAR-RIS-assisted Millimeter wave systems. arXiv preprint arXiv:2207.05344.

He, J., Fakhreddine, A., Vanwynsberghe, C. et al. (2022b). 3D localization with a single partially-connected receiving RIS: positioning error analysis and algorithmic design. arXiv preprint arXiv:2212.02088.

Heath, R.W., González-Prelcic, N., Rangan, S. et al. (2016). An overview of signal processing techniques for millimeter wave MIMO systems. *IEEE Journal of Selected Topics in Signal Processing* 10 (3): 436–453. https://doi.org/10.1109/JSTSP.2016 .2523924.

Hu, S., Rusek, F., and Edfors, O. (2018). Beyond massive MIMO: the potential of positioning with large intelligent surfaces. *IEEE Transactions on Signal Processing* 66 (7): 1761–1774.

Huang, C., Alexandropoulos, G.C., Zappone, A. et al. (2018). Energy efficient multi-user MISO communication using low resolution large intelligent surfaces. *Proceedings of the IEEE GLOBECOM*, Abu Dhabi, UAE, December 2018.

Huang, C., Zappone, A., Alexandropoulos, G.C. et al. (2019). Reconfigurable intelligent surfaces for energy efficiency in wireless communication. *IEEE Transactions on Wireless Communications* 18 (8): 4157–4170.

Islam, M.A., Alexandropoulos, G.C., and Smida, B. (2022). Integrated sensing and communication with millimeter wave full duplex hybrid beamforming. *Proceedings of the IEEE ICC*, Seoul, South Korea, May 2022.

Jamali, V., Alexandropoulos, G.C., Schober, R., and Poor, H.V. (2022). Low-to-zero-overhead IRS reconfiguration: decoupling illumination and channel estimation. *IEEE Communications Letters* 26 (4): 932–936.

Jian, M., Alexandropoulos, G.C., Basar, E. et al. (2022). Reconfigurable intelligent surfaces for wireless communications: overview of hardware designs, channel models, and estimation techniques. *Intelligent and Converged Networks* 3 (1): 1–32.

Jiang, F., Wen, F., Ge, Y. et al. (2021). Beamspace multidimensional ESPRIT approaches for simultaneous localization and communications. *arXiv preprint arXiv:2111.07450*.

Kang, M. and Alouini, M. (2006). Capacity of MIMO Rician channels. *IEEE Transactions on Wireless Communications* 5 (1): 112–122. https://doi.org/10.1109/ TWC.2006.1576535.

Kermoal, J., Schumacher, L., Pedersen, K. et al. (2002). A stochastic MIMO radio channel model with experimental validation. *IEEE Journal on Selected Areas in Communications* 20 (6): 1211–1226. https://doi.org/10.1109/JSAC.2002.801223.

Keykhosravi, K., Keskin, M.F., Seco-Granados, G., and Wymeersch, H. (2021). SISO RIS-enabled joint 3D downlink localization and synchronization. *ICC International Conference on Communications*, 1–6. IEEE.

Keykhosravi, K., Denis, B., Alexandropoulos, G.C. et al. (2022a). Leveraging RIS-enabled smart signal propagation for solving infeasible localization problems. *to appear IEEE Vehicular Technology Magazine*. Available at arXiv preprint arXiv:2204.11538.

Keykhosravi, K., Seco-Granados, G., Alexandropoulos, G.C., and Wymeersch, H. (2022b). RIS-enabled self-localization: leveraging controllable reflections with zero access points. arXiv preprint arXiv:2202.11159.

Kim, H., Granström, K., Gao, L. et al. (2020). 5G mmWave cooperative positioning and mapping using multi-model PHD. *IEEE Transactions on Wireless Communications* 19 (6): 3782–3795.

Kim, H., Chen, H., Keskin, M.F. et al. (2022a). RIS-enabled and access-point-free simultaneous radio localization and mapping. arXiv preprint arXiv:2212.07141.

Kim, H., Granstrom, K., Svensson, L. et al. (2022b). PMBM-based SLAM filters in 5G mmWave vehicular networks. *IEEE Transactions on Vehicular Technology* 71 (8): 8646–8661.

Li, X.R. and Jilkov, V.P. (2003). Survey of maneuvering target tracking. Part I. Dynamic models. *IEEE Transactions on Aerospace and Electronic Systems* 39 (4): 1333–1364.

Liu, Y., Mu, X., Schober, R., and Poor, H.V. (2021a). Simultaneously transmitting and reflecting (STAR)-RISs: a coupled phase-shift model. arXiv preprint arXiv:2110.02374.

Liu, Y., Mu, X., Xu, J. et al. (2021b). STAR: simultaneous transmission and reflection for 360° coverage by intelligent surfaces. *accepted for publication in IEEE Wireless Communications*. Available: arXiv:2103.09104.

Ma, X., Ballal, T., Chen, H. et al. (2021). A maximum-likelihood TDOA localization algorithm using difference-of-convex programming. *IEEE Signal Processing Letters* 28: 309–313.

Mahler, R. (2007). *Statistical Multisource-Multitarget Information Fusion*. Norwood, MA: Artech House. ISBN: 1596930926, 9781596930926

Pan, C., Zhou, G., Zhi, K. et al. (2022). An overview of signal processing techniques for RIS/IRS-aided wireless systems. *IEEE Journal of Selected Topics in Signal Processing* 16 (5): 883–917. https://doi.org/10.1109/JSTSP.2022.3195671.

del Peral-Rosado, J.A., Raulefs, R., López-Salcedo, J.A., and Seco-Granados, G. (2017). Survey of cellular mobile radio localization methods: from 1G to 5G. *IEEE Communication Surveys & Tutorials* 20 (2): 1124–1148.

Rahal, M., Denis, B., Keykhosravi, K. et al. (2022a). Arbitrary beam pattern approximation via RISs with measured element responses. *Proceedings of the IEEE*

Joint European Conference on Networks and Communications & 6G Summit, Grenoble, France, June 2022a.

Rahal, M., Denis, B., Keykhosravi, K. et al. (2022b). Constrained RIS phase profile optimization and time sharing for near-field localization. *2022 IEEE 95th Vehicular Technology Conference:(VTC2022-Spring),* 1–6. IEEE.

Shlezinger, N., Dicker, O., Eldar, Y.C. et al. (2019). Dynamic metasurface antennas for uplink massive MIMO systems. *IEEE Transactions on Communications* 67 (10): 6829–6843.

Strinati, E.C., Alexandropoulos, G.C., Wymeersch, H. et al. (2021). Reconfigurable, intelligent, and sustainable wireless environments for 6G smart connectivity. *IEEE Communications Magazine* 59 (10): 99–105.

Stylianopoulos, K., Shlezinger, N., del Hougne, P., and Alexandropoulos, G.C. (2021). Deep-learning-assisted configuration of reconfigurable intelligent surfaces in dynamic rich-scattering environments. *Proceedings of the IEEE ICASSP,* Singapore, May 2021.

Tasci, R.A., Kilinc, F., Basar, E., and Alexandropoulos, G.C. (2022). A new RIS architecture with a single power amplifier: energy efficiency and error performance analysis. *IEEE Access* 10: 44804–44815.

Vlachos, E., Alexandropoulos, G.C., and Thompson, J. (2019). Wideband MIMO channel estimation for hybrid beamforming millimeter wave systems via random spatial sampling. *IEEE Journal on Selected Topics in Signal Processing* 13 (5): 1136–1150.

Vo, Q.D. and De, P. (2016). A survey of fingerprint-based outdoor localization. *IEEE Communication Surveys & Tutorials* 18 (1): 491–506.

Wang, J., Varshney, N., Gentile, C. et al. (2022). Integrated sensing and communication: enabling techniques, applications, tools and data sets, standardization, and future directions. *IEEE Internet of Things Journal* 9 (23): 23416–23440.

Wen, F., Liu, P., Wei, H. et al. (2018). Joint azimuth, elevation, and delay estimation for 3-D indoor localization. *IEEE Transactions on Vehicular Technology* 67 (5): 4248–4261.

Witrisal, K., Meissner, P., Leitinger, E. et al. (2016). High-accuracy localization for assisted living: 5G systems will turn multipath channels from foe to friend. *IEEE Signal Processing Magazine* 33 (2): 59–70.

Wymeersch, H., He, J., Denis, B. et al. (2020). Radio localization and mapping with reconfigurable intelligent surfaces: challenges, opportunities, and research directions. *IEEE Vehicular Technology Magazine* 15 (4): 52–61.

Wymeersch, H., Shrestha, D., De Lima, C.M. et al. (2021). Integration of communication and sensing in 6G: a joint industrial and academic perspective. *IEEE 32nd Annual International Symposium on Personal, Indoor and Mobile Radio Communications (PIMRC),* 1–7.

Wymeersch, H., Denis, B., Alexandropoulos, G. et al. (2022). Control for RIS-based localisation and sensing. RISE6G project Deliverable D5.1. https://rise-6g.eu/Pages/DELIVERABLES/Livrables.aspx (accessed 14 August 2023).

Xu, J., Liu, Y., Mu, X., and Dobre, O.A. (2021). STAR-RISs: simultaneous transmitting and reflecting reconfigurable intelligent surfaces. *IEEE Communications Letters* 25 (9): 3134–3138. https://doi.org/10.1109/LCOMM.2021.3082214.

Zhang, H., Shlezinger, N., Alamzadeh, I. et al. (2021). Channel estimation with simultaneous reflecting and sensing reconfigurable intelligent metasurfaces. *Proceedings of the IEEE SPAWC*, Lucca, Italy, September 2021.

Zhang, S., Zhang, H., Di, B. et al. (2022). Intelligent omni-surfaces: ubiquitous wireless transmission by reflective-refractive metasurfaces. *IEEE Transactions on Wireless Communications* 21 (1): 219–233.

8

IRS-Aided THz Communications

Boyu Ning and Zhi Chen

National Key Laboratory of Wireless Communications, University of Electronic Science and Technology of China, Chengdu, Sichuan, China

8.1 IRS-Aided THz MIMO System Model

Consider a point-to-point IRS-assisted UM-MIMO communication system as depicted in Fig. 8.1, where the base station (BS), equipped with N_t antennas, transmits $N_s \leq N_t$ data streams to a user, equipped with N_r antennas, with the help of an IRS equipped with N_{IRS} passive elements. In the communication, the BS sends its data message $\mathbf{s} \in \mathbb{C}^{N_s \times 1}$, $\mathbf{s} \sim \mathcal{CN}(\mathbf{0}, \mathbf{I}_{N_s})$ via a precoder $\mathbf{F} \in \mathbb{C}^{N_t \times N_s}$ to the user and the IRS simultaneously. Let $\mathbf{H}_{\text{LoS}} \in \mathbb{C}^{N_r \times N_t}$, $\mathbf{M} \in \mathbb{C}^{N_{\text{IRS}} \times N_t}$, and $\mathbf{N} \in \mathbb{C}^{N_r \times N_{\text{IRS}}}$ denote the channels from the BS to the user, from the BS to the IRS, and from the IRS to the user, respectively. The received signal at the IRS is first phase-shifted by a diagonal reflection matrix $\mathbf{\Theta} = \text{diag}(\beta e^{j\theta_1}, \beta e^{j\theta_2}, \dots, \beta e^{j\theta_{N_{\text{IRS}}}}) \in \mathbb{C}^{N_{\text{IRS}} \times N_{\text{IRS}}}$ and then reflected to the user, where $j = \sqrt{-1}$ is the imaginary unit, $\{\theta_i \in [0, 2\pi)\}_i^{N_{\text{IRS}}}$ are the shifted phases, and $\beta \in [0, 1]$ denote the amplitude of each reflection coefficient. As such, the overall received signal is expressed as

$$\mathbf{y} = \sqrt{\frac{P}{N_s}} (\underbrace{\mathbf{N\Theta MFs}}_{\text{BS-IRS-user}} + \underbrace{\mathbf{H}_{\text{LoS}}\mathbf{Fs}}_{\text{BS-user}}) + \mathbf{n}, \tag{8.1}$$

where P is the total transmitted power and $\|\mathbf{F}\|_F^2 = N_s$. In addition, $\mathbf{n} \sim \mathcal{CN}(\mathbf{0}, \sigma_n^2 \mathbf{I}_{N_r})$ is zero-mean additive Gaussian noise. The aim is to maximize the spectral efficiency by jointly optimizing the precoding matrix $\mathbf{F}$ and the phase shifters $\{\theta_i\}_{i=1}^{N_{\text{IRS}}}$, subject to the power constraint at the BS and the uni-modular constraints on the phase shifters. Let $\mathbf{v} = [e^{j\theta_1}, e^{j\theta_2}, \dots, e^{j\theta_{N_{\text{IRS}}}}]^H$ denote the phase shifter vector at the IRS, i.e., $\mathbf{\Theta} = \beta \cdot \text{diag}(\mathbf{v}^\dagger)$. Define the effective channel in

Intelligent Surfaces Empowered 6G Wireless Network, First Edition.
Edited by Qingqing Wu, Trung Q. Duong, Derrick Wing Kwan Ng, Robert Schober, and Rui Zhang.
© 2024 The Institute of Electrical and Electronics Engineers, Inc. Published 2024 by John Wiley & Sons, Inc.

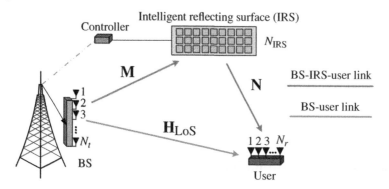

Figure 8.1 A point-to-point IRS-assisted UM-MIMO system.

IRS-assisted UM-MIMO systems as $\mathbf{H}_{\text{eff}} = \mathbf{N\Theta M} + \mathbf{H}_{\text{LoS}}$. Thus, the IRS-assisted joint beamforming problem can be formulated as

$$\max_{\mathbf{F},\mathbf{v}} \log_2 \det \left| \mathbf{I}_{N_b} + \frac{P}{\sigma_n^2 N_s} \mathbf{H}_{\text{eff}} \mathbf{F} \mathbf{F}^H \mathbf{H}_{\text{eff}}^H \right|$$

$$\text{s.t.} \quad \mathbf{H}_{\text{eff}} = \mathbf{N\Theta M} + \mathbf{H}_{\text{LoS}}, \tag{8.2}$$

$$\|\mathbf{F}\|_F^2 = N_s, \quad \mathbf{\Theta} = \beta \cdot \text{diag}(\mathbf{v}^\dagger),$$

$$|\mathbf{v}(i)| = 1, \quad i = 1, 2, \ldots, N_{\text{IRS}},$$

where $\mathbf{v}(i)$ denotes the ith entry of $\mathbf{v}$. Problem (8.2) is quite a hard optimization, as the non-convexity remains on both the objective function and the constraint imposed by IRS's phase shifts. Hitherto, many works have focused on solving it (Ning et al. 2020a,2022; Zhang and Zhang 2020; Wang et al. 2020b).

However, it is practically inefficient to combine the channel estimation and the beamforming designs in UM-MIMO THz systems, as their estimation approaches can hardly establish the beam alignment and their beamforming optimization could result in extremely high implementation complexity in the case of the tremendous number of antenna elements (Chen et al. 2021). Next, we introduce a low-complexity cooperative beam training procedure for THz IRS-assisted systems proposed in (Ning et al. 2020b; Ning et al. 2021), by using diagrams to illustrate its core idea.

8.2 Beam Training Protocol

Assume that the IRS is placed on the same horizontal level as the BS and the user, i.e., without loss of generality, we do not consider the elevation angle of IRS. As shown in Fig. 8.2, the IRS-assisted system consists of six paths angles

Figure 8.2 Path angles in IRS-assisted UM-MIMO systems.

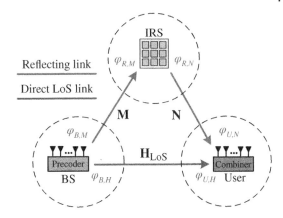

Figure 8.3 Beam modes for active terminal and passive terminal.

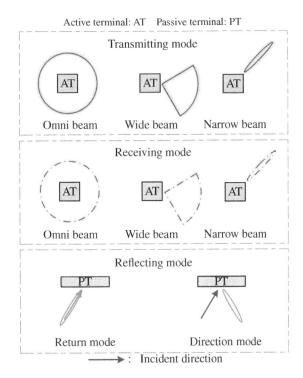

$\varphi_{B,H}$, $\varphi_{U,H}$, $\varphi_{B,M}$, $\varphi_{R,M}$, $\varphi_{R,N}$, and $\varphi_{U,N}$. The beam training aims to find the narrow beams at these angles. For ease of exposition, we first define the training modes for the active terminal (BS and user) and the passive terminal (IRS), respectively. As shown in Fig. 8.3, we use a solid dark gray line to represent the transmit beams, a solid-broken line to represent the receive beams, a solid arrow and a solid light gray line to represent the incoming signal and the reflected signal, respectively. It is worth mentioning that in the return mode of the passive terminal (IRS), the codewords are functions of the angle, i.e., $\{\boldsymbol{\Theta}_{\text{ret}}(\varphi_n^{\text{in}})\}_{n=1}^N$. If the AoA of incoming signals is φ_k^{in}, the codeword $\boldsymbol{\Theta}_{\text{ret}}(\varphi_k^{\text{in}})$ ensures that the AoD of reflected signals is the back direction $\varphi_k^{\text{in}} + \pi$, i.e., $\boldsymbol{\Theta}_{\text{ret}}(\varphi_k^{\text{in}})\mathbf{a}(\varphi_k^{\text{in}}) = \mathbf{a}(\varphi_k^{\text{in}} + \pi)$. The codewords of the directions mode are functions of two angles, i.e., $\{\boldsymbol{\Theta}_{\text{dir}}(\varphi_n^{\text{in}}, \varphi_p^{\text{out}})\}_{n,p=1}^N$.

If the AoA of incoming signals is φ_k^{in}, the codeword $\boldsymbol{\Theta}_{\text{dir}}(\varphi_k^{\text{in}}, \varphi_m^{\text{out}})$ ensures that the AoD of reflected signals is φ_m^{out}, i.e., $\boldsymbol{\Theta}_{\text{dir}}(\varphi_k^{\text{in}}, \varphi_m^{\text{out}})\mathbf{a}(\varphi_k^{\text{in}}) = \mathbf{a}(\varphi_k^{\text{out}})$. Here, we only discuss the effects of these modes and omit the codeword design. Interested readers can learn about the details in Ning et al. (2021).

In the following content, we first present the primary idea of the IRS-assisted joint beam training, albeit with some shortcomings, to draw some basic insights. We divide the overall procedure into three phases to achieve different groups of measurements as illustrated in Fig. 8.4.

Phase 1: Shut down the IRS. Fix the BS to be in an omni-beam transmitting mode and the user sweeps the beam to find the desired path direction. By switching the operation of the BS and the user, we obtain $\varphi_{U,H}$ and $\varphi_{B,H}$.

Phase 2: Keep the user silent and fix the BS to be concurrently in an omni-beam transmitting and receiving mode. Then, the IRS successively sweeps the codewords in return mode, i.e., $\boldsymbol{\Theta}_{\text{ret}}(\varphi_n^{\text{in}})$, which are predefined in time slots and known to all terminals. The best codeword is informed to the BS by determining the time slot with the strongest received signal. By switching the operation of the BS and the user, we obtain $\varphi_{R,M}$ and $\varphi_{R,N}$.

Phase 3: With the obtained $\varphi_{R,M}$ and $\varphi_{R,N}$, we fix IRS to optimally bridge the BS-IRS-user link by direction mode, i.e., $\boldsymbol{\Theta}_{\text{dir}}(\varphi_{R,M}, \varphi_{R,N})$. Fix the BS to be in an omni-beam transmitting mode and the user sweeps the beam to find the desired path direction. By switching the operation of the BS and the user, we obtain $\varphi_{U,N}$ and $\varphi_{B,M}$.

Based on the above, we can find all the path angles in THz IRS-assisted systems (Ning et al. 2020b). However, this strategy suffers from the following main drawbacks.

- Omni-beam may not be effectively detected in THz communication.
- Concurrently transmitting and receiving beams result in interference.
- Sweeping the narrow beams incurs high time complexity.

Focusing on these issues, we now extend this strategy to a more practical procedure. In this procedure, assuming that the path angles $\varphi_{R,M}$ and $\varphi_{R,N}$ at IRS has only N value points, we need to judiciously predefine $2N + 1$ codewords for IRS (the details of codeword design is referred to Ning et al. (2021)), i.e., $\{\boldsymbol{\Theta}_n\}_{n=1}^{2N+1}$, where the optimal phase shifts at IRS is covered by the codewords,

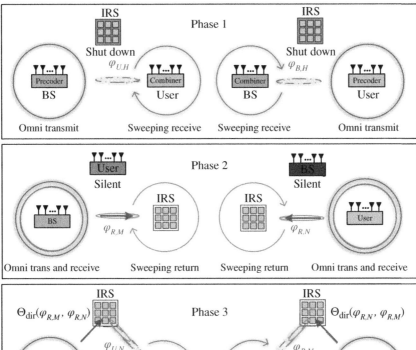

Figure 8.4 Primary idea of the IRS-assisted joint beam training.

i.e., $\Theta_{\text{dir}}(\varphi_{R,M}, \varphi_{R,N}) \in \{\Theta_n\}_{n=1}^{2N+1}$. We show the diagram of the practical procedure in Fig. 8.5.

Phase 1: In Phase 1, we aim to obtain the optimal codeword for IRS. To achieve the beam alignment, we first test 3×3 wide beams in nine successive intervals with BS in the transmitting mode and user k in the receiving mode. In each interval, the IRS successively searches the codewords $\{\Theta_n\}_{n=1}^{2N+1}$. For the IRS, there is only one beam pair that covers both the BS-IRS link and the IRS-user link. During the interval when this beam pair (aligned case) is used, the user will detect an energy pulse in the time slot when IRS uses $\Theta_{\text{dir}}(\varphi_{R,M}, \varphi_{R,N})$. Thus, the user can utilize the pulse slots to identify this optimal codeword for IRS.

Phase 2: In Phase 2, we turn off the IRS and aim to obtain $\varphi_{B,H}$ and $\varphi_{U,H}$ via the following three steps. In step 1, nine wide-beam pairs are tested for alignment.

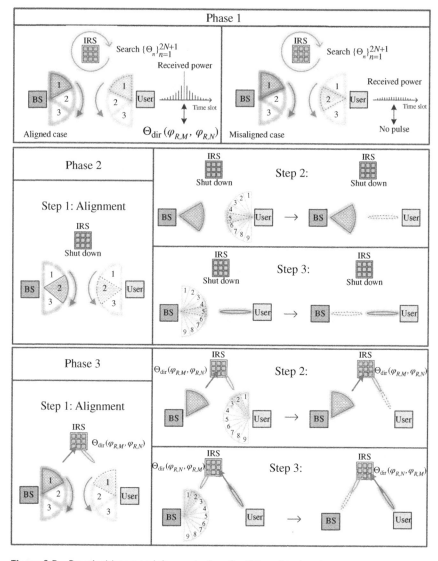

Figure 8.5 Practical beam training procedure for IRS-assisted systems.

The user compares the received energy in nine intervals and determines the aligned pair with the maximum power. The aligned pair is labeled by recording the beams chosen at both sides. In step 2, the BS transmits the labeled wide beam and the user uses a ternary-tree search by narrow beams to obtain $\varphi_{U,H}$. In step 3, the user transmits the labeled narrow beam and BS uses a ternary-tree search to obtain $\varphi_{B,H}$.

Phase 3: In Phase 3, we aim to obtain $\varphi_{B,M}$ and $\varphi_{U,N}$ through three steps similar to Phase 2. We turn on IRS with the obtained optimal codeword, i.e., $\Theta_{\text{dir}}(\varphi_{R,M}, \varphi_{R,N})$. Note that there exist two propagation paths from the BS to the user, i.e., BS-IRS-user path and BS-user path. The BS-user path has been estimated in Phase 2. To estimate the AoA and AoD of the reflecting paths, we can use the ternary-tree search with a small modification. Specifically, in each stage of step 2, the receiver removes the signal component of the BS-user path, by signal processing, when determining the best beam. In step 3, we turn on IRS with $\Theta_{\text{dir}}(\varphi_{R,N}, \varphi_{R,M})$. The user transmits the labeled narrow beam and BS uses a ternary-tree search while removing the signal component of the BS-user path. By this means, we can obtain the best beam pair for the BS-IRS-user path.

Based on the above three phases, all the path angles can be found in IRS-assisted systems, which completes the THz IRS-assisted beam training. We would like to point out that the exhaustive beam training has $N^2 + N^4$ tests in IRS-assisted systems, whereas this training procedure has only $18N + 12\log_3 N - 3$ tests (Ning et al. 2021). Now let us consider a THz IRS-assisted UM-MIMO indoor scenario, in which a BS, a user, and an IRS are at the three vertices of a triangle

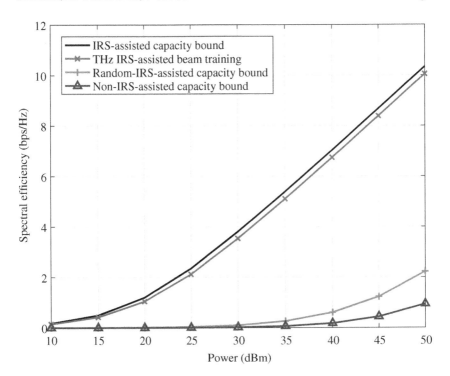

Figure 8.6 The performance of different schemes in THz LoS-blockage case.

with sides of 5 m. The number of BS/user antennas and IRS elements are all 128. The number of narrow beams in the bottom of the training codebook is 243. The operating frequency is set to 140 GHz with background noise power −80 dBm. Figure 8.6 shows the performance of different schemes in THz LoS-blockage case, where $\mathbf{H}_{\text{LoS}} = \mathbf{0}$. In the non-IRS-assisted scheme, we treat IRS as an indoor wall whose first-order ray attenuation is randomly set to between 5.8 and 19.3 dB compared to the LoS (Priebe et al. 2013). It is observed that the performance gains of the IRS-assisted schemes are significant compared to the schemes with random IRS (phase shifts) and without IRS. Besides, the performance of THz IRS-assisted beam training scheme is close to the IRS-assisted capacity bound. Figure 8.7 shows the performance of different schemes in THz non-blockage case, where both the BS-user link and the BS-IRS-user link provide propagation paths. One can see that the performance gained by IRS-assisted schemes is still notable. This is because the IRS can provide additional aperture gains via controllable reflection, so as to increase the received power at the user. These results also validate the effectiveness of THz IRS-assisted beam training in the non-blockage scenario.

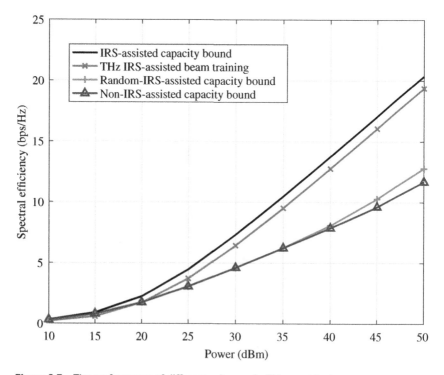

Figure 8.7 The performance of different schemes in THz non-blockage case.

8.3 IRS Prototyping

For the focus of signal energy toward desired targets, THz communication networks raise a huge challenge to hardware design due to the extremely small wave-length scale in THz bands. In this section, the leading-edge THz beam steering approaches at transceivers and IRS are summarized, respectively. Then the digitalized strategy for continuous beam scanning on IRS and a reconfigurable 1-bit reflectarray are respectively introduced in detail, by which an IRS-aided real-time beam tracking transmission was implemented for the first time in THz frequencies.

8.3.1 Active Beam Steering at THz transceivers

THz beamforming at transceivers can be realized by electronic and optical approaches these days. In terms of electronic devices, phase shifters are crucial components for beamforming by phased arrays. Phase modulation can be arranged in the lower frequency to avoid the constrained efficacy of electronic components in THz bands, such as the 370–410 GHz 8×1 uniform linear array shown in Fig. 8.8(a). However, this design cannot allow complex high-order

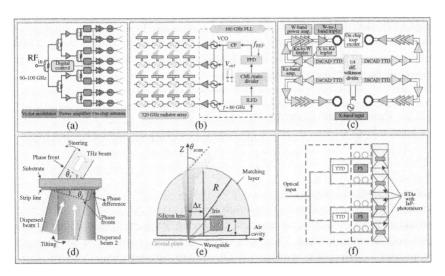

Figure 8.8 Active beam steering devices. Source: (Chen et al. 2021)/IEEE: (a) phased array by phase modulation in lower frequencies; (b) CMOS phased array by phase modulation in THz frequencies; (c) electronic TTD-based antenna array; (d) photoconductive antenna; (e) leaky-wave-lens antenna; and (f) optical TTD-based antenna array.

modulation signals. For direct phase modulation in THz frequencies, Fig. 8.8(b) shows a straightforward way based on coupled harmonic oscillators to engineer a scalable 4×4 uniform planar array in SiGe at 320 GHz, which provides more flexibility for beam control but increases the cost of high hardware complexity. The digitally controlled true-time delay (TTD), an alternative to replace phase shifters, also be adopted by a 280 GHz 2×2 chip-scale dielectric resonator antenna array, as shown in Fig. 8.8(c). The delay of each channel is regulated by a digitally controlled artificial dielectric transmission line. The compensation for signal delay at all frequencies successfully eliminates the beam squint problem in broadband communications.

For optical approaches, there is a virtue that optical devices can avoid the huge phased array structure and costly phase shifters. Such as the photoconductive antenna shown in Fig. 8.8(d), a THz beam can be generated by two laser beams focused onto the surface of the stripline from different directions. By changing the angle between the two pump beams, phase gradients can be formed to steer the directivity of THz beams. Another way is based on optical lens antennas. As shown in Fig. 8.8(e), the THz beam can also be steered by mechanically moving a hemispherical lens located on a leaky waveguide antenna, especially beneficial to THz broadband communications due to the non-dispersive material of the lens, but disadvantaged for high-speed and high-precision beam steering owing to the mechanical movement. Besides, TTD phase shifters are not just applied to electrical ways but also optical approaches to offer stable time delay for wide-band communications. The optical TTD-based chip shown in Fig. 8.8(f) converts optical signals to THz waves by InP photomixers and then radiates THz beams by the 1×4 bow-tie antennas.

8.3.2 Passive Beam Steering on THz IRS

IRS is also referred to as metasurface, which can arbitrarily tailor electromagnetic (EM) properties of incident waves to realize passive beam steering (Pan et al. 2022). The word "passive" indicates that IRS does not radiate EM waves initiatively but still needs to be actively controlled for dynamically reconfigurable ability. Tunable components or materials, as core active parts, are combined with artificial metallic structures to compose composite IRS elements also called hybrid unit cells of metasurfaces. Diodes as common active components are typically embedded in elements to regulate resonant modes by controlling ON/OFF states in the microwave band. However, the scalability of the diodes in the THz regime is hindered by serious parasitics and frequency limits. In recent years, active semiconductor components have gradually broken through the barriers and made a huge leap forward for the THz meta-devices such as via vanadium dioxide, liquid crystals, graphene, high electron mobility transistor (HEMT), and CMOS. In what follows, four representative THz IRSs are introduced.

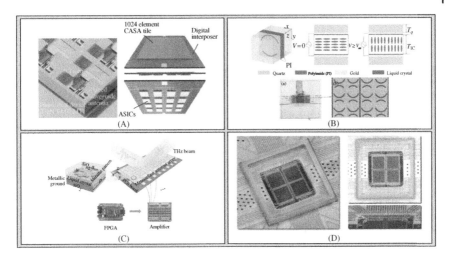

Figure 8.9 Representative THz IRSs: (A) HEMT-based ridged waveguide array; (B) transmissive liquid-crystal metasurface; (C) reflective liquid-crystal metasurface; and (D) CMOS-based programmable metasurface. Source: Jingbo Wu et al., 2020/AIP Publishing LLC/Suresh Venkatesh et al., 2020/Springer Nature.

In 2018, the IRS for dynamic beam steering was proposed for the first time by HRL laboratories. As shown in Fig. 8.9(A), a HEMT-based ridged waveguide array composed of 1024 elements realizes beam shifting between $\pm 20°$ and $\pm 40°$ at 0.22 THz. By through-holes connecting top HEMT electrodes with bottom wiring, this array achieves independent control between each subarray, but through-holes processing still is a great challenge to the current semiconductor fabrications. Subsequent research proposed liquid crystal-based metasurfaces controlled by dynamically changing permittivities of liquid-crystal substrates. As liquid-crystal substrates have no complex microstructure, these metasurfaces naturally avoid serious parasitics in THz bands. Figure 8.9(B) is a transmissive liquid-crystal metasurface that is combined with complementary split resonant rings (SRRs) to attain dynamic beam steering at $\pm 9°$, $\pm 15°$, and $\pm 30°$ at 0.426 THz. Likewise, liquid crystal can also be applied to the reflective metasurface as shown in Fig. 8.9(C), but just achieve three directional beam steering as well. In 2020, Nature Electronics reported a sensational CMOS-based THz programmable metasurface, as shown in Fig. 8.9(D), that simultaneously realized dynamic holographic imaging and beam steering by integrating eight CMOS switches in SRRs. It is worth acknowledging this work is groundbreaking for the development of THz IRS but at the expense of the low-cost advantage of IRS. These THz reconfigurable metasurfaces have experimentally verified their outstanding contributions both in mechanisms and functionalities. However, there is no one until now that realizes high-precision and wide-band beam scanning crucial in future THz communication systems. In the next section, a digitalized approach for continuous beam scanning and a HEMT-based 1-bit reflectarray, possessing the ability for high-accuracy and wide-band beam scanning, is introduced in detail.

8.3.3 Codebook Design for Beam Scanning

The fractional coding strategy is an efficient formalization function (Wang et al. 2020a), just specifying an expected scanning range and reflectarray properties including array scale, element size, and operating frequency, to automatically generate a voluminous set of codebooks for continuous beam scanning.

Taking the 2-bit coding case as an example to illustrate the vice of conventional coding and the virtue of fractional coding, an element size of $d = 200\,\mu m$ and a working frequency of 0.34 THz are assumed firstly. Most of the conventional coding physically allocates the in-phase spacing scale N into integral numbers, which are intrinsically confronting the issue of the Blind Zone. As shown in Fig. 8.10(a), the Blind Zone between $N = 1$: "00 01 10 11" and $N = 2$: "00 00 01 01 10 10 11 11" indicates a 42° angular scope unavailable for scanning. In contrast, the fractional coding method by logically discretizing the spacing scale of supercells is proposed to address this issue that is inherently limited by the fixed physical size of the unit cell.

Figure 8.10(b) shows the general scheme of the logically fractional coding method. In step 1, we assume $L = 1.25d$ as the supercell logical sidelength, which forms a subarray including all four coding states (differentiated by four colors). In step 2, the logical coding sequences will be applied to the physical unit cell one by one. In step 3, two coding states in one unit are recoded by choosing the major filling factor as the unit cell's final state. More specifically, as shown in Fig. 8.10(c), $Code(2)$ is recoded as "01," which is derived from the major filling factor of the combination of "00" and "01" (accounting for 25% and 75%, respectively). Worth noting for the half to half situation, $Code(3)$ can be "01" or "10" because of "01" and "10" both accounting for 50%. Similarly, $Code(4)$ is "10" due to that "10" and "11" account for 75% and 25% in the fourth unit, respectively. A complete coding circulation ends by the fourth state and the null filling factor simultaneously. For instance, $Code(5) = $ "11" and $q(5) = 0$, the periodic coding sequence for $N = 1.25$ is "00 01 10 10 11."

Considering the state-of-the-art, it is currently impossible to divide unit-cell spacing into different portions and reconstruct them into a new coding sequence. We define the filling factor referred to the proportion of the second coding state in a combined unit henceforth a function to the coding state for the ith unit cell has been derived

$$Code(i) = \begin{cases} \left\{ \left(\left[\frac{i}{N} \right] - 1 \right) \Big/ \frac{2\pi}{d\varphi} \right\} \Big/ \frac{d\varphi}{2\pi}, & q < 0.5, \\ \left\{ \left[\frac{i}{N} \right] \Big/ \frac{2\pi}{d\varphi} \right\} \Big/ \frac{d\varphi}{2\pi}, & q \geq 0.5, \end{cases} \tag{8.3}$$

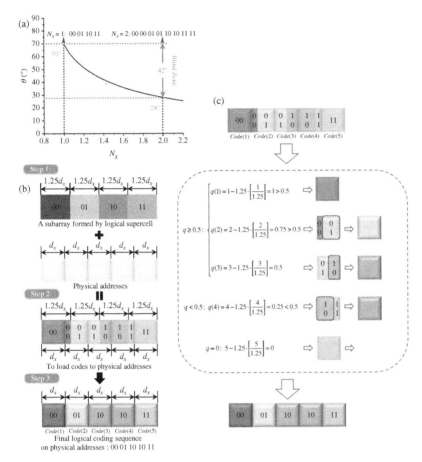

Figure 8.10 The methodology of logically fractional coding: (a) the angular scope of possible main-lobe scanning for the 1D-2 bit case and the Blind Zone; (b) the general scheme of the logically fractional coding method by exemplifying a 1D-2 bit case; and (c) illustration of the numerical coding process.

where $[x]$ is an integer-valued function, $\{x\} = x - [x]$, $d\varphi = \frac{\pi}{2}$, and the outputs decimal "0," "1," "2," and "3" are eventually transferred to binary "00", "01", "10," and "11," accordingly. As a programmable implementation of Fig. 8.10(b) and (c) illustrates the numerical coding process according to Eq. (8.3).

8.3.4 Beam-Scanning Reflectarray

AlGaN/GaN HEMT-based modulators with nanosecond-level response speed and large phase shifting have been demonstrated on the metasurfaces (Zhao et al.

2019; Zhang et al. 2018). According to such advantages, a real-time controlled THz coding reflectarray based on an asymmetrical dipole resonator (ADR) interacting with two-dimensional electron gas (2DEG) is introduced in detail, proposed by University of Electronic Science and Technology of China. The asymmetry of resonant fields can be regulated by the variation of carrier concentration to form a resonant phase compensation for phase-shift modulation. Different from the resonant conversion from inductor–capacitor (LC) to LC modes or long dipole to short dipole modes, the asymmetrical resonant mechanism surmounts the low switch ratio caused by capacitive parasitics. By applying fractional coding, convolutional coding, dual-region coding, and GRS coding to this reflectarray, dynamically multifunctional beam manipulation is successfully attained. In addition, this reflectarray also implement the point-to-point tone-signal transmission toward a moving receiver by real-time beam tracking, further demonstrating the excellent ability for dynamic beam steering and pushing the development of THz IRS to be more closed to next-generation communication networks.

The schematic diagrams of the meta-element and coding reflectarray are shown in Fig. 8.11. From upper to lower in Fig. 8.11(a), the meta-element is composed of a 2DEG-embedded gold pattern, a SiC substrate covered by a nanoscale GaN film, and a backed gold film. The gold pattern consists of a strip and a strip-patch to form a core ADR, a bias wire across the gate of a HEMT, and two ground wires joining the ADR. In Fig. 8.11(b), an AlGaN/GaN heterostructure is nested in the gap between the gold strips to form a 2DEG layer based on

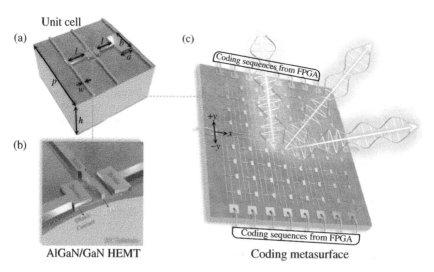

Figure 8.11 Schematic diagrams of the 2DEG-regulated IRS: (a) 2DEG-embedded unit cell; (b) the cross-sectional diagram of the AlGaN/GaN HEMT; and(c) coding reflective IRS.

the effects of spontaneous polarization and piezoelectric polarization within the heterostructure. The source–drain electrodes and gate of the HEMT are respectively constructed by Ohmic contacts and a Schottky contact between the heterostructure and gold pattern. By applying bias voltages to control carrier concentration within 2DEG, large phase modulation can be realized based on asymmetrical resonance regulated by interactions between the ADR and 2DEG. The reflectarray is arranged by the meta-elements in a two-dimensional (2D) plane with an element period p, and two meta-element resonant states with a 180° phase difference are encoded by the binary codes "0" and "1" as a 1-bit coding metasurface. As shown in Fig. 8.11(c), each meta-element shares the bias wires along the +y-direction to compose collectively controlled single-column subarrays fed by an external FPGA. The ground wires of these single-column subarrays are together connected to the common-ground wire across the center of the reflectarray along the x-direction. With the x-axis of symmetry, a mirrored subarray is set toward the −y-direction to enable quasi-2D phase-distribution control. Compared with one-dimensional (1D) column-wise feeding, the dual-region mirrored subarrays provide more flexibility for beam manipulation without redundant circuit coupling.

For the fabricated process, firstly, an AlGaN/GaN epitaxial layer was grown on a SiC substrate by metalorganic chemical vapor deposition (MOCVD). After the standard cleaning process for HEMTs and following the photolithography process, the active region of 2DEG is delimited and outside AlGaN layer was excessive etched by inductively coupled plasma. Next, the source–drain electrodes are formed by photoetching, electron-beam evaporation, lift-off processes, and rapid annealing at 900 °C under an N_2 environment, with Ti/Al/Ni/Au deposition on both ends of the active region as Ohmic contacts. Similarly, for meta-atom patterning, Ni/Au layer was fabricated on the GaN film by similar photoetching, electron-beam evaporation, and lift-off processes. The finished THz reflectarray consists of 64×64 meta-elements with a total area of 13×13 mm^2, in which every 1×32 meta-elements compose one single-column subarray. By the assembly on the FR4-printed circuit board (PCB), the bias- and ground-electrode pads are connected to the external circuits by gold-wire bonds. To avoid onboard interference, we pave a piece of absorbing material around the reflectarray on the PCB.

In the view field from 20° to 60°, a wide-band beam scanning in 0.33–0.4 THz and a 1° scanning accuracy at 0.34 THz are attained in the experiments. Meanwhile, dual-beam manipulation is also demonstrated by dual-region coding and convolutional coding. For diffuse scattering based on GRS coding, the measured maximum reflection is −20 dB, although the scattering ability suffers from suppression due to a high coding-error rate and harsh fault tolerance for quasi-random sequences.

Finally, the THz IRS-assisted point-to-point signal transmission toward different directions is successfully performed by real-time beam tracking. The modulated 1 GHz tone signal kept the good waveform with $V_{pp} \geq 9.8\,\text{mV}$ in the directions from 30° to 60°. In the future, larger-scale and higher-performance THz IRSs can be designed by the proposed designed methods and may be applied in THz high-speed wireless communications, super-resolution imaging systems and other advanced applications owing to the abilities of real-time, high-precision and multifunctional beam manipulation.

8.4 IRS-THz Communication Applications

In this section, we present some appealing scenarios of THz-IRS communications in 6G applications. Benefiting from the unique function of IRS, the THz communication scenarios in 6G will be more diversified.

8.4.1 High Speed Fronthaul/Backhaul

The 6G wireless networks are envisioned to use the dense deployments of cellular cells to meet the unprecedented data rate requirements of hotspots. As envisioned by the IEEE 802.15.3d standard, through beamforming of large scale antennas to provide high gain transmission, THz ultra-high speed wireless communication can be used to flexibly deploy fronthaul (between baseband units and remote radio units) and backhaul (between cells and the core network) links, thereby reducing the complexity and cost issues caused by wired optical fiber connections. More than that, THz communication can also use the widely deployed IRS for joint beamforming to provide high throughput fronthaul/backhaul transmission, as shown in Fig. 8.12(a). The IRS-assisted THz wireless fronthaul/backhaul link not only provides extra aperture gain through passive beamforming at the IRS but also reduces outages by establishing multiple propagation paths.

8.4.2 Cellular Connected Drones

Drones can be regarded as virtual base stations or mobile relays that support flexible deployment to provide adjustable large scale coverage for 6G networks. To fully utilize the potential of drones, 6G is expected to use the THz band to support heavy traffic between drones and BS as well as users. However, the openness of the air space makes THz communications that rely on high power beam to combat fading easily to cause mutual interference, which in turn hurts the transmission performance. As an easy way to adjust the reflection angle of electromagnetic

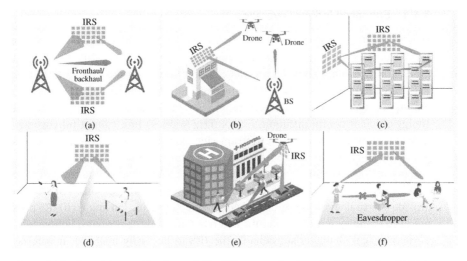

Figure 8.12 Promising applications of THz-IRS communications in six scenarios: (a) high speed fronthaul/backhaul; (b) cellular connected drones; (c) wireless data center; (d) enhanced indoor coverage; (e) vehicular communication; and (f) physical-layer security.

waves, IRS can intelligently adjust the direction of the beam from the BS to provide a variety of drone connections, as shown in Fig. 8.12(b). In urban areas, IRS is generally recommended to be deployed on the exterior wall or roof of a building. Introducing IRS into the airspace BS beamforming scene can improve the interconnection capability of air-ground communication while avoiding multi-drones interference.

8.4.3 Wireless Data Center

As the mobile communications demand for cloud service applications increases steadily, data centers will play a more important role in 6G. With its higher reconfigurability and dynamic operation, THz-enabled wireless data centers have the potential to solve the defects of power consumption, maintenance costs, and space occupied by large cables in wired data centers. However, the point-to-point LoS link established between densely arranged data servers will inevitably be blocked by the server itself or cause link interference. Therefore, as shown in Fig. 8.12(c), the introduction of IRS into the THz wireless data center to establish IRS-assisted THz links will greatly expand the freedom of server interconnection path planning. On the one hand, IRS can provide a variety of link route options for quasi-static data centers. On the other hand, IRS can assist in the establishment of multiple THz backup connections to improve the reliability of data center transmission.

8.4.4 Enhanced Indoor Coverage

Due to the high frequency and short wavelength, the high transmission loss of THz communication results in a short coverage distance, while the limited diffraction capability causes THz transmission to rely on the LoS path. Indoor THz LoS link can be easily blocked by the walls or human bodies, leading to high-speed communication interruption. For the sake of human health and energy-efficient communication, indoor THz wireless link cannot fill coverage holes by simply increasing the transmission power or setting up more access points. As such, it is an arduous task to expand the ubiquitous high-speed indoor coverage in THz communication. IRS has become an innovative and cost-effective THz indoor coverage solution due to the following main reasons: (i) IRS can provide a virtual wireless LoS link by controlling the reflection angle. (ii) IRS does not need complex hardware circuits and has a small thickness with lightweight. These physical characteristics enable the IRS to be easily installed in wireless transmission environments, including walls, ceilings, and furniture. An example is sketched in Fig. 8.12(d), where the IRS assists the THz access point to establish a virtual LoS path behind the obstacle to enhance indoor coverage.

8.4.5 Vehicular Communications

The intelligent transportation system expects the vehicle communication network to provide high data rate, low latency, and reliable communication. For the era of wireless interconnected smart cars, THz communication is a potential support technology for terabit vehicular communications in the future. However, the changeable crowded traffic and dense movement of people will damage the connection stability and alignment speed of the THz beam. To this end, the mobile IRS carried by the drones can follow the traffic flow to assist the THz beam training and tracking process in crowded traffic areas. Figure 8.12(e) shows a schematic diagram of the ultra-high speed THz link used for vehicle communication, in which the IRS carried by the drone can be adjusted to different heights and positions as required. Vehicles can choose cooperative IRS according to obstacle conditions at different locations to ensure high-speed, real time, and stable THz connection.

8.4.6 Physical-Layer Security

As wireless network security relying solely on high-level encryption protocols remains limitations, it is significant to consider physical-layer security in 6G networks. Using massive antenna elements to create highly directional THz beams brings many benefits to the secure physical-layer transmission. However, eavesdroppers located in the sharp sector of the beam coverage can still endanger information security. By jointly active and passive beamforming, the IRS-assisted THz communication systems can concentrate the beam energy to the legitimate user while suppressing the eavesdropper's received power. An illustrative example is sketched in Fig. 8.12(f), in which IRS not only helps the THz beam to bypass the eavesdropper in a reflection path but also intentionally deteriorates the signals power in the direction of the eavesdropper.

Bibliography

Chen, Z., Ning, B., Han, C. et al. (2021). Intelligent reflecting surface assisted terahertz communications toward 6G. *IEEE Wireless Communications* 28 (6): 110–117.

Ning, B., Chen, Z., Chen, W., and Fang, J. (2020a). Beamforming optimization for intelligent reflecting surface assisted MIMO: a sum-path-gain maximization approach. *IEEE Wireless Communications Letters* 9 (7): 1105–1109.

Ning, B., Chen, Z., Chen, W., and Du, Y. (2020b). Channel estimation and transmission for intelligent reflecting surface assisted THz communications. *ICC 2020-2020 IEEE International Conference on Communications (ICC)*, 1–7. IEEE.

Ning, B., Chen, Z., Chen, W. et al. (2021). Terahertz multi-user massive MIMO with intelligent reflecting surface: beam training and hybrid beamforming. *IEEE Transactions on Vehicular Technology* 70 (2): 1376–1393.

Ning, B., Wang, T., Wang, P. et al. (2022). Space-orthogonal scheme for IRSs-aided multi-user MIMO in mmWave/THz communications. *ICC 2022-IEEE International Conference on Communications*, 3478–3483. IEEE.

Pan, Y., Lan, F., Zhang, Y. et al. (2022). Dual-band multifunctional coding metasurface with a mingled anisotropic aperture for polarized manipulation in full space. *Photonics Research* 10 (2): 416–425.

Priebe, S., Kannicht, M., Jacob, M., and Kürner, T. (2013). Ultra broadband indoor channel measurements and calibrated ray tracing propagation modeling at THz frequencies. *Journal of Communications and Networks* 15 (6): 547–558.

Wang, L., Lan, F., Zhang, Y. et al. (2020a). A fractional phase-coding strategy for terahertz beam patterning on digital metasurfaces. *Optics Express* 28 (5): 6395–6407.

Wang, P., Fang, J., Dai, L., and Li, H. (2020b). Joint transceiver and large intelligent surface design for massive MIMO mmWave systems. *IEEE Transactions on Wireless Communications* 20 (2): 1052–1064.

Zhang, S. and Zhang, R. (2020). Capacity characterization for intelligent reflecting surface aided MIMO communication. *IEEE Journal on Selected Areas in Communications* 38 (8): 1823–1838.

Zhang, Y., Zhao, Y., Liang, S. et al. (2018). Large phase modulation of THz wave via an enhanced resonant active hemt metasurface. *Nanophotonics* 8 (1): 153–170.

Zhao, Y., Wang, L., Zhang, Y. et al. (2019). High-speed efficient terahertz modulation based on tunable collective-individual state conversion within an active 3 nm two-dimensional electron gas metasurface. *Nano Letters* 19 (11): 7588–7597.

9

Joint Design of Beamforming, Phase Shifting, and Power Allocation in a Multi-cluster IRS-NOMA Network

Ximing Xie[1], Fang Fang[2], and Zhiguo Ding[1]

[1] *School of Electrical and Electronic Engineering, The University of Manchester, Manchester, UK*
[2] *Department of Electrical and Computer Engineering and the Department of Computer Science, Western University, London, Canada*

9.1 Introduction

Low energy consumption, high spectrum efficiency, and massive multi-device interconnections are considered as key features of the beyond 5G system, which have been attracting more and more research interests (Zhang et al. 2019; Saad et al. 2019; Tariq et al. 2020). The rapidly increasing demand of the fast-growing number of users and Internet of Things (IoT) devices brings challenges to the next-generation wireless communication system. Many techniques, e.g., millimeter wave (Jamali et al. 2020), massive multiple-input and multiple-output (MIMO) (Larsson et al. 2014) and small cell (Bennis et al. 2013) have been proposed and investigated to be extensively adopted in wireless communication area. Non-orthogonal multiple access (NOMA) as a potential technique for the next generation communication has received widespread attention because it can provide high spectral efficiency (Ding et al. 2015; Vaezi et al. 2019). NOMA is more promising than conventional orthogonal multiply access (OMA) techniques such as frequency division multiple access (FDMA), time division multiple access (TDMA), code division multiple access (CDMA), and orthogonal frequency-division multiple access (OFDMA), because it allows one resource block to be shared between different users and devices, which dramatically increases efficiency. To achieve this cooperation among multiple users and devices, successive interference cancellation (SIC) is proposed to cancel interference caused by other NOMA users and devices within the same NOMA network, which can significantly improve the signal-to-interference and noise ratio (SINR) and reception reliability (Saito et al. 2015).

Intelligent Surfaces Empowered 6G Wireless Network, First Edition.
Edited by Qingqing Wu, Trung Q. Duong, Derrick Wing Kwan Ng, Robert Schober, and Rui Zhang.

Recently, the intelligent reflective surface (IRS) has been designed to further enhance the capability of a wireless communication system. Two main advantages have been taken to us from IRS, which are enlarging the communication coverage and improving robustness of signal transmission. Specifically, electromagnetic waves from a base station (BS) can be reflected by an IRS to another destination to extend the cover rage. Moreover, an IRS is equipped with multiple reflecting elements, which have the ability to reconfigure channels by adjusting the phase shift of each element. The quality of received signal at users' end will be greatly improved under this channel modification (Wu and Zhang 2018). The typical architecture of IRS consists of a reflecting panel and a smart controller. There are many reflecting elements and a control circuit on the reflecting panel, which mainly reflects the signal and reconfigures the channel. The smart controller determines the reflection adoption and also communicates with the base station. Furthermore, the base station sends the control signaling to the smart controller and then the smart controller adjusts the phase shift of each reflecting element properly based on the instruction from the base station (Wu et al. 2021).

9.1.1 Previous Works

Extensive research works have been done of NOMA techniques and some of them are combining NOMA with other state-of-the-art techniques, e.g., MIMO and orthogonal time-frequency space (OTFS) modulation (Sun et al. 2018; Fang et al. 2016, 2017; Surabhi et al. 2019). Recently, IRSs have been proposed and designed as a powerful tool to upgrade the performance of an existing wireless network (Wu and Zhang 2019a,2019b; Di Renzo et al. 2019). Among these works, IRSs were proved as a perfect tool in a wireless system, where the channel was intelligently reconfigured.

The IRS-assisted NOMA system has been considered as a promising solution to extensively improve the performance of a communication system because of benefits from NOMA and IRS. Some works studying the combination of NOMA and IRS have been carried out. The recent works (Zhu et al. 2020; Ding et al. 2020) considered a simple scenario, where two NOMA downlink users are served by a single IRS. Specifically, the paper Zhu et al. (2020) aimed to minimize the transmit power of the base station by jointly optimizing beamforming and phase shift of the IRS and an improved quasi-degradation condition was also considered to guarantee NOMA can achieve the capacity region with a high possibility. In Ding et al. (2020), two kinds of phase shift designs were analyzed, which are random phase shifting and coherent phase shifting. The results showed the coherent phase shifting is more adaptive.

Moreover, the scenario consisting of multiple users in an IRS-assisted NOMA network was also investigated in some works (Fu et al. 2021; Liu et al. 2020; Mu et al. 2019; Ding et al. 2020; Ding and Poor 2020). The authors in Fu et al. (2021) minimized the total transmit power at the base station by optimally designing the beamforming vector for each user and the phase shift of IRS. In Liu et al. (2020), a single IRS-assisted downlink NOMA network was considered and reinforcement learning was adopted to design the optimal beamforming vector by minimizing the total transmit power at the base station. The authors of Mu et al. (2019) designed the beamforming vector to achieve the maximal sum rate in a downlink multiple-inputs and single-output (MISO) IRS-assisted NOMA network. The work (Ding et al. 2020) analyzed two kinds of phase shift designs based on given beamforming vectors. However, the work (Ding and Poor 2020) analyzed several beamforming designs in a IRS-NOMA downlink network, which consists of multiple users and multiple IRSs.

Multi-cluster system model is also a typical case, which was widely discussed in many works (Li et al. 2019; Ni et al. 2021). In Li et al. (2019), two types of users were considered, namely the central user and the cell-edge user. Each cluster contains one central user and one cell-edge user. An IRS is deployed to serve all users within this network. Similarly, the authors of this work minimized the total transmit power at the base station by jointly optimizing the beamforming vector – of each user and the phase shift – of the IRS. In Ni et al. (2021), multiple base stations are located into different clusters to serve associated users. Meanwhile, an IRS is deployed to serve all clusters. The sum rate of all clusters was maximized by jointly optimizing power allocation and phase shift.

9.1.2 Motivation and Challenge

All aforementioned works only considered one single IRS. The advantage of adopting one IRS is to save the complicity and the deployment cost. However, the channel condition of each user is different, therefore, one single IRS might not have ability to perfectly reconfigure all channels for every user. The optimality for some users will scarify. Thus, multiple IRSs seem to be more competitive because each IRS's phase shift can be dedicatedly designed according to a specific user or a specific cluster. The work described in this chapter considers a multi-cluster NOMA network and each cluster has an exclusive IRS to fully serve the users within this cluster.

There are some challenges arising from this multi-cluster system. The first one is there are many optimization variables due to multiple IRSs and users. The more optimization variables mean the higher complicity. The second one is that several kinds of variables are coupled together to make the optimization problem non-convex and highly intractable. The last challenge is that the feasibility of the formulated problem is different to be guaranteed. The formulated problem is very sensitive to hyperparameters which means how to wisely choose hyperparameters is very important.

To address these challenges, a new system model consists of multiple users and multiple IRSs is proposed. A few clusters are designed and each of them contains one IRS and two users. A power minimization problem is formulated, which is non-convex and highly intractable. The original problem is approximately transformed into two convex sub-problems and a novel alternating algorithm is proposed to solve them iteratively.

9.2 System Model and Problem Formulation

In this section, a multiuser downlink network is described first. Then, a transmit power minimization problem is formulated based on this system model. The beamforming, phase shift, and power allocation are jointly optimized in the optimization problem.

9.2.1 System Model

As shown in Fig. 9.1, there are multiple clusters in this downlink NOMA network and each cluster contains one IRS and two types of users, namely the central user and the cell-edge user. All users are served by one base station simultaneously. All central users are close to the base station and all cell-edge users are far from the base station. There are K clusters in total, which means this system model consists of $2K$ users and K IRSs. The central user, the cell-edge user, and the IRS in the kth cluster are denoted by CU_k, EU_k, and IRS_k, respectively. For each IRS_k, there are N passive reflecting elements on it. It is assumed that the cell user EU_k can only receive signal from the base station via IRS_k due to the blockage of the direct link. The base station is equipped with M antennas with the assumption that $K \leq M \leq 2K$ and generates K unique beams to serve K clusters. We assume that each cluster is far from others so the inter-cluster interference can be reasonably ignored. Another reason to support this assumption is that the square of the IRS reflecting panel is limited so that only partial electromagnetic waves can be reflected. The energy of the reflected signal will be greatly attenuated if there is severe path loss or fading attenuation. The location of each IRS can be carefully chosen to ensure a strong connection to its associated cell-edge user which cannot directly receive signals from the base station. As such, the IRS may still interfere with the central user within the same cluster, however, it is ignored in this work. The location of each IRS and each user will also affect the total transmit power.

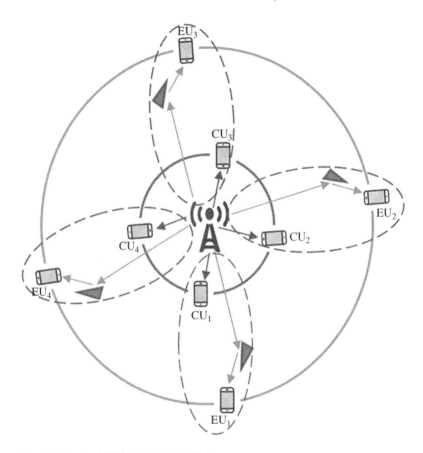

Figure 9.1 An IRS NOMA system model.

We assume the distances between the base station and all the IRSs are the same. In the simulation section, simulation results show the relationship between the performance and BS-IRS distance. This work does not consider the optimal location of each IRS, which is an important research direction in the future.

The unique beam generated by the base station allows all clusters to share the same frequency-time resource block, which greatly improves the spectrum efficiency. The base station broadcasts the superposed signal which is given by

$$\sum_{i=1}^{K} \mathbf{w}_i (\sqrt{\alpha_i} s_{i,c} + \sqrt{1 - \alpha_i} s_{i,e}), \tag{9.1}$$

where $\mathbf{w}_i \in \mathbb{C}^M$ denotes the beam vector for the ith cluster and $i \in 1, 2, \dots, K$. $s_{i,c}$ and $s_{i,e}$ denote the signals to be sent to CU_k and EU_k, respectively, and α_i is the power allocation coefficient of CU_i, thus $1 - \alpha_i$ is the power coefficient of EU_i.

It is assumed that perfect channel state information (CSI) is available to every node, which can be achieved by precisely estimating channel. Therefore, the received signal of CU_k and EU_K can be expressed as follows:

$$y_{k,c} = \underbrace{\mathbf{h}_{k,c}^H \mathbf{w}_k \sqrt{\alpha_k} s_{k,c}}_{\text{signal}} + \underbrace{\mathbf{h}_{k,c}^H \mathbf{w}_k \sqrt{1-\alpha_k} s_{k,e}}_{\text{intra-cluster interference}}$$

$$+ \underbrace{\mathbf{h}_{k,c}^H \sum_{\substack{i=1 \\ i \neq k}}^{K} \mathbf{w}_i (\sqrt{\alpha_i} s_{i,c} + \sqrt{1-\alpha_i} s_{i,e}) + w_{c,k}}_{\text{inter-cluster interference}} \tag{9.2}$$

and

$$y_{k,e} = \underbrace{(\mathbf{h}_{k,e}^H \boldsymbol{\Theta}_k \mathbf{G}_k) \mathbf{w}_k \sqrt{1-\alpha_k} s_{k,e}}_{\text{signal}} + \underbrace{(\mathbf{h}_{k,e}^H \boldsymbol{\Theta}_k \mathbf{G}_k) \mathbf{w}_k \sqrt{\alpha_k} s_{k,c}}_{\text{intra-cluster interference}}$$

$$+ \underbrace{(\mathbf{h}_{k,e}^H \boldsymbol{\Theta}_k \mathbf{G}_k) \sum_{\substack{i=1 \\ i \neq k}}^{K} \mathbf{w}_i (\sqrt{\alpha_i} s_{i,c} + \sqrt{1-\alpha_i} s_{i,e}) + w_{e,k}}_{\text{inter-cluster interference}}, \tag{9.3}$$

where $\mathbf{h}_{k,c} \in \mathbb{C}^{M \times 1}$ denotes the channel vector between the base station and CU_k, $\mathbf{G}_k \in \mathbb{C}^{N \times M}$ denotes the channel matrix between the base station and IRS_k, $w_{c,k} \sim \mathcal{CN}(0, \sigma^2)$ and $w_{e,k} \sim \mathcal{CN}(0, \sigma^2)$ denote the additive white Gaussian noise (AWGN), $\mathbf{h}_{k,e} \in \mathbb{C}^{N \times 1}$ denotes the channel vector between IRS_k and EU_k and $\boldsymbol{\Theta}_k = \text{diag}(\beta e^{j\theta_1^k}, \ldots, \beta e^{j\theta_n^k})$ is the phase-shift matrix of IRS_k, where $\theta_n^k \in [0, 2\pi), n \in \{1, \ldots, N\}$ and $\beta \in [0, 1]$ denotes the phase shift of each reflecting element n and amplitude coefficient on the signal, respectively. In this work, the reflecting coefficient is set to be 1, which means the signal will not loss any energy during reflection. It is worth to point out the $\beta = 1$ may not be the optimal choice in some security sensitive scenarios. However, this work only focuses on general cases, therefore adopting $\beta = 1$ can bring more simplicity. Another concern is that signals may be reflected more than once because there are multiple IRSs existing. The fact that signals reflected twice or more will suffer severe large-scale path loss, hence, we only consider the first-time reflected signal. The SIC strategy adopted in this work is that CU_K only decodes EU_k's signal first and then decodes its own signal by treating inter-cluster signals as interference. Thus, the SINR of signal $s_{k,e}$ observed at CU_k can be expressed as follows:

$$\text{SINR}_{k,c \to e} = \frac{|\mathbf{h}_{k,c}^H \mathbf{w}_k|^2 (1-\alpha_k)}{|\mathbf{h}_{k,c}^H \mathbf{w}_k|^2 \alpha_k + \sum_{\substack{i=1 \\ i \neq k}}^{K} |\mathbf{h}_{k,c}^H \mathbf{w}_i|^2 + \sigma^2}. \tag{9.4}$$

The signal-to-noise ratio (SNR) of CU_k to decode its own signal is given by

$$\text{SINR}_{k,c} = \frac{|\mathbf{h}_{k,c}^H \mathbf{w}_k|^2 \alpha_k}{\sum_{\substack{i=1 \\ i \neq k}}^K |\mathbf{h}_{k,c}^H \mathbf{w}_i|^2 + \sigma^2}. \tag{9.5}$$

The cell-edge user directly decodes its own signal. The SINR of $s_{k,e}$ at EU_k is given by:

$$\text{SINR}_{k,e} = \frac{|\mathbf{h}_{k,e}^H \Theta_k \mathbf{G}_k \mathbf{w}_k|^2 (1 - \alpha_k)}{|\mathbf{h}_{k,e}^H \Theta_k \mathbf{G}_k \mathbf{w}_k|^2 \alpha_k + \sum_{\substack{i=1 \\ i \neq k}}^K |\mathbf{h}_{k,e}^H \Theta_k \mathbf{G}_k \mathbf{w}_i|^2 + \sigma^2}. \tag{9.6}$$

9.2.2 Problem Formulation

In this section, a transmit power minimization problem is formulated consisting of three kinds of optimization variables, which are the beamforming vector $(\mathbf{w}_k, k \in \{1, \dots, K\})$, power allocation coefficients $(\alpha_k, k \in \{1, \dots, K\})$ and phase-shifting matrix $(\Theta_k, k \in \{1, \dots, K\})$. The considered optimization problem can be formulated as follows:

$$\text{P0}: \min_{\alpha, \mathbf{w}, \Theta} \quad \sum_{k=1}^K \|\mathbf{w}_k\|^2 \tag{9.7a}$$

$$\text{s.t. } \log_2(1 + \text{SINR}_{k,c}) \geq R_{k,c}, \quad \forall k \tag{9.7b}$$

$$\log_2(1 + \min(\text{SINR}_{k,e}, \text{SINR}_{k,c \to e})) \geq R_{k,e}, \quad \forall k \tag{9.7c}$$

$$0 \leq \theta_{k,n} \leq 2\pi, \quad \forall k, n \tag{9.7d}$$

$$|\Theta_{k,n,n}| \leq 1, \quad \forall k, n, \tag{9.7e}$$

where the minimum data rate requirement of CU_k and EU_k are denoted by $R_{k,c}$ and $R_{k,e}$, respectively. Constraints (9.7b) and (9.7c) arise from the QoS requirement. Equations (9.7d) and (9.7e) arise from the hardware limit of IRSs. Note that P0 is highly intractable because there are non-convex constraints (9.7b) and (9.7c) existing.

9.3 Alternating Algorithm

In this section, an alternating algorithm is proposed to solve P0 efficiently. The main idea of this algorithm is to divide the original problem into two subproblems and solve them iteritively. In particular, two subproblems, beamforming optimization problem and phase shifting feasibility problem, are iteratively solved instead of directly solving the original problem.

9.3.1 Beamforming Optimization

Given a phase-shift matrix $\boldsymbol{\Theta}$, the beamforming optimization problem can be formulated as follows:

$$\text{P1}:\min_{\alpha,\mathbf{w}} \quad \sum_{k=1}^{K}\|\mathbf{w}_k\|^2 \tag{9.8a}$$

$$\text{s.t.} \quad \log_2(1+\text{SINR}_{k,c}) \geq R_{k,c}, \quad \forall k \tag{9.8b}$$

$$\log_2(1+\text{SINR}_{k,e}) \geq R_{k,e}, \quad \forall k \tag{9.8c}$$

$$\log_2(1+\text{SINR}_{k,c\rightarrow e}) \geq R_{k,e}, \quad \forall k \tag{9.8d}$$

$$0 \leq \alpha_k \leq 1, \quad \forall k. \tag{9.8e}$$

P1 is still a non-convex form. First, constraint (9.8c) can be rewritten as follows:

$$\frac{|e_k^H \mathbf{D}_{k,e} \mathbf{G}_k \mathbf{w}_k|^2 (1-\alpha_k)}{|e_k^H \mathbf{D}_{k,e} \mathbf{G}_k \mathbf{w}_k|^2 \alpha + \sum_{\substack{i=1\\i\neq k}}^{K} |e_k^H \mathbf{D}_{k,e} \mathbf{G}_k \mathbf{w}_i|^2 + \sigma^2} \geq r_{k,e}, \tag{9.9}$$

where $r_{k,e} = 2^{R_{k,e}} - 1$, e_k is an $N \times 1$ vector containing all the diagonal elements of $\boldsymbol{\Theta}_k^H$, and $\mathbf{D}_{k,e}$ is a diagonal matrix, whose main diagonal elements are from the channel vector $\mathbf{h}_{k,e}^H$. Equation (9.9) can be further rewritten as follows:

$$|e_k^H \mathbf{D}_{k,e} \mathbf{G}_k \mathbf{w}_k|^2 (1+r_{k,e})\alpha_k$$
$$\leq |e_k^H \mathbf{D}_{k,e} \mathbf{G}_k \mathbf{w}_k|^2 - \sum_{\substack{i=1\\i\neq k}}^{K} |e_k^H \mathbf{D}_{k,e} \mathbf{G}_k \mathbf{w}_i|^2 r_{k,e} - \sigma^2 r_{k,e}. \tag{9.10}$$

Since each $\boldsymbol{\Theta}_k$ is already given, $e_k^H \mathbf{D}_{k,e} \mathbf{G}_k$ is fixed. For simply notation, we replace $e_k^H \mathbf{D}_{k,e} \mathbf{G}_k$ with $\mathbf{z}_{k,e}^H$ and rewrite (9.10) as follows:

$$\alpha_k |\mathbf{z}_{k,e}^H \mathbf{w}_k|^2 \leq \frac{|\mathbf{z}_{k,e}^H \mathbf{w}_k|^2}{1+r_{k,e}} - \left(\sum_{\substack{i=1\\i\neq k}}^{K} |\mathbf{z}_{k,e}^H \mathbf{w}_i|^2 + \sigma^2\right)\frac{r_{k,e}}{1+r_{k,e}}, \tag{9.11}$$

where $\mathbf{z}_{k,e}^H = e_k^H \mathbf{D}_{k,e} \mathbf{G}_k$. Inspired by semidefinite relaxation (SDR), (9.11) can be equivalently written as follows:

$$\alpha_k \text{Tr}(\mathbf{Z}_{k,e}\mathbf{W}_k) \leq \frac{\text{Tr}(\mathbf{Z}_{k,e}\mathbf{W}_k)}{1+r_{k,e}} - \left(\sum_{\substack{i=1\\i\neq k}}^{K} \text{Tr}(\mathbf{Z}_{k,e}\mathbf{W}_i) + \sigma^2\right)\frac{r_{k,e}}{1+r_{k,e}}, \tag{9.12}$$

$$\mathbf{W}_k \succeq 0, \tag{9.13}$$

$$\text{Rank}(\mathbf{W}_k) = 1, \tag{9.14}$$

where $\mathbf{Z}_{k,e} = \mathbf{z}_{k,e}\mathbf{z}_{k,e}^H$ and $\mathbf{W}_k = \mathbf{w}_k\mathbf{w}_k^H$. It is noted that (9.12) is still non-convex because of the term $\alpha_k \operatorname{Tr}(\mathbf{Z}_{k,e}\mathbf{W}_k)$. By applying the inequality of arithmetic and geometric means, we have the inequality that

$$2\alpha_k \operatorname{Tr}(\mathbf{Z}_{k,e}\mathbf{W}_k) \le (\alpha_k c_k)^2 + \left(\frac{\operatorname{Tr}(\mathbf{Z}_{k,e}\mathbf{W}_k)}{c_k}\right)^2, \tag{9.15}$$

where c_k is a fixed point. Equation (9.15) approximates the non-convex term $\alpha_k \operatorname{Tr}(\mathbf{Z}_{k,e}\mathbf{W}_k)$ with its convex upper bound.

Lemma 9.1 *The fixed point c_k which can tighten this upper bound in the m- iteration can be updated by:*

$$c_k^{(m)} = \sqrt{\frac{\operatorname{Tr}(\mathbf{Z}_{k,e}\mathbf{W}_k^{(m-1)})}{\alpha_k^{(m-1)}}}. \tag{9.16}$$

Proof: A difference function of the original function $2\alpha_k \operatorname{Tr}(\mathbf{Z}_{k,e}\mathbf{W}_k)$ and its approximated upper bound is defined as follows:

$$\mathcal{F}(c_k) = 2\alpha_k \operatorname{Tr}(\mathbf{Z}_{k,e}\mathbf{W}_k) - (\alpha_k c_k)^2 - \left(\frac{\operatorname{Tr}(\mathbf{Z}_{k,e}\mathbf{W}_k)}{c_k}\right)^2. \tag{9.17}$$

When function (9.17) equals to 0, equality (9.15) holds, which tightens the upper bound. From (9.15), it is noted that the maximum value of function $\mathcal{F}(c_k)$ is 0. Since

$$\frac{\partial^2 \mathcal{F}(c_k)}{\partial c_k^2} = -2\alpha_k - \frac{6\operatorname{Tr}(\mathbf{Z}_{k,e}\mathbf{W}_k)}{c_k^4} \le 0, \tag{9.18}$$

when $\alpha_k \ge 0$ and $\operatorname{Tr}(\mathbf{Z}_{k,e}\mathbf{W}_k) \ge 0$, the function $\mathcal{F}(c_k)$ is a concave function with respect to c_k. According to the Karush–Kuhn–Tucker (KKT) conditions, the maximum value of a concave function is obtained by letting the first order derivative equal to 0. Thus, the optimal value of c_k, defined as c_k^*, can be obtained by $\frac{\partial \mathcal{F}(c_k)}{\partial c_k} = 0$, then c_k^* can be given by

$$c_k^* = \sqrt{\frac{\operatorname{Tr}(\mathbf{Z}_{k,e}\mathbf{W}_k)}{\alpha_k}}. \tag{9.19}$$

□

Finally, the non-convex constraint (9.8c) is approximately transformed into a convex constraint, which is given by

$$(\alpha_k c_k)^2 + \left(\frac{\operatorname{Tr}(\mathbf{Z}_{k,e}\mathbf{W}_k)}{c_k}\right)^2 \le 2\frac{\operatorname{Tr}(\mathbf{Z}_{k,e}\mathbf{W}_k)}{1 + r_{k,e}} - 2\left(\sum_{\substack{i=1 \\ i \ne k}}^{K} \operatorname{Tr}(\mathbf{Z}_{k,e}\mathbf{W}_i) + \sigma^2\right)\frac{r_{k,e}}{1 + r_{k,e}}$$

(9.13),   (9.14). $\tag{9.20}$

Similarly, the non-convex constraint (9.8d) can be approximately transformed into

$$(\alpha_k d_k)^2 + \left(\frac{\text{Tr}(\mathbf{H}_{k,c}\mathbf{W}_k)}{d_k}\right)^2 \leq 2\frac{\text{Tr}(\mathbf{H}_{k,c}\mathbf{W}_k)}{1 + r_{k,e}} - 2\left(\sum_{\substack{i=1 \\ i \neq k}}^{K}\text{Tr}(\mathbf{H}_{k,c}\mathbf{W}_i) + \sigma^2\right)\frac{r_{k,e}}{1 + r_{k,e}},$$

(9.13), (9.14), $\qquad\qquad\qquad\qquad\qquad\qquad\qquad\qquad\qquad\qquad\qquad$ (9.21)

where $\mathbf{H}_{k,c} = \mathbf{h}_{k,c}\mathbf{h}_{k,c}^H$, and d_k is a fixed point with a updating rule that

$$d_k^{(m)} = \sqrt{\frac{\text{Tr}(\mathbf{H}_{k,c}\mathbf{W}_k^{(m-1)})}{\alpha_k^{(m-1)}}}. \qquad\qquad\qquad (9.22)$$

We now focus on the non-convex constraint (9.8b), which can be rewritten as follows:

$$\alpha_k\text{Tr}(\mathbf{H}_{k,c}\mathbf{W}_k) \geq \sum_{\substack{i=1 \\ i \neq k}}^{K}\text{Tr}(\mathbf{H}_{k,c}\mathbf{W}_i)r_{k,c} + \sigma^2 r_{k,c}, \qquad (9.23)$$

where $r_{k,c} = 2^{R_{k,c}} - 1$. Simply replacing $\alpha_k\text{Tr}(\mathbf{H}_{k,c}\mathbf{W}_k)$ with its upper bound with the approximation above will not give a convex solution rather than a concave one. To deal with this obstacle, a slack variable t_k is introduced and (9.23) can be transformed to

$$\alpha_k\text{Tr}(\mathbf{H}_{k,c}\mathbf{W}_k) \geq t_k^2. \qquad\qquad\qquad (9.24)$$

$$t_k^2 \geq \sum_{\substack{i=1 \\ i \neq k}}^{K}\text{Tr}(\mathbf{H}_{k,c}\mathbf{W}_i)r_{k,c} + \sigma^2 r_{k,c}. \qquad (9.25)$$

Equation (9.24) is equivalent to a convex form, which is

$$\begin{bmatrix} \alpha_k & t_k \\ t_k & \text{Tr}(\mathbf{H}_{k,c}\mathbf{W}_i) \end{bmatrix} \succeq 0, \qquad\qquad (9.26)$$

because of the Schur complement theory. The successive convex approximation (SCA) is applied to (9.25) to replace the quadratic term t_k^2 with its first-order Taylor series. Then, the approximated convex form of (9.25) is given by

$$t_{k,0}^2 + 2t_{k,0}(t_k - t_{k,0}) \geq \sum_{\substack{i=1 \\ i \neq k}}^{K}\text{Tr}(\mathbf{H}_{k,c}\mathbf{W}_i)r_{k,c} + \sigma^2 r_{k,c}, \qquad (9.27)$$

where $t_{k,0}$ is a fixed point introduced by SCA. The updating rule of $t_{k,0}$ at the mth iteration is given by $t_{k,0}^{(m)} = t_k^{(m-1)}$.

The last non-convex constraint is the rank one constraint (9.14). By applying SDR. the rank one constraint is temporally omitted to make the whole problem convex. However, the Gaussian randomization may be necessary when rank of the optimal solution of transformed semidefinite programming (SDP) problem is not 1. P1 is eventually transformed to

$$
\text{P2}: \min_{\alpha, \mathbf{W}, t} \quad \sum_{k=1}^{K} \text{Tr}(\mathbf{W}_k) \tag{9.28a}
$$

$$
\text{s.t.} \quad (\alpha_k c_k)^2 + \left(\frac{\text{Tr}(\mathbf{Z}_{k,e} \mathbf{W}_k)}{c_k} \right)^2
$$

$$
\leq 2 \frac{\text{Tr}(\mathbf{Z}_{k,e} \mathbf{W}_k)}{1 + r_{k,e}} - 2 \left(\sum_{\substack{i=1 \\ i \neq k}}^{K} \text{Tr}(\mathbf{Z}_{k,e} \mathbf{W}_i) + \sigma^2 \right) \frac{r_{k,e}}{1 + r_{k,e}}, \quad \forall k \tag{9.28b}
$$

$$
(\alpha_k d_k)^2 + \left(\frac{\text{Tr}(\mathbf{H}_{k,c} \mathbf{W}_k)}{d_k} \right)^2
$$

$$
\leq 2 \frac{\text{Tr}(\mathbf{H}_{k,c} \mathbf{W}_k)}{1 + r_{k,e}} - 2 \left(\sum_{\substack{i=1 \\ i \neq k}}^{K} \text{Tr}(\mathbf{H}_{k,c} \mathbf{W}_i) + \sigma^2 \right) \frac{r_{k,e}}{1 + r_{k,e}}, \quad \forall k \tag{9.28c}
$$

$$
\begin{bmatrix} \alpha_k & t_k \\ t_k & \text{Tr}(\mathbf{H}_{k,c} \mathbf{W}_i) \end{bmatrix} \geq 0, \quad \forall k \tag{9.28d}
$$

$$
t_{k,0}^2 + 2t_{k,0}(t_k - t_{k,0}) \geq \sum_{\substack{i=1 \\ i \neq k}}^{K} \text{Tr}(\mathbf{H}_{k,c} \mathbf{W}_i) r_{k,c} + \sigma^2 r_{k,c}, \quad \forall k \tag{9.28e}
$$

$$
0 \leq \alpha_k \leq 1, \quad \forall k. \tag{9.28f}
$$

Randomly choosing three fixed points c_k, d_k and $t_{k,0}$, $\forall k$ gives a large chance to make P2 infeasible. How to wisely initialize these fixed points is important. An auxiliary variable q is introduced to relax all the constants. The initial point search problem can be formulated as follows:

$$
\text{P3}: \min_{\alpha, \mathbf{W}, t, q} \quad q \tag{9.29a}
$$

$$
\text{s.t.} \quad (\alpha_k c_k)^2 + \left(\frac{\text{Tr}(\mathbf{Z}_{k,e} \mathbf{W}_k)}{c_k} \right)^2 - q
$$

$$
\leq 2 \frac{\text{Tr}(\mathbf{Z}_{k,e} \mathbf{W}_k)}{1 + r_{k,e}} - 2 \left(\sum_{\substack{i=1 \\ i \neq k}}^{K} \text{Tr}(\mathbf{Z}_{k,e} \mathbf{W}_i) + \sigma^2 \right) \frac{r_{k,e}}{1 + r_{k,e}}, \quad \forall k \tag{9.29b}
$$

$$(\alpha_k d_k)^2 + \left(\frac{\text{Tr}(\mathbf{H}_{k,c}\mathbf{W}_k)}{d_k}\right)^2 - q$$

$$\leq 2\frac{\text{Tr}(\mathbf{H}_{k,c}\mathbf{W}_k)}{1+r_{k,e}} - 2\left(\sum_{\substack{i=1 \\ i \neq k}}^{K} \text{Tr}(\mathbf{H}_{k,c}\mathbf{W}_i) + \sigma^2\right)\frac{r_{k,e}}{1+r_{k,e}}, \quad \forall k \qquad (9.29c)$$

$$\begin{bmatrix} \alpha_k & t_k \\ t_k & \text{Tr}(\mathbf{H}_{k,c}\mathbf{W}_i) \end{bmatrix} \geq 0, \quad \forall k \qquad (9.29d)$$

$$t_{k,0}^2 + 2t_{k,0}(t_k - t_{k,0}) + q \geq \sum_{\substack{i=1 \\ i \neq k}}^{K} \text{Tr}(\mathbf{H}_{k,c}\mathbf{W}_i)r_{k,c} + \sigma^2 r_{k,c}, \quad \forall k \qquad (9.29e)$$

$$0 \leq \alpha_k \leq 1, \quad \forall k \qquad (9.29f)$$

$$q \geq 0. \qquad (9.29g)$$

When q equals to 0, all the constraints in P3 are exactly the same as those in P2 and the value of c_k, d_k, and $t_{k,0}$ when $q = 0$ can be the initial choice. Finally, the beamforming optimization problem is solved.

9.3.2 Phase-Shift Feasibility

Since the objective function has no relationship with Θ, the aim is to find a feasible phase shift matrix. Note that Θ only exists in some constraints, thus, the phase-shift feasibility problem can be formulated as follows:

$$\text{P4} : \text{find} \quad \Theta \qquad (9.30a)$$

$$\text{s.t.} \quad \log_2(1 + \text{SINR}_{k,e}) \geq R_{k,e}, \quad \forall k \qquad (9.30b)$$

$$0 \leq \theta_{k,n} \leq 2\pi, \quad \forall k, n \qquad (9.30c)$$

$$|\Theta_{k,n,n}| = 1, \quad \forall k, n. \qquad (9.30d)$$

It is straightforward to find out that the non-convexity arises from constraint (9.30b). The first step is to transform this non-convex constraint to be a convex constraint. Thus, (9.30b) can be rewritten as follows:

$$|\mathbf{h}_{k,e}^H \Gamma_{\mathbf{p}_k} \mathbf{e}_k|^2(1 + r_{k,e})\alpha_k \leq |\mathbf{h}_{k,e}^H \Gamma_{\mathbf{p}_k} \mathbf{e}_k|^2 - \sum_{\substack{i=1 \\ i \neq k}}^{K} |\mathbf{h}_{k,e}^H \Gamma_{\mathbf{p}_i} \mathbf{e}_k|^2 - \sigma^2 r_{k,e}, \qquad (9.31)$$

where $\Gamma_{\mathbf{p}_i}$ is a diagonal matrix whose main diagonal elements are from $\mathbf{p}_i = \mathbf{G}_k\mathbf{w}_i$ and $\mathbf{e}_k$ is the phase-shift vector. However, with $\mathbf{W}_k$, $\alpha_k, \forall k$ already obtained from the beamforming optimization problem, constraint (9.31) is a quartic form with respect to $\mathbf{e}_k$. For simplicity, we substitute $\mathbf{h}_{k,e}^H \Gamma_{\mathbf{p}_i}$ with $\mathbf{r}_{k,e}^{iH}$. It is known that a quartic form can be equivalently transformed to a linear form with a rank-one constraint. Thus, (9.31) can be expressed as follows:

$$\text{Tr}(\mathbf{R}_{k,e}^k \mathbf{V}_k)(1 + r_{k,e})\alpha_k$$

$$\leq \text{Tr}(\mathbf{R}_{k,e}^k \mathbf{V}_k) - \sum_{\substack{i=1 \\ i \neq k}}^{K} \text{Tr}(\mathbf{R}_{k,e}^i \mathbf{V}_i) - \sigma^2 r_{k,e}, \tag{9.32}$$

$$\mathbf{V}_k \succeq 0, \tag{9.33}$$

$$\text{Rank}(\mathbf{V}_k) = 1, \tag{9.34}$$

where $\mathbf{R}_{k,e}^i = \mathbf{r}_{k,e}^i \mathbf{r}_{k,e}^{iH}$ and $\mathbf{V}_i = \mathbf{e}_i \mathbf{e}_i^H$. Given $\mathbf{w}_k, \alpha_k, \forall k$, (9.32) is an affine constraint. The rank-one constraint makes the whole problem intractable, thus SDR is adopted again to remove this rank-one constraint. Then, P4 can be transformed as follows:

$$\text{P5 : find} \quad \mathbf{V}_k, \quad \forall k \tag{9.35a}$$

$$\text{s.t.} \quad (9.32), \quad \forall k$$

$$\mathbf{V}_k \succeq 0, \quad \forall k \tag{9.35b}$$

$$\mathbf{V}_{k,n,n} = 1, \quad \forall k, n. \tag{9.35c}$$

P5 is a convex problem, which can be solved by CVX efficiently. Since the rank-one constraint is removed, the optimal solution of P5 may not be the optimal solution of P4. Therefore, Gaussian randomization will be applied to achieve a suboptimal solution for P4.

9.3.3 Algorithm Design

The detail of the proposed alternating algorithm are illustrated in Algorithm 9.1, where P2 and P5 are alternately solved until the convergence metric is satisfied.

Algorithm 9.1 The Proposed Alternating Algorithm

1: **Initialize:** $\epsilon = 0.001, j = 0$.

2: **while** $\sum_{k=1}^{K} \text{Tr}(\mathbf{W}_k^{*(j)}) - \sum_{k=1}^{K} \text{Tr}(\mathbf{W}_k^{*(j)}) \geq \epsilon$ **do**

3: Searching initial fixed feasible point $\{c_k^{*(j)}, d_k^{*(j)}, t_{k,0}^{*(j)}\}$, $\forall k$ by solving P3.

4: Update $\{\mathbf{W}_k^{*(j)}, \mathbf{w}_k^{*(j)}, \alpha_k^{*(j)}\}$, $\forall k$ by solving P2.

5: Update $\mathbf{V}_k^{*(j)}$, $\forall k$ by solving P5 with $\{\mathbf{w}_k^{*(j)}, \alpha_k^{*(j)}\}$, $\forall k$

6: Update phase-shift vector $\mathbf{e}_k^{*(j)}$, $\forall k$ by decomposing $\mathbf{V}_k^{*(j)}$, $\forall k$ based on Gaussian Randomization method.

7: $j = j + 1$

8: **end while**

9: **Output** $\{\mathbf{w}_k^{*(j)}, \alpha_k^{*(j)}, \mathbf{e}_k^{*(j)}\}$, $\forall k$.

9.4 Simulation Result

In this section, the simulation result of the proposed algorithm is provided. In these simulations, channel coefficients are generated by

$$\mathbf{h}_{k,e} = \frac{\mathbf{h}_{k,e}^*}{\sqrt{d_0^{\alpha_0}}} \quad \mathbf{G}_k = \frac{\mathbf{G}_k^*}{\sqrt{d_1^{\alpha_1}}} \quad \mathbf{h}_{k,c} = \frac{\mathbf{h}_{k,c}^*}{\sqrt{d_2^{\alpha_2}}}, \tag{9.36}$$

where $k = 1, 2, \ldots, K$, $\mathbf{h}_{k,e}^*$ and $\mathbf{h}_{k,c}^*$ are complex Rayleigh channel coefficients. $\mathbf{G}_k^*$ is complex Rician channel coefficient generated by

$$\mathbf{f} = \sqrt{\frac{\kappa}{1+\kappa}} \mathbf{f}^{\text{LoS}} + \sqrt{\frac{1}{1+\kappa}} \mathbf{f}^{\text{nLoS}}, \tag{9.37}$$

where κ is the Rician factor, $\mathbf{f}^{\text{LoS}}$ is the line-of-sight (LoS) component and $\mathbf{f}^{\text{nLoS}}$ is the non-LoS (nLoS) component following the Rayleigh distribution. $d_0 = 10\,\text{m}$, $d_1 = 50\,\text{m}$ and $d_2 = 10\,\text{m}$ stand for distances between each two nodes. The loss exponents α_0, α_1, and α_2 are set to be 1.8, 1.8, and 2, respectively. The noise power adopted in this work is $-80\,\text{dBm}$. The number of clusters is 4. Some benchmarks, e.g., NOMA scheme with grid search, NOMA scheme with random phase shift and OMA scheme are also provided.

In Fig. 9.2, the transmit power at the base station is as a function of the number of antennas equipped by the base station. We set the number of reflecting elements

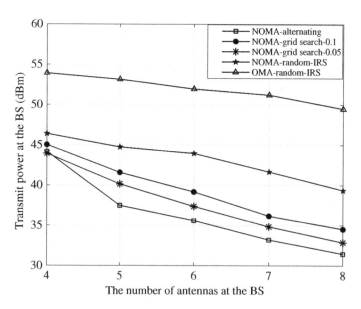

Figure 9.2 The transmit power versus the number of reflecting elements at the IRS.

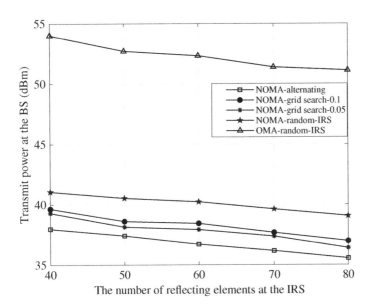

Figure 9.3 The transmit power versus the number of reflecting elements at the IRS.

at each IRS as $N = 32$, and the minimum date rate requirement of all users are chosen as 1 bps/Hz. The figure shows all algorithms have a better performance when the base station is equipped more antennas. The alternating algorithm outperforms than other benchmarks.

In Fig. 9.3, the transmit power at the base station is as a function of the number of reflecting elements on each IRS. We set the number of antennas at the base station as $M = 6$ and the minimum date rate requirement of all users is chosen 1 bps/Hz. The figure shows the IRS with more reflecting elements can save more transmit power. The alternating algorithm also outperforms than other benchmarks.

In Fig. 9.4, the transmit power at the base station is as a function of the minimum data rate requirement of the central user. We set the number of antennas at the base station as $M = 6$, the number of reflecting elements on each IRS as $N = 32$ and the date rate requirement of cell-edge users as 1 bps/Hz. The figure shows if the minimum data rate requirement of each central user is increasing, the transmit power at the base station will be consumed more. The alternating algorithm also outperforms than other benchmarks.

In Fig. 9.5, the transmit power at the base station is as a function of the distance between each IRS and the base station. We set the number of antennas at the base station as $M = 6$, the number of reflecting elements on each IRS as $N = 32$ and the minimum date rate requirement of all users as 1 bps/Hz. The figure shows the long distance brings large channel fading, which will cause more transmit power consumption. Among all algorithms, alternating algorithms are the best.

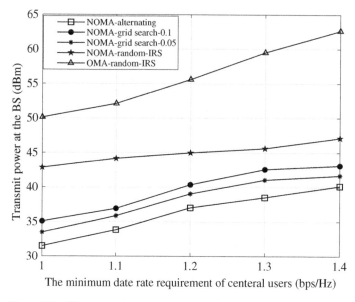

Figure 9.4 The transmit power versus the minimum date rate of the central users.

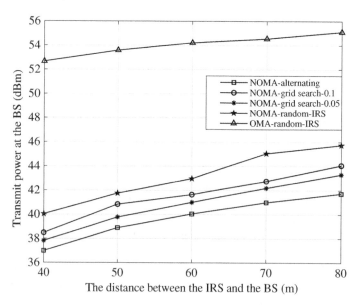

Figure 9.5 The transmit power versus the distance between the IRS and the BS.

9.5 Conclusion

The joint optimization of beamforming, power allocation, and IRS phase shift in a NOMA-IRS-assisted multi-cluster network was investigated in this work. By introducing inequality approximation, SCA and SDR, an alternating algorithm was proposed to minimize the transmit power by iteratively solving beamforming optimization and phase shifting feasibility until the algorithm converges. Furthermore, an initial point search algorithm was proposed to guarantee the feasibility of the beamforming optimization subproblem. In future research, the scenario that the IRS reconfigures the imperfect channel will be studied and the inter-cluster interference caused by IRS will also be considered.

Bibliography

Bennis, M., Simsek, M., Czylwik, A. et al. (2013). When cellular meets WiFi in wireless small cell networks. *IEEE Communications Magazine* 51 (6): 44–50.

Di Renzo, M., Debbah, M., Phan-Huy, D.-T. et al. (2019). Smart radio environments empowered by reconfigurable AI meta-surfaces: an idea whose time has come. *EURASIP Journal on Wireless Communications and Networking* 2019 (1): 1–20.

Ding, Z. and Poor, H.V. (2020). A simple design of IRS-NOMA transmission. *IEEE Communications Letters* 24 (5): 1119–1123.

Ding, Z., Adachi, F., and Poor, H.V. (2015). The application of MIMO to non-orthogonal multiple access. *IEEE Transactions on Wireless Communications* 15 (1): 537–552.

Ding, Z., Schober, R., and Poor, H.V. (2020). On the impact of phase shifting designs on IRS-NOMA. *IEEE Wireless Communications Letters* 9 (10): 1596–1600.

Fang, F., Zhang, H., Cheng, J., and Leung, V.C.M. (2016). Energy-efficient resource allocation for downlink non-orthogonal multiple access network. *IEEE Transactions on Communications* 64 (9): 3722–3732.

Fang, F., Zhang, H., Cheng, J. et al. (2017). Joint user scheduling and power allocation optimization for energy-efficient NOMA systems with imperfect CSI. *IEEE Journal on Selected Areas in Communications* 35 (12): 2874–2885.

Fu, M., Zhou, Y., Shi, Y., and Letaief, K.B. (2021). Reconfigurable intelligent surface empowered downlink non-orthogonal multiple access. *IEEE Transactions on Communications* 69 (6): 3802–3817.

Jamali, V., Tulino, A.M., Fischer, G. et al. (2020). Intelligent surface-aided transmitter architectures for millimeter-wave ultra massive MIMO systems. *IEEE Open Journal of the Communications Society* 2: 144–167.

Larsson, E.G., Edfors, O., Tufvesson, F., and Marzetta, T.L. (2014). Massive mimo for next generation wireless systems. *IEEE Communications Magazine* 52 (2): 186–195.

Li, Y., Jiang, M., Zhang, Q., and Qin, J. (2019). Joint beamforming design in multi-cluster MISO NOMA intelligent reflecting surface-aided downlink communication networks. *arXiv preprint arXiv:1909.06972*.

Liu, X., Liu, Y., Chen, Y., and Poor, H.V. (2020). RIS enhanced massive non-orthogonal multiple access networks: deployment and passive beamforming design. *IEEE Journal on Selected Areas in Communications* 39 (4): 1057–1071.

Mu, X., Liu, Y., Guo, L. et al. (2019). Exploiting intelligent reflecting surfaces in multi-antenna aided NOMA systems. *arXiv preprint arXiv:1910.13636*, pages 0090–6778.

Ni, W., Liu, X., Liu, Y. et al. (2021). Resource allocation for multi-cell IRS-aided NOMA networks. *IEEE Transactions on Wireless Communications* 20 (7): 4253–4268.

Saad, W., Bennis, M., and Chen, M. (2019). A vision of 6G wireless systems: applications, trends, technologies, and open research problems. *IEEE Network* 34 (3): 134–142.

Saito, K., Benjebbour, A., Kishiyama, Y. et al. (2015). Performance and design of SIC receiver for downlink NOMA with open-loop SU-MIMO. *2015 IEEE International Conference on Communication Workshop (ICCW)*, 1161–1165. IEEE.

Sun, X., Yang, N., Yan, S. et al. (2018). Joint beamforming and power allocation in downlink NOMA multiuser MIMO networks. *IEEE Transactions on Wireless Communications* 17 (8): 5367–5381.

Surabhi, G.D., Augustine, R.M., and Chockalingam, A. (2019). On the diversity of uncoded OTFS modulation in doubly-dispersive channels. *IEEE Transactions on Wireless Communications* 18 (6): 3049–3063.

Tariq, F., Khandaker, M.R.A., Wong, K.-K. et al. (2020). A speculative study on 6G. *IEEE Wireless Communications* 27 (4): 118–125.

Vaezi, M., Ding, Z., and Poor, H.V. (2019). *Multiple Access Techniques for 5G Wireless Networks and Beyond*, vol. 159. Springer.

Wu, Q. and Zhang, R. (2018). Intelligent reflecting surface enhanced wireless network: joint active and passive beamforming design. *2018 IEEE Global Communications Conference (GLOBECOM)*, 1–6. IEEE.

Wu, Q. and Zhang, R. (2019a). Intelligent reflecting surface enhanced wireless network via joint active and passive beamforming. *IEEE Transactions on Wireless Communications* 18 (11): 5394–5409.

Wu, Q. and Zhang, R. (2019b). Towards smart and reconfigurable environment: intelligent reflecting surface aided wireless network. *IEEE Communications Magazine* 58 (1): 106–112.

Wu, Q., Zhang, S., Zheng, B. et al. (2021). Intelligent reflecting surface-aided wireless communications: a tutorial. *IEEE Transactions on Communications* 69 (5): 3313–3351.

Zhang, Z., Xiao, Y., Ma, Z. et al. (2019). 6G wireless networks: vision, requirements, architecture, and key technologies. *IEEE Vehicular Technology Magazine* 14 (3): 28–41.

Zhu, J., Huang, Y., Wang, J. et al. (2020). Power efficient IRS-assisted NOMA. *IEEE Transactions on Communications* 69 (2): 900–913.

10

IRS-Aided Mobile Edge Computing: From Optimization to Learning

Xiaoyan Hu[1], Kai-Kit Wong[2], Christos Masouros[2], and Shi Jin[3]

[1]*School of Information and Communications Engineering, Xi'an Jiaotong University, Xi'an, China*
[2]*Department of Electronic and Electrical Engineering, University College London, London, UK*
[3]*School of Information Science and Engineering, Southeast University, Nanjing, China*

10.1   Introduction

The proliferation of Internet-of-Things (IoT) and mobile devices together with the increasing data rates provided by 5G and beyond technologies, have given rise to massive connectivity communications. Accompanied by a wide range of emerging computation-intensive latency-critical applications and services, e.g., virtual reality and automatic driving, etc., the computing and processing demands for user equipment (UEs) are growing unprecedentedly.

In order to liberate the resource-limited UEs from heavy computation workloads and provide them with high-performance low-latency computing services, mobile edge computing (MEC) promotes to use cloud computing capabilities at the edge of mobile networks through integrating MEC servers at the wireless access points (APs) (Mao et al. 2017). Hence, UEs' computation-intensive tasks can be offloaded and completed at the adjacent APs with less cost, energy and time. Extensive works have contributed to the performance enhancement of applying MEC in various wireless networks, either improving the energy efficiency or reducing the execution latency through jointly optimizing the radio and computational resources (Sardellitti et al. 2015; You et al. 2016; Hu et al. 2018,2019a,b; Tran and Pompili 2018).

In order to avoid the communication performance degradation caused by random link blockage and further enhance the uplink offloading performance of the resource-limited UEs, recently great attention has been drawn to the technology of intelligent reflecting surface (IRS), due to its advantages of low cost, easy deployment, fine-grained passive beamforming, and directional signal enhancement or interference nulling (Wu and Zhang 2019a; Basar et al. 2019). Through controlling

Intelligent Surfaces Empowered 6G Wireless Network, First Edition.
Edited by Qingqing Wu, Trung Q. Duong, Derrick Wing Kwan Ng, Robert Schober, and Rui Zhang.

the reflecting elements on the surface, IRSs can be reconfigured to provide a more favorable propagation environment for communications. Clearly, leveraging IRSs into MEC systems is a cost-effective and environment-friendly way to facilitate UEs' computation offloading. Several pioneer IRS-aided MEC works have been done to explore the potential benefits of utilizing IRSs in MEC systems (Bai et al. 2020; Hua et al. 2021).

For IRS-aided MEC systems, the resource allocation issues are quite challenging considering joint radio/computation resource scheduling and joint active/passive beamforming, leading to typically non-convex optimization problems with coupled optimization variables. Hence, iterative algorithms such as block coordinate descending (BCD) or alternating optimization algorithms are usually necessary for the related resource allocation. While iterative algorithms may be capable of providing near-optimal solutions even with guaranteed convergence, they have very high computational complexity which usually require long execution time and thus may hinder their utilization in practical networks. To tackle this issue, deep learning architectures provide a promising way to achieve lightweight online implementation via offline training (Goodfellow et al. 2016; Oshea and Hoydis 2017). Note that deep learning methods have been investigated in some MEC systems to simplify optimization algorithm or fulfill online implementations (Dong et al. 2019; Huang et al. 2020). As per the above literature, deep learning approaches are promising to offer low-complexity solutions for traditional MEC-related systems or IRS-aided downlink communications. However, the potentials of deep learning methods in simplifying the optimization algorithms of complex IRS-aided MEC systems have not been fully explored in the existing literature.

In this chapter, we mainly overview the resource allocation issues for multi-user IRS-aided MEC systems with multiplexing computation offloading, including both joint radio/computational resource scheduling and joint active/passive beamforming, and the relevant system model is presented in Section 10.2. Our focus is on the evolution from the more classical optimization-based solutions to the learning-based solutions for improving the efficiency of the IRS-aided MEC, highlighting the benefits and limitations of each. In terms of optimization, we first overview the iterative BCD algorithm to maximize the system computation efficiency with guaranteed convergence in Section 10.3, and then in Section 10.4 we demonstrate two deep learning architectures to facilitate online implementations with significantly reduced complexity. Section 10.5 provides the simulation results, and we conclude this chapter in Section 10.6.

10.2 System Model and Objective

Let us consider a typical IRS-aided MEC system as shown in Fig. 10.1, which consists of N single-antenna ground UEs, one IRS with K reflecting elements, and

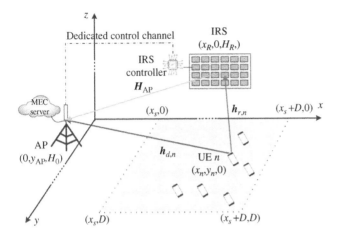

Figure 10.1 An illustration of the typical IRS-aided MEC architecture. Source: (Hu et al. 2021)/IEEE, where a K-element IRS assists the multiplexing computation offloading of N UEs. The phase shifts of IRS elements can be adaptively adjusted to refine signal propagations.

one M-antenna AP. The IRS can be flexibly installed on the surrounding buildings, and it is under the control of the AP through a wireless controller to dynamically adjust the phase shift of each reflecting element. By choosing a desirable location for the IRS, it is possible to achieve line-of-sight (LoS) connections between IRS and AP as well as UEs within a certain area.

Each UE $n \in \mathcal{N} = \{1, 2, \ldots, N\}$ has intensive computation task-input bits to be dealt with, but with a limited energy budget dedicated for completing these task bits, denoted as E_n in Joules (J). Partial offloading is leveraged where UEs' task-input bits can be arbitrarily divided to facilitate parallel operations at UEs for local computing and offloading to the AP for remote computing. Hence, the accounted computation energy consumption of each UE includes both used for local computing and computation offloading. The grid-powered AP is co-located with a powerful MEC server for helping UEs compute their offloaded tasks and it is also capable of downloading UEs' computation results, both in negligible time. We use C_n to represent the amount of required computing resource, i.e., the number of CPU cycles, for completing 1-bit of UE n's input data. A typical objective in the considered scenario is to maximize the total completed task-input bits (TCTB) of all UEs each with a limited energy budget during a given time slot T. It has been shown that this is equivalent to maximizing the system computation efficiency in both energy and time, which reflects the fundamental requirements for practical MEC systems with or without the assistance of IRS (Bi and Zhang 2018; Zhou and Hu 2020; Hu et al. 2020; Bai et al. 2020; Hua et al. 2021).

The objective can be solved by first introducing a partition parameter $a_n \in [0, 1]$ for UE $n \in \mathcal{N}$, and $a_n E_n$ (J) of energy will be used for computation offloading while $(1 - a_n)E_n$ (J) of energy will be used for local computing. In this case, the transmit power of UE n for computation offloading is given as

$$p_n = \frac{a_n E_n}{T} \triangleq a_n \tilde{E}_n, \quad \forall n \in \mathcal{N}. \tag{10.1}$$

Let s_n denote the task-input data-bearing signal transmitted by UE $n \in \mathcal{N}$ for computation offloading with $|s_n| = 1$. Note that all the UEs with offloading requirements transmit their signals simultaneously in a multiplexing way within the given time slot, and thus one can express the corresponding received signal $\mathbf{y} \in \mathbb{C}^{M \times 1}$ at the AP as (Wu and Zhang 2019b)

$$\mathbf{y} = \sum_{n=1}^{N} (\mathbf{H}_{AP} \mathbf{\Phi} \mathbf{h}_{r,n} + \mathbf{h}_{d,n}) \sqrt{p_n} s_n + \mathbf{n}, \tag{10.2}$$

where $\mathbf{h}_{d,n} \in \mathbb{C}^{M \times 1}$ is the direct link between UE n and the AP, $\mathbf{h}_{r,n} \in \mathbb{C}^{K \times 1}$ indicates the relay channel between UE n and the IRS, and $\mathbf{H}_{AP} \in \mathbb{C}^{M \times K}$ represents the channel between IRS and the AP. It is typically assumed that all the channels are quasi-static within the given time slot. Additionally, $\mathbf{\Phi} = \text{diag}\{\boldsymbol{\phi}\}$ indicates the reflection-coefficient matrix of the IRS, where $\boldsymbol{\phi} = [\phi_1, \ldots, \phi_K]^T$ and $\phi_k = e^{j\theta_k}$ being the phase shift of the kth reflecting element of the IRS with $\theta_k \in [0, 2\pi]$ for $k \in \mathcal{K} = \{1, 2, \ldots, K\}$. Also, $\mathbf{n} \sim \mathcal{CN}(0, \sigma^2 \mathbf{I}_M)$ is the additive white Gaussian noise (AWGN) at the AP with σ^2 being the noise power. The linear beamforming strategy is then adopted at the AP to decode UEs' transmit signals, and $\mathbf{w}_n \in \mathbb{C}^{M \times 1}$ is the specific receive beamforming vector for UE n. Thus, the estimated signal for UE n can be given as $\hat{s}_n = \mathbf{w}_n^H \mathbf{y}$. Based on the analysis above, we can obtain the uplink signal-to-interference-plus-noise ratio (SINR) for offloading UE n's tasks as

$$\gamma_n(\mathbf{a}, \mathbf{w}_n, \boldsymbol{\phi}) = \frac{a_n \tilde{E}_n |\mathbf{w}_n^H (\mathbf{H}_{AP} \mathbf{\Phi} \mathbf{h}_{r,n} + \mathbf{h}_{d,n})|^2}{\sum_{i=1, i \neq n}^{N} a_i \tilde{E}_i |\mathbf{w}_n^H (\mathbf{H}_{AP} \mathbf{\Phi} \mathbf{h}_{r,i} + \mathbf{h}_{d,i})|^2 + \sigma^2 \|\mathbf{w}_n^H\|^2}, \quad \forall n \in \mathcal{N}, \tag{10.3}$$

where $\mathbf{a} = [a_1, \ldots, a_N]$. Given B as the bandwidth shared with all UEs, then the completed task-input bits (CTB) of UE n through computation offloading can be expressed as

$$R_n^{\text{off}}(\mathbf{a}, \mathbf{w}_n, \boldsymbol{\phi}) = BT \log_2(1 + \gamma_n(\mathbf{a}, \mathbf{w}_n, \boldsymbol{\phi})), \quad \forall n \in \mathcal{N}. \tag{10.4}$$

As for the case of local computing, the relevant literature (Zhang et al. 2013) has shown that an energy-efficient solution at the UEs is given by leveraging the dynamic voltage and frequency scaling (DVFS). Thus, the computation energy consumption of UE $n \in \mathcal{N}$ can be expressed as $(1 - a_n)E_n = T\kappa_n f_n^3$, where κ_n is the effective capacitance coefficient and f_n is the CPU frequency of UE n. Hence, the CTB of UE n for local computing can be given as

$$R_n^{\text{loc}}(a_n) = \frac{f_n T}{C_n} = \frac{T}{C_n} \sqrt[3]{\frac{(1-a_n)\tilde{E}_n}{T\kappa_n}}, \quad \forall n \in \mathcal{N}. \tag{10.5}$$

As mentioned before, the TCTB maximization is of practical meaning for improving the computation efficiency of IRS-aided MEC systems. This typically necessitates jointly designing the passive reflection coefficients in ϕ, the active receive beamforming vectors in $\mathbf{W} = [\mathbf{w}_1, \ldots, \mathbf{w}_N]$, and the energy partition parameters in $\mathbf{a}$. In the rest of this chapter, we overview this joint design for IRS-aided MEC from optimization to learning techniques.

10.3 Optimization-Based Approaches to IRS-Aided MEC

Based on the system model and objective given in Section 10.2, one can formulate the TCTB maximization problem as problem (P0) given below

$$(\text{P0}): \max_{\mathbf{a},\mathbf{W},\phi} \sum_{n=1}^{N} \left(R_n^{\text{off}}(\mathbf{a}, \mathbf{w}_n, \phi) + R_n^{\text{loc}}(a_n) \right) \tag{10.6a}$$

$$\text{s.t.} \quad a_n \in [0,1], \quad \forall n \in \mathcal{N}, \tag{10.6b}$$

$$|\phi_k| = 1, \quad \forall k \in \mathcal{K}, \tag{10.6c}$$

which is a non-convex optimization problem and it is difficult to find the global optimal solution due to the strong coupling among ϕ, $\mathbf{W}$, and $\mathbf{a}$. This is a typical bottleneck in such IRS-aided MEC scenarios (Bai et al. 2020; Hua et al. 2021), and one way to overcome this is through a three-step BCD optimization algorithm so as to effectively separate the coupling among the optimization variables and solve the problem iteratively with guaranteed convergence.

10.3.1 IRS Reflecting Coefficients Design

The χth ($\chi = 1, 2, \ldots$) iteration of the BCD algorithm first involves designing the IRS reflecting coefficients ϕ with given $\mathbf{W} = \mathbf{W}_{\chi-1}$ and $\mathbf{a} = \mathbf{a}_{\chi-1}$. The IRS's reflecting coefficients design problem (P1) can be given as

$$(\text{P1}): \max_{\phi} \sum_{n=1}^{N} \log_2(1 + \gamma_n(\phi)) \tag{10.7a}$$

$$\text{s.t.} \quad |\phi_k| = 1, \quad \forall k \in \mathcal{K}, \tag{10.7b}$$

which is still non-convex and difficult to deal with directly. Based on $\gamma_n(\boldsymbol{\phi})$ in (10.3) for $n \in \mathcal{N}$, one can re-express $|\mathbf{w}_n^H(\mathbf{H}_{AP}\boldsymbol{\Phi}\mathbf{h}_{r,i} + \mathbf{h}_{d,i})|^2$ as

$$|\mathbf{w}_n^H(\mathbf{H}_{AP}\boldsymbol{\Phi}\mathbf{h}_{r,i} + \mathbf{h}_{d,i})|^2 = |\mathbf{w}_n^H\mathbf{H}_{AP}\,\mathrm{diag}(\mathbf{h}_{r,i})\boldsymbol{\phi} + \mathbf{w}_n^H\mathbf{h}_{d,i}|^2 \tag{10.8}$$
$$= |\mathbf{h}_{r,n,i}^{IRS}\boldsymbol{\phi} + h_{d,n,i}|^2, \quad \forall n, \quad i \in \mathcal{N}.$$

By defining a matrix $\mathbf{Q}_{n,i} \in \mathbb{C}^{(K+1)\times(K+1)}$ as

$$\mathbf{Q}_{n,i} = \begin{bmatrix} (\mathbf{h}_{r,n,i}^{IRS})^H \mathbf{h}_{r,n,i}^{IRS}, & (\mathbf{h}_{r,n,i}^{IRS})^H h_{d,n,i} \\ h_{d,n,i}^H \mathbf{h}_{r,n,i}^{IRS}, & 0 \end{bmatrix}, \quad \forall n \in \mathcal{N}, \tag{10.9}$$

and a vector $\tilde{\boldsymbol{\phi}} = [\phi_1, \dots, \phi_K, \xi]^T \in \mathbb{C}^{(K+1)\times 1}$ with an auxiliary scalar ξ, we can then re-express $|\mathbf{h}_{r,n,i}^{IRS}\boldsymbol{\phi} + h_{d,n,i}|^2 = \tilde{\boldsymbol{\phi}}^H\mathbf{Q}_{n,i}\tilde{\boldsymbol{\phi}} + |h_{d,n,i}|^2 = \mathrm{Tr}(\mathbf{Q}_{n,i}\boldsymbol{\Psi}) + |h_{d,n,i}|^2$, where $\boldsymbol{\Psi} = \tilde{\boldsymbol{\phi}}\tilde{\boldsymbol{\phi}}^H \in \mathbb{C}^{(K+1)\times(K+1)}$ is a positive semidefinite matrix (PSD) related to the IRS reflecting coefficients. Note that each added item in the objective function, i.e., $\log_2(1 + \gamma_n(\boldsymbol{\phi}))$, can be re-written as

$$\log_2(1 + \gamma_n(\boldsymbol{\phi})) = \log_2(1 + \gamma_n(\tilde{\boldsymbol{\phi}})) \tag{10.10}$$
$$= \log_2\left(\sum_{j=1}^N p_j(\mathrm{Tr}(\mathbf{Q}_{n,j}\boldsymbol{\Psi}) + |h_{d,n,j}|^2) + \sigma^2\|\mathbf{w}_n^H\|^2\right)$$
$$- \log_2\left(\sum_{i=1, i\neq n}^N p_i(\mathrm{Tr}(\mathbf{Q}_{n,i}\boldsymbol{\Psi}) + |h_{d,n,i}|^2) + \sigma^2\|\mathbf{w}_n^H\|^2\right)$$
$$\triangleq F_{1,n}(\boldsymbol{\Psi}) - F_{2,n}(\boldsymbol{\Psi}), \quad \forall n \in \mathcal{N},$$

where $F_{1,n}(\boldsymbol{\Psi})$ and $F_{2,n}(\boldsymbol{\Psi})$ are two concave functions w.r.t. $\boldsymbol{\Psi}$. Hence, the problem (P1) can be equivalently transformed into problem ($\tilde{P}$1)

$$(\tilde{P}1): \max_{\boldsymbol{\Psi} \succeq 0} \sum_{n=1}^N \left(F_{1,n}(\boldsymbol{\Psi}) - F_{2,n}\right)(\boldsymbol{\Psi}) \tag{10.11a}$$

$$\text{s.t.} \quad \boldsymbol{\Psi}_{k,k} = 1, \quad \forall k = 1, 2, \dots, K+1, \tag{10.11b}$$

$$\mathrm{rank}(\boldsymbol{\Psi}) = 1. \tag{10.11c}$$

Even though the objective function in (10.11a) and the rank-one constraint (10.11b) make problem ($\tilde{P}$1) non-convex, it is easy to note that the objective function is a sum of differences of concave functions. Next, we will show that the difference of convex functions (DC) programming (An and Tao 2005) can be leveraged to effectively address the issues of the objective function and the rank-one constraint.

As for the objective function, in the $(l + 1)$th ($l = 0, 1, \dots$) iteration of the DC programming, the second concave item, i.e., $F_{2,n}(\boldsymbol{\Psi})$ for $n \in \mathcal{N}$, can be approximated by its linear upper bound at the point $\boldsymbol{\Psi}^{(l)}$ (the solution obtained from the previous lth iteration), which is given as

$$F_{2,n}(\boldsymbol{\Psi}) \leq \hat{F}_{2,n}(\boldsymbol{\Psi}; \boldsymbol{\Psi}^{(l)}) = F_{2,n}(\boldsymbol{\Psi}^{(l)}) \tag{10.12}$$

$$+ \frac{\sum_{i=1,i\neq n}^{N} p_i \left\langle (\boldsymbol{\Psi} - \boldsymbol{\Psi}^{(l)}), \nabla_{\boldsymbol{\Psi}} \operatorname{Tr}(\mathbf{Q}_{n,i}\boldsymbol{\Psi})|_{\boldsymbol{\Psi}=\boldsymbol{\Psi}^{(l)}} \right\rangle}{\ln 2 \left(\sum_{i=1,i\neq n}^{N} p_i (\operatorname{Tr}(\mathbf{Q}_{n,i}\boldsymbol{\Psi}^{(l)}) + |h_{d,n,i}|^2) + \sigma^2 \|\mathbf{w}_n^H\|^2 \right)},$$

where $\nabla_{\boldsymbol{\Psi}} \operatorname{Tr}(\mathbf{Q}_{n,i}\boldsymbol{\Psi})|_{\boldsymbol{\Psi}=\boldsymbol{\Psi}^{(l)}}$ denotes the Jacobian matrix of $\operatorname{Tr}(\mathbf{Q}_{n,i}\boldsymbol{\Psi})$ w.r.t. $\boldsymbol{\Psi}$ at the point $\boldsymbol{\Psi}^{(l)}$, and it is easy to note that the equality holds when $\boldsymbol{\Psi} = \boldsymbol{\Psi}^{(l)}$.

As for the rank-one constraint (10.11c), it can be equivalently transformed into the following form $\operatorname{Tr}(\boldsymbol{\Psi}) - \|\boldsymbol{\Psi}\|_s = 0$ where $\|\boldsymbol{\Psi}\|_s$ denotes the spectral norm of the PSD matrix $\boldsymbol{\Psi}$. It is noticeable that $\operatorname{Tr}(\boldsymbol{\Psi}) = \sum_{k=1}^{K+1} \rho_k(\boldsymbol{\Psi})$ and $\|\boldsymbol{\Psi}\|_s = \rho_1(\boldsymbol{\Psi})$, where $\rho_k(\boldsymbol{\Psi})$ indicates the kth largest singular value of $\boldsymbol{\Psi}$. Hence, the equality of $\operatorname{Tr}(\boldsymbol{\Psi}) = \|\boldsymbol{\Psi}\|_s$ holds when the rank-one constraint is satisfied with $\rho_1(\boldsymbol{\Psi}) > 0$ and $\rho_k(\boldsymbol{\Psi}) = 0$ for $k = 2, \ldots, K+1$, and vice versa. Similarly, in the $(l+1)$th iteration of the DC programming, a linear lower-bound of the convex item $\|\boldsymbol{\Psi}\|_s$ at the point $\boldsymbol{\Psi}^{(l)}$ can be expressed as

$$\|\boldsymbol{\Psi}\|_s \geq \|\boldsymbol{\Psi}^{(l)}\|_s + \left\langle (\boldsymbol{\Psi} - \boldsymbol{\Psi}^{(l)}), \partial_{\boldsymbol{\Psi}} \|\boldsymbol{\Psi}\|_s|_{\boldsymbol{\Psi}=\boldsymbol{\Psi}^{(l)}} \right\rangle \triangleq \Upsilon(\boldsymbol{\Psi}; \boldsymbol{\Psi}^{(l)}), \tag{10.13}$$

where $\partial_{\boldsymbol{\Psi}} \|\boldsymbol{\Psi}\|_s|_{\boldsymbol{\Psi}=\boldsymbol{\Psi}^{(l)}}$ is a subgradient of the spectral norm $\|\boldsymbol{\Psi}\|_s$ w.r.t. $\boldsymbol{\Psi}$ at the point $\boldsymbol{\Psi}^{(l)}$, and the equality holds when $\boldsymbol{\Psi} = \boldsymbol{\Psi}^{(l)}$. Note that one subgradient of $\|\boldsymbol{\Psi}\|_s$ at point $\boldsymbol{\Psi}^{(l)}$ can be computed as $\mathbf{z}_1 \mathbf{z}_1^H$, where $\mathbf{z}_1$ is the vector corresponding to the largest singular value of $\boldsymbol{\Psi}^{(l)}$ (Yang et al. 2020).

With the obtained linear lower bound of $\|\boldsymbol{\Psi}\|_s$ in (10.13), we can generate an approximate rank-one constraint of (10.11c), given as $\operatorname{Tr}(\boldsymbol{\Psi}) - \Upsilon(\boldsymbol{\Psi}; \boldsymbol{\Psi}^{(l)}) \leq \varepsilon_{\boldsymbol{\Psi}}$, where $\varepsilon_{\boldsymbol{\Psi}}$ is a positive threshold with very small value. The approximated rank-one constraint can guarantee that $0 \leq \operatorname{Tr}(\boldsymbol{\Psi}) - \|\boldsymbol{\Psi}\|_s \leq \operatorname{Tr}(\boldsymbol{\Psi}) - \Upsilon(\boldsymbol{\Psi}; \boldsymbol{\Psi}^{(l)}) \leq \varepsilon_{\boldsymbol{\Psi}}$, and the rank-one constraint can be approached with an arbitrary accuracy by setting $\varepsilon_{\boldsymbol{\Psi}}$ infinitely close to zero. Hence, an approximation problem of (P̃1) at the $(l+1)$th iteration can be expressed as

$$(\text{P1.1}): \max_{\boldsymbol{\Psi} \geq 0} \sum_{n=1}^{N} \left(F_{1,n}(\boldsymbol{\Psi}) - \hat{F}_{2,n} \right)(\boldsymbol{\Psi}; \boldsymbol{\Psi}^{(l)}) \tag{10.14a}$$

$$\text{s.t.} \quad \boldsymbol{\Psi}_{k,k} = 1, \quad \forall k = 1, 2, \ldots, K+1, \tag{10.14b}$$

$$\operatorname{Tr}(\boldsymbol{\Psi}) - \Upsilon(\boldsymbol{\Psi}; \boldsymbol{\Psi}^{(l)}) \leq \varepsilon_{\boldsymbol{\Psi}}, \tag{10.14c}$$

which is a convex optimization problem and can be readily solved by the existing convex solvers such as CVX, and the optimal solution can be obtained as $\boldsymbol{\Psi}^{(l+1)}$. Through choosing $\boldsymbol{\Psi}^{(0)} = \boldsymbol{\Psi}_{\chi-1} = \tilde{\boldsymbol{\phi}}_{\chi-1} \tilde{\boldsymbol{\phi}}_{\chi-1}^H$, it is easy to prove that the feasibility of problem (P1.1) in each iteration l can always be guaranteed since $\boldsymbol{\Psi}^{(l-1)}$ is always a feasible solution.

Lemma 10.1 *The objective function of problem (P1) in (10.11a) monotonically increases with the iteration index l as*

$$F_{1,n}(\Psi^{(l+1)}) - F_{2,n}(\Psi^{(l+1)}) \tag{10.15}$$

$$\overset{(a)}{\geq} F_{1,n}(\Psi^{(l+1)}) - \hat{F}_{2,n}(\Psi^{(l+1)}; \Psi^{(l)})$$

$$\overset{(b)}{\geq} F_{1,n}(\Psi^{(l)}) - \hat{F}_{2,n}(\Psi^{(l)}; \Psi^{(l)}) = F_{1,n}(\Psi^{(l)}) - F_{2,n}(\Psi^{(l)}), \quad \forall n \in \mathcal{N},$$

where (a) comes from the inequality (10.12) and (b) holds since $\Psi^{(l)}$ is a feasible solution while $\Psi^{(l+1)}$ is the optimal solution of problem (P1.1) in (10.14).

The final solution of Ψ converged at the $(l+1)$th iteration for DC programming is the solution of the BCD algorithm at the χth iteration, i.e., $\Psi_\chi = \Psi^{(l+1)}$. Then, we can retrieve $\tilde{\phi}_\chi$ by decomposing $\Psi_\chi = \tilde{\phi}_\chi \tilde{\phi}_\chi^H$ with denoting $\tilde{\phi}_\chi = [\phi_{\chi,0}, \xi_{\chi,0}]^T$, and thus it is easy to obtain the IRS reflecting coefficient vector at the χth iteration of the BCD algorithm as $\phi_\chi = \phi_{\chi,0}/\xi_{\chi,0}$ and accordingly $\Phi_\chi = \text{diag}\{\phi_\chi\}$. In order to facilitate the following analysis, we define the effective UE-AP channels with given Φ (or ϕ) as

$$\mathbf{h}_n(\Phi) = \mathbf{H}_{AP}\Phi\mathbf{h}_{r,n} + \mathbf{h}_{d,n}, \quad \forall n \in \mathcal{N}. \tag{10.16}$$

10.3.2 Receive Beamforming Design

With given $\mathbf{a} = \mathbf{a}_{\chi-1}$ and $\Phi = \Phi_\chi$, the sub-problem for optimizing AP's receive beamforming matrix $\mathbf{W}$ can be expressed as the following problem (P2)

$$(\text{P2}) : \max_{\mathbf{W}} \sum_{n=1}^{N} R_n^{\text{off}}(\mathbf{w}_n), \tag{10.17}$$

which can be equivalently solved by addressing N parallel sub-problems

$$(\text{P2.1}) : \max_{\mathbf{w_n}} \gamma_n(\mathbf{w}_n) = \frac{\mathbf{w}_n^H \Theta_n \mathbf{w}_n}{\mathbf{w}_n^H \Theta_{-n} \mathbf{w}_n}, \quad n \in \mathcal{N}, \tag{10.18}$$

where $\Theta_n = p_n \mathbf{h}_n(\mathbf{h}_n)^H$ and $\Theta_{-n} = \sum_{i=1, i \neq n}^{N} p_i \mathbf{h}_i(\mathbf{h}_i)^H + \sigma^2 \mathbf{I}_M$.

Lemma 10.2 *It is easy to note that problem (P2.1) in (10.18) is a generalized eigenvector problem, and its optimal solution $\mathbf{w}_n^*$ should be the eigenvector corresponds to the largest eigenvalue of the matrix $(\Theta_{-n})^{-1}\Theta_n$. Hence, the optimal $\mathbf{w}_n^*$ of problem (P2.1) for $n \in \mathcal{N}$ can be given as*

$$\mathbf{w}_n^* = \text{eigvec}\left\{ \max\left\{ \text{eig}\{(\Theta_{-n})^{-1}\Theta_n\} \right\} \right\}. \tag{10.19}$$

We then denote the receive beamforming matrix obtained at the χth iteration of the BCD algorithm as $\mathbf{W}_\chi = [\mathbf{w}_1^*, \ldots, \mathbf{w}_N^*]$.

10.3.3 Energy Partition Optimization

Here, the sub-problem (P3) for optimizing the energy partition parameters in $\mathbf{a}$ with given $\boldsymbol{\Phi} = \boldsymbol{\Phi}_\chi$ and $\mathbf{W} = \mathbf{W}_\chi$ is considered, which is given below

$$(\text{P3}): \max_{\mathbf{a}} \sum_{n=1}^{N} \left(R_n^{\text{off}}(\mathbf{a}) + R_n^{\text{loc}}(a_n) \right) \tag{10.20a}$$

$$\text{s.t. } a_n \in [0,1], \quad \forall n \in \mathcal{N}. \tag{10.20b}$$

Note that problem (P3) is non-convex because of the non-concave items $\{R_n^{\text{off}}(\mathbf{a})\}_{n\in\mathcal{N}}$ in the objective function (10.20a). Actually, $R_n^{\text{off}}(\mathbf{a})$ for $n \in \mathcal{N}$ can be re-expressed as the difference of two concave functions as follows

$$R_n^{\text{off}}(\mathbf{a}) \triangleq R_{n,1}^{\text{off}}(\mathbf{a}) - R_{n,2}^{\text{off}}(\mathbf{a}_{-n}) \tag{10.21}$$

$$= BT\log_2\left(\sum_{j=1}^{N} a_j \tilde{E}_j |\mathbf{w}_n^H \mathbf{h}_j|^2 + \sigma^2 \|\mathbf{w}_n^H\|^2 \right)$$

$$- BT\log_2\left(\sum_{i=1,i\neq n}^{N} a_i \tilde{E}_i |\mathbf{w}_n^H \mathbf{h}_i|^2 + \sigma^2 \|\mathbf{w}_n^H\|^2 \right),$$

where $\mathbf{a}_{-n} = [a_1, \ldots, a_{n-1}, a_{n+1}, \ldots, a_N]$. Then, the problem (P3) can also be solved with the DC programming method. Assuming $\mathbf{a}^{(m)}$ is the solution obtained at the mth ($m = 0, 1, \ldots$) iteration of the DC programming, a linear upper bound of $R_{n,2}^{\text{off}}(\mathbf{a}_{-n})$ at the point $\mathbf{a}^{(m)}$ can be obtained through the first-order Taylor series expansion as

$$R_{n,2}^{\text{off}}(\mathbf{a}_{-n}) \leq \hat{R}_{n,2}^{\text{off}}(\mathbf{a}_{-n}; \mathbf{a}_{-n}^{(m)}) \tag{10.22}$$

$$= R_{n,2}^{\text{off}}(\mathbf{a}_{-n}^{(m)}) + \sum_{i=1,i\neq n}^{N} R_{n,2,i}^{\text{off}\prime}(\mathbf{a}_{-n}^{(m)}) * (a_i - a_i^{(m)}),$$

where $R_{n,2,i}^{\text{off}\prime}(\mathbf{a}_{-n}^{(m)})$ is the first-order derivative of $R_{n,2}^{\text{off}}(\mathbf{a}_{-n})$ w.r.t. a_i at the point $\mathbf{a}_{-n}^{(m)}$, and the equality holds when $\mathbf{a}_{-n} = \mathbf{a}_{-n}^{(m)}$. At the $(m+1)$th iteration of DC programming, we aim at maximizing the following approximation problem

$$(\text{P3.1}): \max_{\mathbf{a}} \sum_{n=1}^{N} \left(R_{n,1}^{\text{off}}(\mathbf{a}) - \hat{R}_{n,2}^{\text{off}}(\mathbf{a}_{-n}; \mathbf{a}_{-n}^{(m)}) + R_n^{\text{loc}}(a_n) \right) \tag{10.23a}$$

$$\text{s.t. } a_n \in [0,1], \quad \forall n \in \mathcal{N}, \tag{10.23b}$$

which is a convex problem and can be easily solved by CVX, and thus the optimal solution, i.e., $\mathbf{a}^{(m+1)}$, can be finally obtained.

Lemma 10.3 *The objective function of problem (P3) in* (10.20a) *is monotonic increasing w.r.t. the iteration index m as*

$$R_n^{\text{off}}(\mathbf{a}^{(m+1)}) + R_n^{\text{loc}}(a_n^{(m+1)}) \tag{10.24}$$

$$\geq R_{n,1}^{\text{off}}(\mathbf{a}^{(m+1)}) - \hat{R}_{n,2}^{\text{off}}(\mathbf{a}_{-n}^{(m+1)}; \mathbf{a}_{-n}^{(m)}) + R_n^{\text{loc}}(a_n^{(m+1)})$$

$$\geq R_{n,1}^{\text{off}}(\mathbf{a}^{(m)}) - \hat{R}_{n,2}^{\text{off}}(\mathbf{a}_{-n}^{(m)}; \mathbf{a}_{-n}^{(m)}) + R_n^{\text{loc}}(a_n^{(m)}) = R_n^{\text{off}}(\mathbf{a}^{(m)}) + R_n^{\text{loc}}(a_n^{(m)}).$$

We can obtain the final solution of $\mathbf{a}$ at the χth iteration of the BCD algorithm when the DC programming converges at the $(m + 1)$th iteration, which is denoted as $\mathbf{a}_\chi = [a_1^{(m+1)}, \ldots, a_N^{(m+1)}]$.

10.3.4 Convergence and Complexity

Based on the three-step BCD optimization algorithm overviewed previously, it is easy to verify that the original TCTB maximization problem (P0) in (10.6) can be effectively solved with guaranteed convergence (Hu et al. 2021).

In (Hu et al. 2021), it was shown that the above optimization is computationally demanding, as it commonly involves interior point solvers for convex optimization (Wang et al. 2014). Indeed, the total computational complexity of the utilized three-step BCD algorithm can be given as $O(L(L_1 K^6 + L_3 N^{3.5}))$, where L_1, L_3 and L represent the iteration numbers for solving problem (P1), (P3), and that of the BCD algorithm, respectively. It is easy to observe that the computational complexity of the BCD algorithm increases dramatically with the number of IRS reflecting elements and the number of UEs.

10.4 Deep Learning Approaches to IRS-Aided MEC

Although the overviewed BCD optimization algorithm can achieve effective solutions with guaranteed convergence, its high computational complexity may hinder it from being applied in real-time applications, which is a major bottleneck for most of the iterative optimization algorithms proposed in state-of-the-art works (Bai et al. 2020; Hua et al. 2021). One way to overcome this drawback is leveraging the deep learning approaches, not only due to the fact that deep neural networks (DNNs) are regarded as universal function approximators but also because deep learning is well known as a promising way to achieve effective online implementations (Goodfellow et al. 2016; Oshea and Hoydis 2017). Hence, in this section, we explore the potentials of deep learning approaches in obtaining effective solutions to the original problem (P0).

As mentioned before, deep learning methods have great potential in reducing the computational complexity of iterative optimization approaches (Dong et al. 2019; Huang et al. 2020). Following up the BCD optimization design overviewed above, we show that learning-based approaches are capable of emulating the BCD optimization algorithm with high accuracy and low implementation overheads

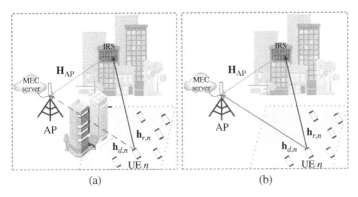

Figure 10.2 Two typical IRS-aided edge computation offloading scenarios. Source: (Hu et al. 2021)/IEEE: (a) Scenario without LoS direct links between UEs and AP; (b) Scenario with strong LoS direct links between UEs and AP.

via well-designed DNN architectures. Two typical IRS-aided edge computation offloading scenarios in terms of whether LoS direct links exist between UEs and AP are shown in Fig. 10.2, and two deep learning architectures of (Hu et al. 2021) based on channel state information (CSI) and only UEs' locations (CSI-free) are overviewed in Sections 10.4.1 and 10.4.2, respectively.

10.4.1 CSI-Based Learning Architecture

It is well known that CSI is crucial for achieving reliable communications which is commonly required for resource scheduling via either optimization algorithms (Bai et al. 2020; Hua et al. 2021) or leaning methods (Dong et al. 2019; Huang et al. 2020). In this section, a CSI-based deep learning architecture is given as shown in Fig. 10.3, to obtain the solutions of $\{\boldsymbol{\phi}, \mathbf{a}, \mathbf{W}\}$, which is applicable to both scenarios (a) and (b) without and with LoS direct links between UEs and AP. The real and imaginary parts of the channel coefficients $\{\mathbf{h}_{d,n}\}$, $\{\mathbf{h}_{r,n}\}$ and $\mathbf{H}_{AP}$ constitute the input feature of the constructed DNN-CSI, represented by the input vector $\mathbf{x}$ with a dimension of $I = 2(MN + KN + MK)$. In contrast, the normalized angles of the IRS reflecting coefficients $\boldsymbol{\phi}$, denoted as $\tilde{\theta} = \theta/2\pi$, and the energy partition parameters in $\mathbf{a}$ constitute the output vector $\mathbf{y} = [\tilde{\theta}_1, \dots, \tilde{\theta}_K, a_1, \dots, a_N]$ with the dimension of $(K + N)$. Note that all the elements of the output vector are within the range of $[0, 1]$, and thus one can use the sigmoid function, i.e., $\text{Sigmoid}(z) = \frac{1}{1+e^{-z}}$ as the output activation function. With the final output $\boldsymbol{\phi}$ and $\mathbf{a}$ of the DNN-CSI, the optimal receive beamforming vector for each UE $n \in \mathcal{N}$, i.e., $\mathbf{w}_n$, can be readily obtained according to Lemma 10.2 in Section 10.3.2.

One way of implementing the constructed learning architecture is through feed-forward DNNs and here we overview one well-designed example that consists of

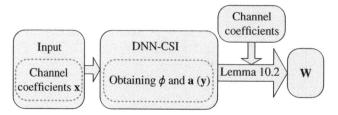

Figure 10.3 The architecture for obtaining the solutions of $\{\phi, \mathbf{a}, \mathbf{W}\}$ with the CSI-based DNN-CSI.

an I-dimensional input layer, 5 normal hidden dense layers respectively with 1024, 512, 256, 128, 128 neurons, and a $(K + N)$-dimensional output layer, which are the key functional layers of DNN-CSI. The exponential linear units (ELU) is used as the activation functions of the hidden layers with

$$
\text{ELU}(z) = \begin{cases} z, & \text{if } z > 0, \\ \alpha(\exp(z) - 1), & \text{otherwise, } z \le 0. \end{cases} \tag{10.25}
$$

It should be noted that we add the Batch-Normalization layers and Dropout layers between two normal dense layers to accelerate the training speed, avoid gradients vanishing, as well as prevent overfitting of the DNN. To be specific, this fully connected feedforward DNN-CSI is with 10% of random dropout of neurons for the hidden layer 1 to hidden layer 4 and 5% of random dropout for the hidden layer 5 during each training epoch, so as to avoid overfitting.

10.4.2 Location-Only Learning Architecture

For CSI-based deep learning architecture, it is required to obtain CSI in advance via pilot channel estimation for time division duplex (TDD) systems. Due to the random characteristics of wireless channels, it is quite difficult to estimate CSI accurately especially considering pilot contamination. The use of DNN offers a unique opportunity to develop techniques that can remove CSI estimation, such as ViterbiNet (Shlezinger et al. 2020) and ensemble learning (Zhang et al. 2022) are designed for CSI-Free symbol detection.

Here for the IRS-aided MEC scenario (b), the channel coefficients and the solutions of $\{\phi, \mathbf{a}, \mathbf{W}\}$ are highly correlated to the locations of UEs, i.e., $\{(x_n, y_n)\}_{n \in \mathcal{N}}$. Hence, one may use UEs' locations as the input feature of DNNs to obtain $\{\phi, \mathbf{a}, \mathbf{W}\}$, which provides a promising way to circumvent CSI estimation prior task offloading and achieve lighter-weight implementations. In fact, UEs' location information is less random and its dimension size is much smaller than that of the CSI, which is much easier and more convenient to obtain in

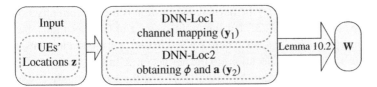

Figure 10.4 The architecture for obtaining the solutions of $\{\phi, \mathbf{a}, \mathbf{W}\}$ with the location-only (CSI-free) DNN-Loc1 and DNN-Loc2.

practice.[1] In Fig. 10.4, we present a location-only (CSI-free) deep learning architecture, where the prior knowledge in Lemma 10.2 indicating the relationship between $\{\mathbf{W}\}$ with $\{\phi, \mathbf{a}\}$ and the channel coefficients, is fully considered to reduce training overheads. Here, two DNNs are constructed, where DNN-Loc1 aims at calculating the channel mapping between UEs' locations and channel coefficients with $2N$-dimensional input feature $\mathbf{z}$ and I-dimensional output feature $\mathbf{y}_1$ while DNN-Loc2 focuses on obtaining $\{\phi, \mathbf{a}\}$ with input feature $\mathbf{z}$ and $(K + N)$-dimensional output feature $\mathbf{y}_2$. Then, the optimal receive beamforming matrix $\mathbf{W}$ can be easily calculated via Lemma 10.2.

The constructed DNN-Loc1 and DNN-Loc2 are also feedforward networks, where both have 5 normal hidden dense layers. There are respectively 512, 512, 256, 128, 256 neurons for the five hidden layers of DNN-Loc1, and 512, 256, 128, 64, 32 neurons for layers of DNN-Loc2. Similarly, the layers of Batch-Normalization and Dropout are also utilized for DNN-Loc1 and DNN-Loc2 with same dropout policy of DNN-CSI. Here, the sigmoid activation function is not only leveraged at the output layer of DNN-Loc2 for obtaining $\{\phi, \mathbf{a}\}$ but also at that of DNN-Loc1 for channel mapping where the output data samples are scaled into the range of $[0, 1]$ with the MinMaxScaler in Tensorflow.

10.4.3 Input Feature Uncertainty

In Sections 10.4.1 and 10.4.2, we assume that the input features to the constructed DNNs, i.e., the input vector $\mathbf{x}$ of CSI in Fig. 10.3 and the input vector $\mathbf{z}$ of UEs' locations in Fig. 10.4, are perfectly known, which is quite ideal. However, the obtained CSI and UEs' locations are usually imperfect in practice due to the deviation of channel estimation and GPS/Wi-Fi localization.

In this section, we focus on a more practical scenario where the input features of CSI and UEs' locations are corrupted with uncertainty. For the input vector $\mathbf{x}$ of CSI, the corresponding corrupted counterpart can be modeled as $\hat{\mathbf{x}} = \mathbf{x} + \triangle\mathbf{x}$,

---

1 In our considered scenarios, we assume that each UE is installed with an advanced global positioning system (GPS) module for outdoor localization and is capable to apply the Wi-Fi round-trip time technology and standards for indoor localization with high accuracy.

where $\triangle \mathbf{x} \sim \mathcal{N}(0, \sigma_{\triangle \mathbf{x}}^2)$ following the normal distribution is the random offset of the achieved CSI to the perfect CSI. For the input vector $\mathbf{z}$ of UEs' locations, the corrupted counterpart is $\hat{\mathbf{z}} = \mathbf{z} + \triangle \mathbf{z}$, where $\triangle \mathbf{z} \sim \mathcal{N}(0, \sigma_{\triangle \mathbf{z}}^2)$ is the random offset (in meter) of the achieved UEs' locations to the perfect ones. In this practical case with uncertain input features, the CSI-based DNN and the location-only DNNs are trained and tested based on the corrupted input information of CSI and UEs' locations, so as to verify the effectiveness and robustness of the two constructed deep learning architectures.

10.4.4 Comparison Between the CSI-Based and CSI-Free Learning Architectures

The training (including validation) and testing of the constructed DNNs can easily be implemented based on the platforms of Tensorflow and Keras via supervised learning, and the related parameters are given in Table 10.1. More details about the DNN setting can be found in (Hu et al. 2021).

Furthermore, in Table 10.2, we present the values of trainable parameters, training time, testing time, and average inference time of the constructed DNN-CSI, DNN-Loc1, DNN-Loc2 for the case with $M = 8$, $N = 8$, $K = 24$ in both situations with perfect and imperfect input features. It is shown that the time overhead for training and testing the constructed DNNs with imperfect input features is slightly larger than that with perfect input features due to the fact that more complicated input-output mappings need to be figured out. Note that the total required training parameters of the two location-only DNNs are considerably less (nearly half) than that of the DNN-CSI. Also, the training, testing, and inference overhead can be significantly reduced by leveraging the location-only data-driven method, which is verified by the much less required training, testing, and average inference time of DNN-Loc1, DNN-Loc2 shown in Table 10.2 that are only around 60% of those for DNN-CSI in both cases. Hence, it is of great benefit to leverage the location-only

Table 10.1 Parameters related to training and testing of the DNNs.

Parameters	Values
Number of training samples	200,000
Number of testing samples	10,000
Batch size	128
Number of epoches	1000
Initial learning rate	0.001
Validation split	0.2

Table 10.2 Processing time of the overviewed algorithms.

Parameters	DNN-CSI	DNN-Loc1	DNN-Loc2
Trainable parameters	1,632,288	702,208	186,304
Training samples	$(\mathbf{x}, \mathbf{y})$	$(\mathbf{z}, \mathbf{y}_1)$	$(\mathbf{z}, \mathbf{y}_2)$
Training time	5.7426 h	3.3504 h	1.3979 h
Testing time	0.3883 s	0.2418 s	0.1025 s
Average inference time	38.83 µs	24.18 µs	10.25 µs
Training samples	$(\hat{\mathbf{x}}, \mathbf{y})$	$(\hat{\mathbf{z}}, \mathbf{y}_1)$	$(\hat{\mathbf{z}}, \mathbf{y}_2)$
Training time	6.0345 h	3.5123 h	1.4862 h
Testing time	0.4015 s	0.2568 s	0.1156 s
Average inference time	40.15 µs	25.68 µs	11.56 µs
BCD running time		28.7 s	

deep learning architecture in situations where it can provide satisfactory inference solutions, such as in the scenario with strong LoS direct links between UEs and AP.

10.4.5 Complexity Reduction via Learning

In Table 10.2, the average running time of the BCD algorithm for one realization is also given for comparison, i.e., 28.7 seconds/s, nearly 10^6 times to those of the two deep learning methods that the CSI-based and the location-only DNNs only require 38.83 microseconds/µs (40.15 µs) and 24.18 µs (25.68 µs) with perfect (imperfect) input features. This result effectively validates the ability of the constructed deep learning architectures in reducing the computational complexity/running time for providing lightweight online inference solutions.

In Fig. 10.5, we further show more details of the average running time for one realization of the overviewed BCD optimization algorithm, the CSI-based and location-only deep learning architectures versus the number of UEs (N), in both scenarios with NLoS and LoS direct links. It is noticeable that the average running time, i.e., the average inference time, of both two deep learning architectures are always quite small in all scenarios (measured in microsecond-µs) and increase slightly with N, especially compared with those of the BCD algorithm (measured in second-s) that increase quite considerably with N. Specifically, no matter with or without input uncertainty, the average running time of the CSI-based learning approach is nearly millionth and the location-only learning method is usually less than millionth of that required by the BCD optimization solution,

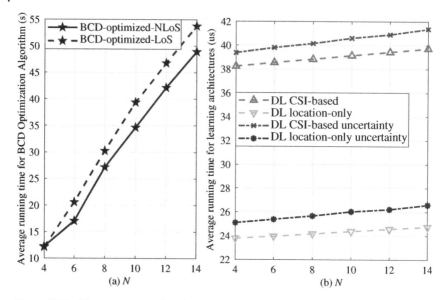

Figure 10.5 The average running time versus the number of UEs (*N*): (a) optimization-based solutions and (b) learning-based solutions.

which indicates that lightweight online implementations are available by the data-driven architectures via periodical training.

10.5 Comparative Evaluation Results

In this section, simulation results are given to verify the effectiveness and performance improvement of the overviewed BCD optimization algorithm as well as the CSI-based and location-only deep learning architectures. In addition, the effectiveness and robustness of the two considered deep learning methods to the corrupted input features are also validated by simulations.

A three-dimensional (3D) Euclidean coordinate system is adopted with the locations of the AP as $(0, y_{AP}, H_0)$, the IRS as $(x_R, 0, H_R)$ and UE $n \in \mathcal{N}$ as $(x_n, y_n, 0)$, all measured in meters (m) as shown in Fig. 10.1. The IRS employs with a uniform rectangular array (URA) of $K = K_x K_z$ reflecting elements, while the M-antenna AP is equipped with a uniform linear array (ULA). We assume that N ground UEs are randomly distributed in a square serving area with four vertices at horizontal locations of $(x_s, 0)$, $(x_s + D, 0)$, (x_s, D), and $(x_s + D, D)$. The Rician fading channel model is adopted to account for both LoS and non-LoS (NLoS) components of all channels. More details about the channel setting can be found

Table 10.3 Simulation parameters.

Symbol	Value	Symbol	Value
x_s, D	20 m, 40 m	$(0, y_{AP}, H_0)$	$(0,20,5)$ m
$(x_R, 0, H_R)$	$(40,0,20)$ m	T	5 s
N, M	8, 8,	$K = K_x K_z$	$24 = 8 \times 3$
E_n $(n \in \mathcal{N})$	10 J	C_n $(n \in \mathcal{N})$	200 cycles/bit
κ_n $(n \in \mathcal{N})$	10^{-28}	B	40 MHz
σ^2	-60 dBm	$\sigma_{\triangle x}, \sigma_{\triangle z}$	0.001,1

in (Hu et al. 2021). The other basic simulation parameters are listed in Table 10.3 unless specified otherwise.

10.5.1 Scenario Without LoS Direct Links

In this section, the LoS paths of UEs' direct links are blocked as shown in the scenario (a) of Fig. 10.2, which is quite common in urban areas. Numerical results for the overviewed BCD optimization solution ("BCD-optimized solution"), the CSI-based deep learning solution ("deep learning [DL] CSI-based") and its counterpart with input uncertainty ("DL CSI-based uncertainty") are presented in comparison with three other existing solutions, where the "Direct Offloading-No IRS" scheme corresponds to the case without deploying IRS, the "ZF receive beamforming" scheme considers the ZF beamforming for detecting UEs' signals, and the "Equal energy allocation" scheme is operated by equally allocating the UEs' energy for local computing and computation offloading.

In Fig. 10.6, we first show the TCTB of all the considered schemes w.r.t. the UEs' uniform energy budget, i.e., $E = E_n$ for $n \in \mathcal{N}$. From this figure, we can observe that the TCTB curves of all the schemes increase with E, which coincides with the intuition that more computation task-input bits can be completed if the UEs are endowed with more energy. It is clear to see that significant performance improvement can be achieved by the BCD-optimized solution, verifying the great benefits of deploying the aided IRS, also jointly optimizing the IRS coefficients, the receive beamforming, and the UEs' energy allocation. It is confirmed that the BCD algorithm provides 26% improvement in TCTB over the direct offloading method without the assistance of IRS. More importantly, the CSI-based deep learning method can achieve a performance very close to the BCD optimization solution in both cases with perfect and imperfect CSI, clearly demonstrating that the CSI-based deep learning architecture can effectively emulate the BCD optimization algorithm, with a much reduced online complexity and high robustness.

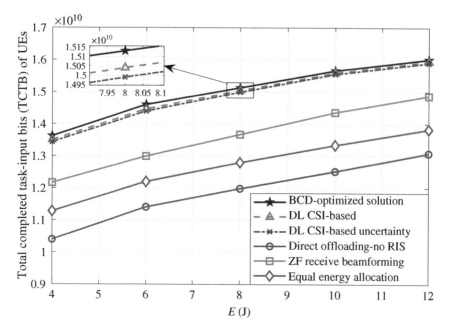

Figure 10.6 The TCTB of UEs versus the uniform energy budget $E = E_n, n \in \mathcal{N}$.

Figure 10.7 describes the effects of the number of UEs, i.e., N, on the system performance of TCTB. Here, the effectiveness of the CSI-based deep learning architecture is further verified in the scenarios with different number of users considering both perfect and imperfect input CSI, which also demonstrates its robustness and generalizability. It is easy to note that the BCD-optimized solution as well as the CSI-based deep learning schemes with and without input uncertainty have strong robustness in dealing with the cases when serving more UEs than the number of the AP's antennas. These cases are particularly relevant to massive connectivity scenarios. Instead of degrading the performance like the compared three methods, the overviewed solutions in this chapter are able to provide even better performance as N becoming larger than M through effectively designing the receive beamforming and UE's energy allocation.

10.5.2 Scenario with Strong LoS Direct Links

In this section, we focus on implementing the mentioned schemes in the scenario where strong LoS direct links exist for UEs in the considered serving area, which is exactly the scenario (b) of Fig. 10.2. This scenario is practically relevant when considering the suburb districts. In this scenario, the location-only deep learning architecture ("DL location-only") can be leveraged to effectively mimic the

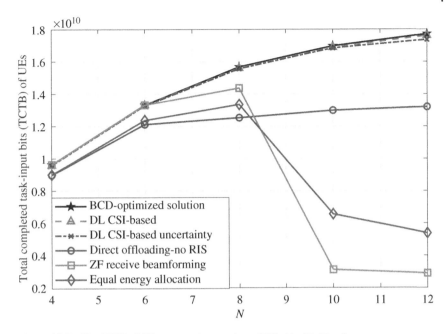

Figure 10.7 The TCTB of UEs versus the number of UEs N with $M = 8$.

mapping of the BCD algorithm. In addition, the performance of its counterpart solution with uncertain input of UEs' locations denoted as "DL location-only uncertainty" is also given in this section.

The TCTB performance of all the considered schemes versus the UEs' uniform energy budget E is shown in Fig. 10.8. Obviously, both the CSI-based and the location-only deep learning methods can achieve excellent system performance, in both cases with and without input uncertainty. Even though the DL location-only solution is slightly worse than the DL CSI-based solution, it is far more superior than the other compared approaches. Note that the TCTB gap between these two DL schemes becomes slightly smaller in the case with uncertain input features. More importantly, the UEs' locations are quite easier to obtain compared with the related CSI, which makes it more flexible to achieve online implementation through the location-only deep learning architecture. Also from this figure, we can clearly see that the zero-forcing (ZF) receive beamforming scheme is even worse than the scheme without IRS and the scheme of Equal energy allocation, due to the fact that the effective channels of different UEs may be highly correlated in the considered scenario with LoS direct links. This phenomenon further indicates the importance of effectively designing the AP's receive beamforming and managing the UEs' energy budgets.

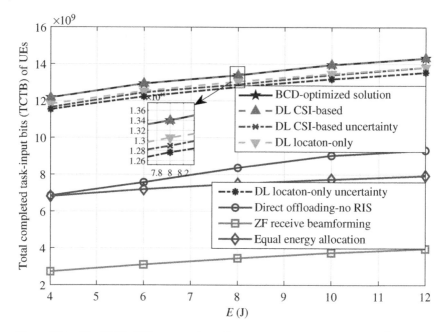

Figure 10.8 The TCTB of UEs versus the uniform energy budget $E = E_n, n \in \mathcal{N}$.

10.6 Conclusions

In this chapter, a typical IRS-aided MEC system with multiplexing computation offloading has been investigated, where the IRS constructively reflects UEs' offloaded input-data-bearing signals to improve UEs' computation efficiency. A number of design approaches from optimization to learning were presented to solve the TCTB maximization problem. First, we overviewed an optimization-based approach in the form of a three-step BCD algorithm with guaranteed convergence, and then two deep learning architectures based on CSI and UEs' locations are constructed to mimic the mapping of the BCD algorithm with a considerable complexity reduction. The simulation results have confirmed that significant performance improvement can be achieved by utilizing the BCD algorithm comparing with some other existing schemes. For both cases with perfect and imperfect input features, the CSI-based learning architecture can always approach the performance of the BCD algorithm, while the more practical location-only learning architecture can provide satisfactory performance when strong LoS direct links exist between UEs and AP.

Bibliography

An, L.T.H. and Tao, P.D. (2005). The DC (difference of convex functions) programming and DCA revisited with DC models of real world nonconvex optimization problems. *Annals of Operations Research* 133 (1): 23–46.

Bai, T., Pan, C., Deng, Y. et al. (2020). Latency minimization for intelligent reflecting surface aided mobile edge computing. *IEEE Journal on Selected Areas in Communications* 38 (11): 2666–2682.

Basar, E., Di Renzo, M., De Rosny, J. et al. (2019). Wireless communications through reconfigurable intelligent surfaces. *IEEE Access* 7: 116753–116773.

Bi, S. and Zhang, Y.J. (2018). Computation rate maximization for wireless powered mobile-edge computing with binary computation offloading. *IEEE Transactions on Wireless Communications* 17 (6): 4177–4190.

Dong, R., She, C., Hardjawana, W. et al. (2019). Deep learning for hybrid 5G services in mobile edge computing systems: Learn from a digital twin. *IEEE Transactions on Wireless Communications* 18 (10): 4692–4707.

Goodfellow, I., Bengio, Y., and Courville, A. (2016). *Deep learning*. MIT press.

Hu, X., Wong, K.-K., and Yang, K. (2018). Wireless powered cooperation-assisted mobile edge computing. *IEEE Transactions on Wireless Communications* 17 (4): 2375–2388.

Hu, X., Wang, L., Wong, K.-K. et al. (2019a). Edge and central cloud computing: A perfect pairing for high energy efficiency and low-latency. *IEEE Transactions on Wireless Communications* 19 (2): 1070–1083.

Hu, X., Wong, K.-K., Yang, K., and Zheng, Z. (2019b). UAV-assisted relaying and edge computing: Scheduling and trajectory optimization. *IEEE Transactions on Wireless Communications* 18 (10): 4738–4752.

Hu, X., Wong, K.-K., and Zhang, Y. (2020). Wireless-powered edge computing with cooperative UAV: Task, time scheduling and trajectory design. *IEEE Transactions on Wireless Communications* 19 (12): 8083–8098.

Hu, X., Masouros, C., and Wong, K.-K. (2021). Reconfigurable intelligent surface aided mobile edge computing: From optimization-based to location-only learning-based solutions. *IEEE Transactions on Communications* 69 (6): 3709–3725.

Hua, S., Zhou, Y., Yang, K. et al. (2021). Reconfigurable intelligent surface for green edge inference. *IEEE Transactions on Green Communications and Networking* 5 (2): 964–979.

Huang, C., Mo, R., and Yuen, C. (2020). Reconfigurable intelligent surface assisted multiuser MISO systems exploiting deep reinforcement learning. *IEEE Journal on Selected Areas in Communications* 38 (8): 1839–1850.

Mao, Y., You, C., Zhang, J. et al. (2017). A survey on mobile edge computing: the communication perspective. *IEEE Communication Surveys & Tutorials* 19 (4): 2322–2358.

Oshea, T. and Hoydis, J. (2017). An introduction to deep learning for the physical layer. *IEEE Transactions on Cognitive Communications and Networking* 3 (4): 563–575.

Sardellitti, S., Scutari, G., and Barbarossa, S. (2015). Joint optimization of radio and computational resources for multicell mobile-edge computing. *IEEE Transactions on Signal and Information Processing over Networks* 1 (2): 89–103.

Shlezinger, N., Farsad, N., Eldar, Y.C., and Goldsmith, A.J. (2020). ViterbiNet: a deep learning based Viterbi algorithm for symbol detection. *IEEE Transactions on Wireless Communications* 19 (5): 3319–3331.

Tran, T.X. and Pompili, D. (2018). Joint task offloading and resource allocation for multi-server mobile-edge computing networks. *IEEE Transactions on Vehicular Technology* 68 (1): 856–868.

Wang, K.-Y., So, A.M.-C., Chang, T.-H. et al. (2014). Outage constrained robust transmit optimization for multiuser MISO downlinks: Tractable approximations by conic optimization. *IEEE Transactions on Signal Processing* 62 (21): 5690–5705.

Wu, Q. and Zhang, R. (2019a). Towards smart and reconfigurable environment: Intelligent reflecting surface aided wireless network. *IEEE Communications Magazine* 58 (1): 106–112.

Wu, Q. and Zhang, R. (2019b). Intelligent reflecting surface enhanced wireless network via joint active and passive beamforming. *IEEE Transactions on Wireless Communications* 18 (11): 5394–5409.

Yang, K., Jiang, T., Shi, Y., and Ding, Z. (2020). Federated learning via over-the-air computation. *IEEE Transactions on Wireless Communications* 19 (3): 2022–2035.

You, C., Huang, K., Chae, H., and Kim, B.-H. (2016). Energy-efficient resource allocation for mobile-edge computation offloading. *IEEE Transactions on Wireless Communications* 16 (3): 1397–1411.

Zhang, W., Wen, Y., Guan, K. et al. (2013). Energy-optimal mobile cloud computing under stochastic wireless channel. *IEEE Transactions on Wireless Communications* 12 (9): 4569–4581.

Zhang, J., Masouros, C., and Huang, Y. (2022). CSI-free geometric symbol detection via semi-supervised learning and ensemble learning. *IEEE Transactions on Communications* 70 (11): 7265–7278.

Zhou, F. and Hu, R.Q. (2020). Computation efficiency maximization in wireless-powered mobile edge computing networks. *IEEE Transactions on Wireless Communications* 19 (5): 3170–3184.

11

Interference Nulling Using Reconfigurable Intelligent Surface

Tao Jiang[1], Foad Sohrabi[2], and Wei Yu[1]

[1] *The Edward S. Rogers Sr. Department of Electrical and Computer Engineering, University of Toronto, Toronto, Ontario, Canada*
[2] *Nokia Bell Labs, Murray Hill, NJ, USA*

11.1 Introduction

Reconfigurable intelligent surface (RIS) is a class of emerging devices (Di Renzo et al. 2019; Wu et al. 2021) that are capable of intelligently reconfiguring the wireless propagation channel by altering the phases of the reflected radio signals in a controlled manner. The RIS is typically made of many passive reconfigurable elements. It has very low energy consumption and can be easily integrated into the existing wireless communication systems. It provides a low-cost way of adaptively reengineering the radio propagation channel and is envisioned as a key technology for the next generation of wireless communication networks.

Most of the literature, (e.g., Huang et al. 2019; Wu and Zhang 2019), investigate the passive beamforming capability of the RIS. The idea is to reconfigure the passive elements of the RIS so that the incoming electromagnetic radiation is refocused with a narrow beam toward the intended receiver so that the signal-to-noise ratio (SNR) of the overall communication link is improved.

This chapter presents a new potential use of the RIS for a wireless environment with multiple transmitter-receiver pairs. The idea is to reconfigure the passive elements of the RIS so that the interferences between these multiple links can be reduced or even eliminated. In other words, the RIS is used to reengineer a multiuser wireless communication channel so that all the links become effectively interference-free.

Consider a scenario in which K single-antenna transceiver pairs communicate independently while utilizing the same set of time and frequency resources.

Intelligent Surfaces Empowered 6G Wireless Network, First Edition.
Edited by Qingqing Wu, Trung Q. Duong, Derrick Wing Kwan Ng, Robert Schober, and Rui Zhang.
© 2024 The Institute of Electrical and Electronics Engineers, Inc. Published 2024 by John Wiley & Sons, Inc.

When the transmitters and the receivers are located close to each other, the interferences between these independent transmissions are typically the main channel impairments. The main insight of this chapter is that as long as the direct channels between the transmitters and the receivers are not too strong as compared to the reflected path through the RIS, it is possible to reconfigure the RIS in such a way that the interferences are completely eliminated in the overall effective channels. This is possible if the RIS has a sufficiently large number of elements – in the order of $O(K^2)$.

In conventional wireless communication deployment scenarios, interference is managed by scheduling nearby transmit-receive links into different time or frequency resource blocks. This leads to traditional deployment strategies such as frequency reuse but is not the most spectrally efficient. Modern wireless systems are often designed with full frequency reuse. To deal with interference, modern wireless systems utilize spatial domain techniques based on multiple-input multiple-output (MIMO) technology (e.g., zero-forcing beamforming) to spatially separate the communication links. The main result of this chapter is that instead of relying on multiple antennas either at the transmitter or at the receiver to separate the interfering transmission pairs, it is possible to use a sufficiently large intelligent reflector to modify the $K \times K$ wireless channel so that it becomes interference-free, even for the case in which all the transmitters and the receivers are equipped with a single antenna only.

From a theoretical perspective, it is possible to show that under a rectangular array model of RIS, if the number of elements is sufficiently large and if the direct channels between the transceivers are blocked and the reflection channels between the transceivers and the RIS have line-of-sight, then it is always possible to tune the phase shifts of the RIS so that the effective overall channel is interference-free. From an algorithmic perspective, when the channel state information (CSI) is fully available, it is possible to use an alternating projection algorithm to efficiently find an interference-nulling solution numerically. Such an algorithm can be used either with or without the direct channels between the transceivers being present. Numerical results show that the alternating projection algorithm can find an interference-nulling solution if the number of RIS elements is slightly greater than $2K(K-1)$ and the direct links are weak.

The above results rely on the availability of perfect CSI. However, obtaining an accurate estimate of the CSI in an RIS system would in general require the number of pilots to scale with $O(KN)$, where N is the number of RIS elements. This is a huge overhead, because N is typically large. Poor channel estimation can lead to a solution with high interference power. The second part of the chapter describes a learning-based method for finding the RIS configuration that can be easily incorporated into the conventional channel estimation and optimization framework with much reduced pilot training overhead. Specifically, using a few received pilots, the scheme first uses a linear minimum mean square estimator (LMMSE) to estimate the channel, then utilizes a deep neural network (DNN) to learn an initial point for the alternating projection algorithm from the estimated channel. This proposed method is based on the observation that the RIS optimization problem is highly nonconvex and different initial points can lead to solutions with different interference powers. As compared to the conventional methods using random initial points for the alternating projection algorithm, simulations show that a DNN can learn a better initial point that can lead to a solution with significantly lower interference power at the same pilot length.

The interference cancelation capability of RIS has been investigated for device-to-device (D2D) communications in the cellular network (Fu et al. 2021; Chen et al. 2021; Ji et al. 2022; Abrardo et al. 2021). In that setting, the RIS is used to mitigate interference from the D2D communications to the cellular communications. Most existing works also assume that perfect CSI is available when designing the reflection coefficients. This chapter studies the interference-nulling capability of RIS in a K-user interference channel and explicitly studies the impact of imperfect CSI.

11.2 System Model

Consider an RIS-assisted K-user interference environment, in which K single-antenna transceiver pairs communicate simultaneously using the same frequency and time resources. As shown in Fig. 11.1, an RIS is deployed between the transmitters and receivers. The idea is to tune the phase shifts of the RIS to create an interference-free communication network. We assume a block-fading channel model where the channels remain constant in one coherence block but change independently across different coherence blocks. Let $t_j \in \mathbb{C}^N$ denote the channel from the jth transmitter to the RIS, and $r_k^\top \in \mathbb{C}^{1 \times N}$ denote the channel from the RIS to the kth receiver. The RIS is modeled as an array of phase shifters. Let $\omega_i \in (0, 2\pi]$ be the phase shift of the ith element of the RIS, then the reflection coefficients of the RIS can be denoted by $\theta = [e^{j\omega_1}, \ldots, e^{j\omega_N}]^\top \in \mathbb{C}^N$. The direct channel from the jth transmitter to the kth receiver without the RIS reflection is denoted as $d_{kj} \in \mathbb{C}$. The received signal at the kth receiver can be represented as

$$y_k = \sum_{j=1}^{K} \left(r_k^\top \operatorname{diag}(\theta) t_j + d_{kj} \right) x_j + n_k, \tag{11.1}$$

RIS reflection coefficients $\boldsymbol{\theta} \in \mathbb{C}^N$

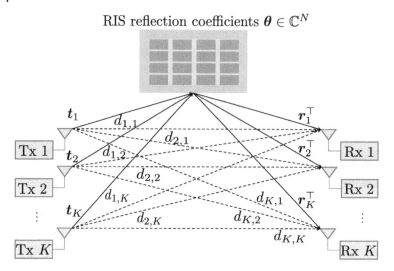

Figure 11.1 System model of an RIS-assisted K-user interference environment.

where $x_j \in \mathbb{C}$ is the transmitted signal from the jth transmitter and intended to the jth receiver, and $n_k \sim \mathcal{CN}(0, \sigma^2)$ is the additive Gaussian noise. The transmitted signal is subject to a power constraint, i.e., $\mathbb{E}\left[|x_j|^2\right] = p_j$, where p_j is the transmit power of the jth transmitter.

With these notations, the theoretical achievable rate of the kth transceiver pair can now be written as

$$R_k = \log_2\left(1 + \frac{p_k|\boldsymbol{r}_k^{\mathsf{T}}\operatorname{diag}(\boldsymbol{\theta})\boldsymbol{t}_k + d_{k,k}|^2}{\sum_{j\neq k}p_j|\boldsymbol{r}_k^{\mathsf{T}}\operatorname{diag}(\boldsymbol{\theta})\boldsymbol{t}_j + d_{k,j}|^2 + \sigma^2}\right). \tag{11.2}$$

The chapter focuses on designing the reflection coefficients $\boldsymbol{\theta}$ to create K interference-free channels for the corresponding K transceiver pairs. This achieves the maximum degrees-of-freedom K for the system.

11.3 Interference Nulling via RIS

To achieve the maximum degrees-of-freedom of the overall K-user system, we need to find a configuration of the RIS reflection coefficients $\boldsymbol{\theta}$ such that all the interference is nulled and the desired links have positive effective channel strength, i.e.,

$$\boldsymbol{r}_k^{\mathsf{T}}\operatorname{diag}(\boldsymbol{\theta})\boldsymbol{t}_j + d_{k,j} = 0, \quad \forall j \neq k, \ \forall k = 1, \dots, K, \tag{11.3a}$$

$$\boldsymbol{r}_k^{\mathsf{T}}\operatorname{diag}(\boldsymbol{\theta})\boldsymbol{t}_k + d_{k,k} \neq 0, \quad \forall k = 1, \dots, K. \tag{11.3b}$$

Let $a_{k,j} \triangleq \text{diag}(t_j) r_k$ denote the cascaded channel from transmitter j to receiver k, the conditions in (11.3) can be rewritten in the following form:

$$a_{k,j}^\top \theta + d_{k,j} = 0, \quad \forall j \neq k, \ \forall k = 1, 2, \ldots, K, \tag{11.4a}$$

$$a_{k,k}^\top \theta + d_{k,k} \neq 0, \quad \forall k = 1, 2, \ldots, K. \tag{11.4b}$$

Since the channel realizations are random, the conditions in (11.4b) hold almost surely. Thus, we focus on seeking a feasible solution θ to the conditions in (11.4a). Note that the RIS coefficients are subject to the constraint $\theta = [e^{j\omega_1}, \ldots, e^{j\omega_N}]$, so finding a feasible set of phase shifts $\{\omega_1, \ldots, \omega_N\}$ is a highly nontrivial problem.

11.3.1 Feasibility of Interference Nulling

The conditions in (11.4a) are not always feasible. If the RIS reflection coefficients are treated as unconstrained variables, the number of complex variables N should be greater than the number of complex linear equations $K(K-1)$ to guarantee that the set of Eq. (11.4a) has a solution. This would be the case if both the amplitude and phase can be controlled arbitrarily for each RIS element. If only the phase of the RIS can be configured, that is, the RIS coefficient vector is subject to the constraint $\theta = [e^{j\omega_1}, \ldots, e^{j\omega_N}]$, the feasibility problem becomes more complicated. By simply counting the number of equations and the number of variables in the real field, we can see that there are N real variables ω_i in θ and $2K(K-1)$ nonlinear real equations of ω_i's in (11.4a). Therefore, intuitively we would need $N \geq 2K(K-1)$ to ensure that there exists a feasible solution to (11.4a).

However, it is not easy to establish the above results rigorously since the equations of the phase shifts ω_i's are nonlinear. Further, even if the existence of the interference-nulling solutions can be guaranteed, finding a solution is still computationally challenging. Nevertheless, later in this chapter, we show that the intuition of requiring $N \geq 2K(K-1)$ is correct and present a computationally efficient alternating projection algorithm, which can be used to find an interference nulling solution when N is slightly larger than $2K(K-1)$. Before going into the algorithm, we first present a theoretical result that provides a sufficient condition for the feasibility of (11.4a) for a specific RIS model and specific deployment scenario. This theoretical result suggests that for an RIS-assisted network with negligible direct channels and line-of-sight transceiver-RIS channels, the interference can be completely nulled out by a uniform rectangular RIS with a sufficiently large number of reflective elements.

Specifically, consider an $N_1 \times N_2$ uniform rectangular array RIS, which has N_1 elements per row (horizontal direction) and N_2 elements per column (vertical direction). Denote $\varrho \in [-\frac{\pi}{2}, \frac{\pi}{2}]$, $\phi \in [-\frac{\pi}{2}, \frac{\pi}{2}]$ as the azimuth angle and the elevation angle of arrival, respectively. We can write the nth element of the RIS steering vector as follows (Björnson and Sanguinetti 2021):

$$[\psi(\varrho, \phi)]_n = e^{j\frac{2\pi}{\lambda}[i_1(n)b_1 \sin(\varrho)\cos(\phi)+i_2(n)b_2 \sin(\phi)]}, \tag{11.5}$$

where b_1 and b_2 are the horizontal and vertical spacings between the RIS elements, and $i_1(n) = \text{mod}(n - 1, N_1)$ and $i_2(n) = \lfloor (n - 1)/N_2 \rfloor$ denote the horizontal index and vertical index of element n, respectively. The channel vector between the RIS and the kth transmitter can then be written as

$$t_k = \beta_k^t \psi(\varrho_k^t, \phi_k^t), \tag{11.6}$$

where β_k^t denotes the pathloss between the kth transmitter and the RIS, and ϱ_k^t and ϕ_k^t are the corresponding azimuth and elevation angles. Similarly, we represent the channel between RIS and the kth receiver as $r_k = \beta_k^r \psi(\varrho_k^r, \phi_k^r)$.

When the direct channels can be ignored, the interference nulling conditions (11.4) become

$$a_{k,j}^{\mathsf{T}}\theta = 0, \quad \forall k = 1, 2, \ldots, K, \ \forall j \neq k, \tag{11.7a}$$

$$a_{k,k}^{\mathsf{T}}\theta \neq 0, \quad \forall k = 1, 2, \ldots, K, \tag{11.7b}$$

where $\theta = [e^{j\omega_1}, \ldots, e^{j\omega_N}]$. In order to find a sufficient condition for the feasibility of (11.7), we use the following lemma, which ensures the existence of a polynomial with unit modulus coefficients for given roots on the complex unit circle.

Lemma 11.1 *(Newman and Giroux (1990))* *Given $z_1, \ldots, z_n$ on the complex unit circle $C = \{x \in \mathbb{C} : |x| = 1\}$, there exists a polynomial f of degree $\sum_{i=1}^{n} 4^{i-1}$ with unit modulus coefficients such that the points $z_1, \ldots, z_n$ are the only zeros of f on C.*

Now, we establish the following sufficient condition for the feasibility of interference nulling criterion (11.7).

Theorem 11.1 *For an $N_1 \times N_2$ rectangular RIS, assuming that the channels between the RIS and the users are given by (11.6), there exists a feasible solution to the interference nulling conditions in (11.7) if either $N_1 = \sum_{k=1}^{K(K-1)} 4^{k-1}$ or $N_2 = \sum_{k=1}^{K(K-1)} 4^{k-1}$.*

Proof: We begin with the case $N_2 = 1$, which corresponds to the uniform linear array case. In this case, the channel between the RIS and transmitter j can be written as

$$t_j = \beta_j^t [1, \ldots, e^{j\frac{2\pi b_2(N_1-1)}{\lambda} \sin(\phi_j^t)}]^{\mathsf{T}}, \tag{11.8}$$

and the channel between the RIS and receiver k can be written as

$$r_k = \beta_k^r [1, \ldots, e^{j\frac{2\pi b_2(N_1-1)}{\lambda} \sin(\phi_k^r)}]^{\mathsf{T}}. \tag{11.9}$$

Thus, we have

$$\boldsymbol{a}_{k,j} = \beta_k^r \beta_j^t [1, \ldots, e^{j\frac{2\pi b_2(N_1-1)}{\lambda}(\sin(\phi_j^t)+\sin(\phi_k^r))}]^\top \tag{11.10a}$$

$$\triangleq \beta_k^r \beta_j^t [1, z_{k,j}, z_{k,j}^2, \ldots, z_{k,j}^{N_1-1}]^\top, \tag{11.10b}$$

where $z_{k,j} = e^{j\frac{2\pi b_2}{\lambda}(\sin(\phi_j^t)+\sin(\phi_k^r))}$.

Let $f(z)$ denote a polynomial of degree N_1 with unit modulus coefficients as follows:

$$f(z) = \theta_1 + \theta_2 z + \theta_3 z^2 + \cdots + \theta_{N_1} z^{N_1-1}, \tag{11.11}$$

where $|\theta_i| = 1$, $\forall i$. Observe that the interference nulling conditions in (11.7a) and (11.7b) can now be expressed as below with the coefficients of $f(z)$ playing the role of RIS phase shifts:

$$f(z_{k,j}) = 0, \quad \forall k = 1, \ldots, K, \ \forall j \neq k \tag{11.12a}$$

$$f(z_{k,k}) \neq 0, \quad \forall k = 1, \ldots, K. \tag{11.12b}$$

Thus, the original problem of finding RIS phase shifts for interference nulling is now transformed into the problem of finding a polynomial with unit modulus coefficients as in (11.11) such that $z_{k,j}$'s are the roots, while the polynomial does not vanish at $z_{k,k}$'s. By Lemma 11.1, we know such a polynomial exists if $N_1 = \sum_{k=1}^{K(K-1)} 4^{k-1}$.

To generalize the above result to the rectangular array case with $N_2 > 1$, we make the argument that a rectangular uniform array can be viewed as N_2 rows of the uniform linear array of size $N_1 \times 1$. Thus, if we set all the rows of the RIS to have the same reflection coefficients that null all the interference, which can be achieved if $N_1 = \sum_{k=1}^{K(K-1)} 4^{k-1}$, the entire rectangular array would also achieve the interference nulling condition. Mathematically, this means that the same reflection coefficients are used N_2 times to produce a zero-forcing solution for all N_2 sub-vectors of $\boldsymbol{\psi}(\varrho, \phi)$ as in (11.5).

Finally, if we exchange the roles of N_1 and N_2, it is easily seen that the interference nulling conditions are also achievable if $N_2 = \sum_{k=1}^{K(K-1)} 4^{k-1}$. This completes the proof. □

Theorem 11.1 is a theoretical result showing that an interference nulling solution must exist if the number of RIS elements N scales as an exponential function of K. But in practice, we empirically find that by using an alternating projection algorithm, a feasible solution can be found as long as the number of RIS elements N is only slightly larger than the number of nonlinear equations $2K(K-1)$.

We remark that if the direct channel $d_{k,j}$ is present, the feasibility condition becomes more strict since it is necessary to have

$$\|\boldsymbol{a}_{k,j}\|_1 \geq |d_{k,j}|, \quad \forall k = 1, \ldots, K, \, j \neq k, \tag{11.13}$$

as otherwise even aligning all the strength of the cascaded channel would not be sufficient to cancel out the interference in the direct channel $d_{k,j}$. It can be observed that as the strength of the direct channel increases, the number of RIS elements N also needs to increase in order to make interference nulling feasible.

11.3.2 Alternating Projection Algorithm

This section presents an alternating projection algorithm for finding an interference nulling solution to (11.4). The proposed algorithm is applicable regardless of the RIS channel model and regardless of whether the direct links are present.

Specifically, we formulate the problem of finding an interference-nulling solution as the following feasibility problem:

$$\text{find} \quad \boldsymbol{\theta} \tag{11.14a}$$

$$\text{subject to} \quad \boldsymbol{a}_{k,j}^{\mathsf{T}}\boldsymbol{\theta} + d_{k,j} = 0, \quad k = 1, \ldots, K, \forall j \neq k, \tag{11.14b}$$

$$|\theta_i| = 1, \quad i = 1, \ldots, N. \tag{11.14c}$$

For ease of presentation, let $\boldsymbol{A}_k \in \mathbb{C}^{N \times (K-1)}$ denote the collection of all the interference channels to the kth receiver through the reflection of the RIS, i.e.,

$$\boldsymbol{A}_k = [\boldsymbol{a}_{k,1}, \ldots, \boldsymbol{a}_{k,k-1}, \boldsymbol{a}_{k,k+1}, \ldots, \boldsymbol{a}_{k,K}]. \tag{11.15}$$

Let $\boldsymbol{d}_k \in \mathbb{C}^{K-1}$ denote all the interference channels to the kth receiver through the direct links, i.e.,

$$\boldsymbol{d}_k = [d_{k,1}, \ldots, d_{k,k-1}, d_{k,k+1}, \ldots, d_{k,K}]^{\mathsf{T}}. \tag{11.16}$$

To further simplify the notation, let

$$\boldsymbol{A} = [\boldsymbol{A}_1, \ldots, \boldsymbol{A}_K], \tag{11.17a}$$

$$\boldsymbol{d} = [\boldsymbol{d}_1^{\mathsf{T}}, \ldots, \boldsymbol{d}_K^{\mathsf{T}}]^{\mathsf{T}}, \tag{11.17b}$$

where $\boldsymbol{A} \in \mathbb{C}^{N \times (K-1)K}$ and $\boldsymbol{d} \in \mathbb{C}^{(K-1)K}$ contain all the interference channels through the RIS reflection and the direct link, respectively.

Then, (11.14) can be rewritten more compactly as

$$\text{find} \quad \boldsymbol{\theta} \tag{11.18a}$$

$$\text{subject to} \quad \boldsymbol{A}^{\mathsf{T}}\boldsymbol{\theta} + \boldsymbol{d} = \boldsymbol{0}, \tag{11.18b}$$

$$|\theta_i| = 1, \quad i = 1, \ldots, N. \tag{11.18c}$$

Problem (11.18) is computationally challenging to solve due to the nonconvex unit modulus constraints.

To solve this problem efficiently, this chapter presents an alternating projection algorithm that transforms the feasibility problem (11.18) into a problem of finding a point in the intersection of two sets. For ease of presentation, we define the following two constraint sets

$$S_1 = \left\{ \theta \ : \ A^{\mathsf{T}}\theta + d = 0 \right\} \tag{11.19a}$$

$$S_2 = \left\{ \theta \ : \ |\theta_i| = 1, \ i = 1, \dots, N \right\}, \tag{11.19b}$$

and rewrite problem (11.18) equivalently as

$$
\begin{aligned}
&\text{find} && \theta \\
&\text{subject to} && \theta \in S_1 \cap S_2.
\end{aligned}
\tag{11.20}
$$

The feasibility problem (11.18) now becomes that of finding a point in the intersection of S_1 and S_2. To this end, given an initial point $\theta^{(0)}$, the alternating projection algorithm alternatively projects onto S_1 and S_2 as follows:

$$\tilde{\theta}^{(t)} = \Pi_{S_1}(\theta^{(t)}), \tag{11.21a}$$

$$\theta^{(t+1)} = \Pi_{S_2}(\tilde{\theta}^{(t)}), \tag{11.21b}$$

where the operator $\Pi_S(\theta)$ denotes the Euclidean projection of θ onto S, defined as the solution to the following problem:

$$
\begin{aligned}
&\underset{x}{\text{minimize}} && \|\theta - x\|_2^2 \\
&\text{subject to} && x \in S.
\end{aligned}
\tag{11.22}
$$

Fortunately, the projections onto the sets S_1 and S_2 both have simple analytical expressions, as given by (Parikh and Boyd 2014):

$$\Pi_{S_1}(\theta) = \theta - A^*(A^{\mathsf{T}}A^*)^{-1}(A^{\mathsf{T}}\theta + d), \tag{11.23a}$$

$$\Pi_{S_2}(\theta) = \theta/|\theta|, \tag{11.23b}$$

where A^* is complex conjugate of A. In (11.23a) the channel matrix A is assumed to be full column rank (otherwise the matrix A can be replaced by a new matrix A' constructed from the basis of the column space of A). In (11.23b), if some elements of the vector θ are zero, these elements can be projected to any random point on the complex unit circle. The alternating projection algorithm for solving problem (11.20) is summarized in Algorithm 11.1.

The local convergence of the alternating projection algorithm is established in (Jiang and Yu 2022). That is, if the intersection of the sets S_1 and S_2 is nonempty, Algorithm 11.1 is guaranteed to converge to an intersection point of S_1 and S_2 from an initial point sufficiently close to the intersection point.

Algorithm 11.1: Alternating Projection for Solving (11.18)

Input: Initial point $\theta \in \mathbb{C}^N$. Channel matrix A.

Initialization: $\theta^{(0)} = \theta$

for $t = 0, 1, 2, \ldots$ **do**

$\quad \tilde{\theta}^{(t)} = \Pi_{S_1}(\theta^{(t)})$

$\quad \theta^{(t+1)} = \Pi_{S_2}(\tilde{\theta}^{(t)})$

$\quad$ **if** *stopping criterion is satisfied* **then**

$\quad \quad |$ break

$\quad$ **end**

end

Output: $\theta^{(t+1)}$

The computational complexity of the proposed algorithm is dominated by the step (11.23a). This step requires time complexity $O(N^2K^2 + K^6 + K^4N)$ for computing $A^*(A^\top A^*)^{-1}A^\top$. But, this matrix can be precalculated and reused in each iteration, so the overall time complexity of the Algorithm 11.1 is $O(N^2K^2 + K^6 + K^4N + tN^2)$, where t is the number of iterations required for convergence.

11.3.3 Simulation Results

To illustrate the performance of the proposed algorithm, we report the simulation results in (Jiang and Yu 2022). The results in Figs. 11.2 and 11.3 correspond to the scenario without direct links and the results in Figs. 11.4 and 11.5 correspond to the scenario with direct links.

In the simulations, an interference-nulling solution is said to be found if the maximum interference-to-signal ratio across all transceiver pairs is below -36 dB. In Fig. 11.2, we plot the empirical probability of finding an interference nulling solution for different numbers of RIS elements N and transceiver pairs K. It can be seen from Fig. 11.2 that there is a phase-transition phenomenon as N increases for each fixed K. The interference nulling probability is almost 0 if N is smaller than some threshold, while the probability increases to 1 dramatically if N exceeds the threshold. The threshold that the phase transition occurs is about $2K(K-1)$ as K varies.

To observe the phase-transition location more precisely, Fig. 11.3 shows the value of N below which there is a "0% success" rate for finding an interference-nulling solution and the points above which there is empirically a "95% success" rate for finding an interference-nulling solution. We also plot the line $N = 2K(K-1)$, which precisely matches the "0% success" line. This implies that $N \geq 2K(K-1)$ is a necessary condition for the feasibility of interference

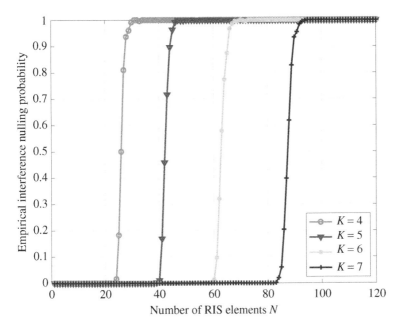

Figure 11.2 Empirical interference nulling probability versus number of RIS elements in the scenario without direct links.

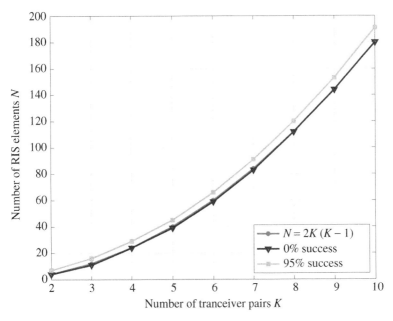

Figure 11.3 Number of RIS elements versus number of transceiver pairs K in the scenario without direct links.

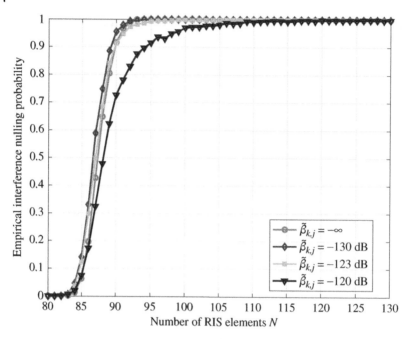

Figure 11.4 Empirical interference nulling probability versus number of RIS elements in the scenario with direct links for a system with $K = 7$ transceiver pairs.

nulling. Both Figs. 11.2 and 11.3 show that the proposed alternating projection algorithm can find an interference nulling solution with high probability if the number of RIS elements N is slightly above the threshold $2K(K − 1)$. It is also observed in simulations that the phase-transition location is not sensitive to the channel model.

We present the results of the case with direct links in Figs. 11.4 and 11.5. In Fig. 11.4, the number of transceiver pairs $K = 7$. The path loss of the cascaded channels is about $−122.6$ dB. As the path loss of the direct link $\tilde{\beta}_{k,j}$ becomes stronger, the phase transition location shifts toward a larger N as shown in Fig. 11.4. This implies that more RIS elements are needed to cancel out the interference in the direct link.

From (11.13), we know that the relative strength of the cascaded channels and the direct channels can affect the interference nulling probability. So in Fig. 11.5, we plot the empirical interference nulling probability for $K = 8$ for various values of the ratio between the strength of the direct link and the cascaded channels, i.e.,

$$\eta = \max_{k,j} \frac{|d_{k,j}|}{\|\boldsymbol{a}_{k,j}\|_1}. \tag{11.24}$$

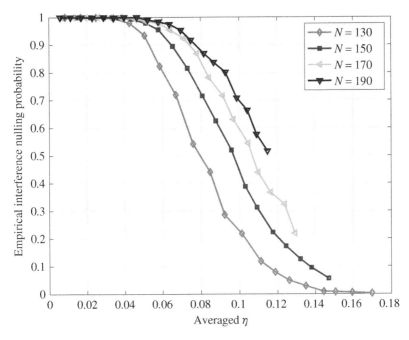

Figure 11.5 Empirical interference nulling probability versus direct-to-reflective link strength η for a system with $K = 8$ transceiver pairs.

As can be seen from Fig. 11.5, the empirical interference nulling probability decreases as the parameter η increases. To maintain high interference nulling probability as η increases, we need to increase the number of RIS elements.

11.4 Learning to Minimize Interference

The discussions so far in this chapter all assume that the perfect CSI is available for configuring the RIS. This represents the ultimate interference-nulling capability of the RIS system. But in practice the perfect CSI is never available; the CSI needs to be estimated using pilots with inherent channel estimation error. In this section, we consider the case where a pilot stage is used to obtain the CSI. We assume that the pilots are transmitted from the transmitter to the receiver and the received pilots are collected at a central node in order to design the reflection coefficients of the RIS. For simplicity, we ignore the direct links in this section but the proposed method can be readily extended to the scenario where the direct links exist.

We assume that the pilots are sent in a time orthogonal manner. Suppose that KL samples are reserved for the pilot stage in each coherence block. Each transmitter occupies L samples and sends a random sequence of L pilots while the other transmitters remain silent. More specifically, the jth transmitter sends a sequence of pilots $x_j(1), \ldots, x_j(L)$, and the received pilots at the kth receiver are given by

$$y_{kj}(\ell) = a_{kj}^\mathsf{T} \theta(\ell) x_j(\ell) + n_{kj}(\ell), \quad \ell = 1, \ldots, L, \tag{11.25}$$

where $\theta(\ell)$ is the reflection coefficients in the ℓth pilot transmission slot (typically randomly chosen), $n_{kj} \sim \mathcal{CN}(0, \sigma_2^2)$ is the additive noise and we set the pilot symbol $x_j(\ell) = 1$ without loss of generality.

Let $Y_j \in \mathbb{C}^{K \times L}$ denote the matrix whose entry in the kth row and ℓth column is $y_{kj}(\ell)$, we have

$$Y_j = \tilde{A}_j^\mathsf{T} \Theta + N_j, \tag{11.26}$$

where $\tilde{A}_j = [a_{1j}, \ldots, a_{Kj}]$ contains the cascaded channel from transmitter j to all the K receivers and $\Theta = [\theta(1), \ldots, \theta(L)]$. Concatenating Y_j's into a matrix $Y = [Y_1^\mathsf{T}, \ldots, Y_K^\mathsf{T}]^\mathsf{T}$, we can write

$$Y = \tilde{A}^\mathsf{T} \Theta + N, \tag{11.27}$$

where $\tilde{A} = [\tilde{A}_1, \ldots, \tilde{A}_K]$ and $N = [N_1^\mathsf{T}, \ldots, N_K^\mathsf{T}]^\mathsf{T}$. Given the received pilots Y and the RIS reflection coefficients matrix Θ, a linear minimum mean square error (LMMSE) estimator for $\tilde{A}$ takes the following form (Jiang et al. 2021):

$$\hat{A} = (Y - \mathbb{E}[Y]) \mathbb{E}\left[(Y - \mathbb{E}[Y])^\mathsf{H} (Y - \mathbb{E}[Y]) \right]^{-1}$$
$$\mathbb{E}\left[(Y - \mathbb{E}[Y])^\mathsf{H} (A - \mathbb{E}[A]) \right] + \mathbb{E}[A]. \tag{11.28}$$

Using the estimated channel $\hat{A}$, we can run the alternating projection algorithm (i.e., Algorithm 11.1) to obtain an interference nulling solution. The overall algorithmic framework is illustrated in Fig. 11.6.

Obtaining a good interference nulling solution, however, requires a very accurate estimate of the channel, which in turn requires significant pilot overhead. If the estimated channel is not sufficiently accurate, the alternating projection algorithm often returns an interference-nulling solution that still results in a relatively high level of interference. This is because the alternating projection algorithm is very sensitive to the channel estimation error.

To reduce the pilots training overhead, it is possible to use machine-learning approaches to directly map the received pilots to the reflection coefficients (Özdogan and Björnson 2020; Jiang et al. 2021; Zhang et al. 2022; Sohrabi et al. 2022). Through end-to-end learning, the neural network can be trained to exploit the received pilots more efficiently for optimizing the final task objective (Yu et al. 2022). However, applying such an idea to the considered interference-nulling problem is suitable only if the target interference power is not too close to zero.

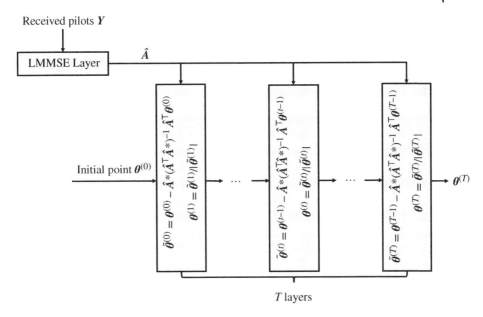

Figure 11.6 Conventional method runs the alternating projection algorithm from a random initial point.

If the target interference power level is too low, an enormous amount of training data and a huge DNN would be required to obtain an accurate zero-forcing solution.

For the task of learning a highly accurate solution to an optimization problem, algorithm unfolding-based neural network architecture can often be used to achieve good performance with reasonably sized training sets (Monga et al. 2021; Chen et al. 2022). In the algorithm unfolding-based DNN, each layer mimics an iteration of an particular algorithm with some parameters being made trainable. This ensures that the DNN is capable of achieving at least as good performance as the original algorithm and can potentially obtain performance gain by tuning the trainable parameters to make the algorithm tailored to the specific data distribution in the training stage.

In this section, we propose an unfolding-based DNN architecture to learn the interference-nulling solution from received pilots. But the alternating projection algorithm described earlier in the chapter is actually parameter-free, so we propose to include trainable parameters *before* the alternating projection algorithm for producing the initial point (instead of having trainable parameters within each iteration). This is also justified by the fact that the considered optimization problem is highly nonconvex, so different initial points lead to solutions with different performances. The hope is that the DNN can learn a better initial point from the training instances of the problem than the traditional random initialization strategy.

11.4.1 Learning to Initialize

To reduce the pilot training overhead and make the algorithm robust in the imperfect CSI scenario, this chapter proposes a learning approach that can find a better interference-nulling solution than the conventional approach. This is achieved by using a neural network to learn a good initial point for the subsequent alternating projection algorithm.

The overall framework of the proposed learning method is illustrated in Fig. 11.7. The central node first collects the received pilots in the pilot stage and uses an LMMSE layer described in (11.28) to extract an estimation of the interference channel matrix $\hat{A}$. The matrix $\hat{A}$ is used to form the alternating projection steps as well as the input to the DNN initialization layer. The DNN initialization layer then outputs an initial point $\theta^{(0)}$ as the input to the subsequent alternating projection algorithm, which consists of T layers. The neural network is trained end-to-end to minimize the loss function $-\mathbb{E}[\|A^\top \theta^{(T)}\|_2^2]$, where $\theta^{(T)}$ is the solution after T layer iterations. In the training stage, the DNN aims to learn to produce an initial point that leads to a solution with lower interference power when evaluated on the true channel matrix A.

As shown in Fig. 11.7, the proposed learning-based method is quite similar to the conventional approach in Fig. 11.6 except that an additional DNN layer outputs the initial points for the subsequently alternating projection algorithm. So the complexities of these two approaches are almost the same. In the following simulations, we can see that the DNN initialization layer can learn a more robust initial point than the conventional scheme with random initialization.

11.4.2 Simulation Results

To illustrate the performance of the proposed DNN, we use a simulation setting with the number of RIS elements $N = 64$ and the number of transceivers $K = 4$. The channels between the RIS and the transceivers are assumed to be multipath channels of the form $t = \frac{1}{\sqrt{L_p}} \sum_{i=1}^{L_p} \alpha_i \psi(\varrho_i, \phi_i)$, where $L_p = 3$ is the number of paths, $\alpha_i \sim \mathcal{CN}(0, 1)$ is the fading coefficient and $\psi(\varrho_i, \phi_i)$ is the steering vector of RIS with azimuth and elevation angles of arrival (ϱ_i, ϕ_i). The SNR in the pilot stage is 25 dB. The number of iterations of the alternating projection algorithm is 30.

In Fig. 11.8, we compare the performance between the conventional channel estimation method followed by alternating projection (AP) with random initialization and the proposed method with DNN initialization. Assuming a

Received pilots Y

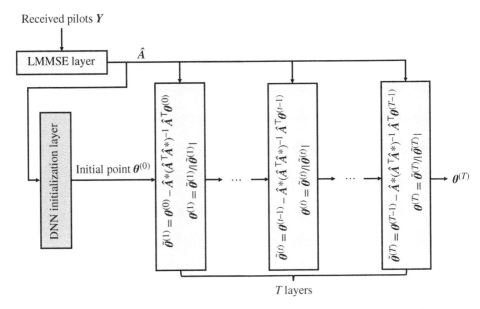

T layers

Figure 11.7 Proposed approach uses a deep neural network to learn a good initial point.

transmit power of 23 dBm and a pathloss of -108 dB for all $K = 4$ links, the total interference power across the four users is plotted for the two cases. As can be seen from Fig. 11.8, the interference powers of both methods decrease as the pilot length L increases, but the proposed approach achieves a much lower interference level than the conventional method with random initialization. This implies that the neural network learns a better initial point for the downstream alternating projection algorithm.

To investigate where the performance gain comes from, we conduct a simulation where many random initial points are used to run the alternating projection algorithm based on the estimated interference channel $\hat{A}$. The system setting is as follows, $N = 64$, $K = 4$, and $L = 70$. After the algorithm converges, we compute the interference power based on the true interference channel A. We report the minimum total interference power across the solutions obtained from different initial points in Fig. 11.9. It can be seen that as the number of random initial points increases, the total interference power for the alternating projection algorithm decreases rapidly. This is because the optimization problem (11.20) is highly nonconvex and the alternating projection algorithm starting from different initial points can converge to different solutions, some of which have lower interference power than others. Interestingly, the proposed learning-based approach can make use of the training data to learn a better initial point. From Fig. 11.9, we can see

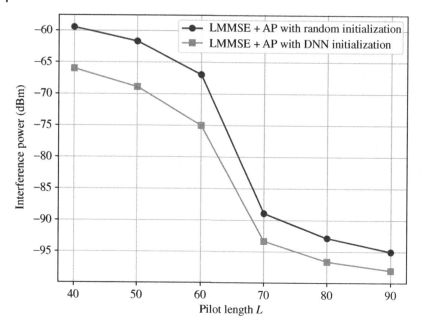

Figure 11.8 Performance comparison for a system with $N = 64$ RIS elements and $K = 4$ transceiver pairs.

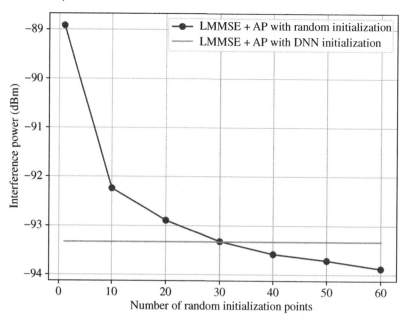

Figure 11.9 Performance comparison for a system with $N = 64$ RIS elements, $K = 4$ transceiver pairs and $L = 70$ pilots.

that the learning approach can achieve similar performance as the alternating projection algorithm with 30 different random initial points.

11.5 Conclusions

This chapter shows that the RIS can be used not only to adaptively reflect the signals, thereby enhancing the direct channels, but also to reduce interference in a multiuser transmission environment. From a theoretical perspective, it can be shown that a rectangular-array RIS is capable of nulling all the interference of a K-user interference channel if the number of RIS elements is sufficiently large when the channels between the RIS and the transceivers are line-of-sight and the direct links can be ignored. From an algorithmic perspective, this chapter presents an alternating projection algorithm that can efficiently find an interference-nulling solution for the general channel models with direct links. Numerically, it is found that the alternating projection algorithm can produce an interference-nulling solution as long as the number of RIS elements is slightly greater than $2K(K-1)$ and the direct links are not too strong. Moreover, for the scenario where CSI needs to be estimated using pilots, unlike the traditional approach of using random initial points for the alternating projection algorithm, the chapter shows that a DNN can be trained to learn a better initial point from the estimated CSI. This new approach can result in a much lower interference level at the same number of pilots. In summary, the chapter shows that reducing or nulling interference is a promising use case for RIS in future-generation wireless networks.

Bibliography

Abrardo, A., Dardari, D., Di Renzo, M., and Qian, X. (2021). MIMO interference channels assisted by reconfigurable intelligent surfaces: mutual coupling aware sum-rate optimization based on a mutual impedance channel model. *IEEE Wireless Communications Letters* 10 (12): 2624–2628. https://doi.org/10.1109/LWC.2021.3109017.

Björnson, E. and Sanguinetti, L. (2021). Rayleigh fading modeling and channel hardening for reconfigurable intelligent surfaces. *IEEE Wireless Communications Letters* 10 (4): 830–834. https://doi.org/10.1109/LWC.2020.3046107.

Chen, Y., Ai, B., Zhang, H. et al. (2021). Reconfigurable intelligent surface assisted device-to-device communications. *IEEE Transactions on Wireless Communications* 20 (5): 2792–2804. https://doi.org/10.1109/TWC.2020.3044302.

Chen, T., Chen, X., Chen, W. et al. (2022). Learning to optimize: a primer and a benchmark. *Journal of Machine Learning Research* 23 (189): 1–59.

Di Renzo, M., Zappone, A., Debbah, M. et al. (2019). Smart radio environments empowered by reconfigurable intelligent surfaces: how it works, state of research, and road ahead. *IEEE Journal on Selected Areas in Communications* 38 (11): 2450–2525. https://doi.org/10.1109/JSAC.2020.3007211.

Fu, M., Zhou, Y., and Shi, Y. (2021). Reconfigurable intelligent surface for interference alignment in MIMO device-to-device networks. *IEEE International Conference on Communications Workshops (ICC Workshops)*, 1–6, Montreal, QC, Canada, June 2021. https://doi.org/10.1109/ICCWorkshops50388.2021.9473762.

Huang, C., Zappone, A., Alexandropoulos, G.C. et al. (2019). Reconfigurable intelligent surfaces for energy efficiency in wireless communication. *IEEE Transactions on Wireless Communications* 18 (8): 4157–4170. https://doi.org/10 .1109/TWC.2019.2922609.

Ji, Z., Qin, Z., and Parini, C.G. (2022). Reconfigurable intelligent surface aided cellular networks with device-to-device users. *IEEE Transactions on Communications* 70 (3): 1808–1819. https://doi.org/10.1109/TCOMM.2022.3145570.

Jiang, T. and Yu, W. (2022). Interference nulling using reconfigurable intelligent surface. *IEEE Journal on Selected Areas in Communications* 40 (5): 1392–1406. https://doi.org/10.1109/JSAC.2022.3143220.

Jiang, T., Cheng, H.V., and Yu, W. (2021). Learning to reflect and to beamform for intelligent reflecting surface with implicit channel estimation. *IEEE Journal on Selected Areas in Communications* 39 (7): 1931–1945. https://doi.org/10.1109/JSAC .2021.3078502.

Monga, V., Li, Y., and Eldar, Y.C. (2021). Algorithm unrolling: interpretable, efficient deep learning for signal and image processing. *IEEE Signal Processing Magazine* 38 (2): 18–44. https://doi.org/10.1109/MSP.2020.3016905.

Newman, D.J. and Giroux, A. (1990). Properties on the unit circle of polynomials with unimodular coefficients. *Proceedings of the American Mathematical Society* 109 (1): 113–116. https://doi.org/10.2307/2048369.

Özdogan, Ö. and Björnson, E. (2020). Deep learning-based phase reconfiguration for intelligent reflecting surfaces. *Asilomar Conference on Signals, Systems & Computers*, 707–711, Pacific Grove, CA, USA, November 2020. https://doi.org/10 .1109/IEEECONF51394.2020.9443516.

Parikh, N. and Boyd, S. (2014). Proximal algorithms. *Foundations and Trends® in Optimization* 1 (3): 127–239. https://doi.org/10.1561/2400000003.

Sohrabi, F., Jiang, T., Cui, W., and Yu, W. (2022). Active sensing for communications by learning. *IEEE Journal on Selected Areas in Communications* 40 (6): 1780–1794. https://doi.org/10.1109/JSAC.2022.3155496.

Wu, Q. and Zhang, R. (2019). Intelligent reflecting surface enhanced wireless network via joint active and passive beamforming. *IEEE Transactions on Wireless Communications* 18 (11): 5394–5409. https://doi.org/10.1109/TWC.2019.2936025.

Wu, Q., Zhang, S., Zheng, B. et al. (2021). Intelligent reflecting surface-aided wireless communications: a tutorial. *IEEE Transactions on Communications* 69 (5): 3313–3351. https://doi.org/10.1109/TCOMM.2021.3051897.

Yu, W., Sohrabi, F., and Jiang, T. (2022). Role of deep learning in wireless communications. *IEEE BITS the Information Theory Magazine* 2 (2): 56–72. https://doi.org/10.1109/MBITS.2022.3212978.

Zhang, Z., Jiang, T., and Yu, W. (2022). Learning based user scheduling in reconfigurable intelligent surface assisted multiuser downlink. *IEEE Journal on Selected Topics in Signal Processing* 16 (5): 1026–1039. https://doi.org/10.1109/JSTSP.2022.3178213.

12

Blind Beamforming for IRS Without Channel Estimation

Kaiming Shen[1] and Zhi-Quan Luo[1,2]

[1]*School of Science and Engineering, The Chinese University of Hong Kong (Shenzhen), Shenzhen, China*
[2]*Shenzhen Research Institute of Big Data, Shenzhen, China*

12.1   Introduction

Passive beamforming, i.e., coordinating phase shifts across the reflective elements, lies at the core of the intelligent reflecting surface (IRS) technology. Despite the intrinsic difference between passive beamforming and the conventional active beamforming for the MIMO channels, many existing works on IRS still follow the traditional model-driven paradigm of first estimating channels and then optimizing phase shifts, thereby readily applying the classical beamforming tools such as semidefinite relaxation (SDR) (Luo et al. 2010), weighted minimum mean square error (WMMSE) (Shi et al. 2011), and fractional programming (FP) (Shen and Yu 2018).

However, channel acquisition for an IRS-aided system could pose formidable challenges in engineering practice, mainly in the following three respects:

i. Each reflected channel alone can be easily overwhelmed by the other channels and noise, so it can be difficult to obtain precise estimation.
ii. One has to modify the current networking protocol (e.g., frame structure) to enable channel estimation for IRS. Besides, some channel estimating algorithms entail the full information of the received signal, but this is not supported by many communication chips on the market.
iii. Channel estimation for the reflected channels may incur huge time complexity if the IRS consists of a large number reflective elements.

In order to address the above issues, the past few years have seen a surge of research interests in *blind beamforming* without relying on any channel knowledge. Both (Psomas and Krikidis 2021) and (Nadeem et al. 2021) advocate a random rotation strategy for passive beamforming in the absence of the instantaneous

Intelligent Surfaces Empowered 6G Wireless Network, First Edition.
Edited by Qingqing Wu, Trung Q. Duong, Derrick Wing Kwan Ng, Robert Schober, and Rui Zhang.

channel information. Another line of studies (You et al. 2020; Ning et al. 2021; Wang et al. 2022; Wang and Zhang 2021) suggest simply trying out all possible directions of the reflected beam; this procedure does not require any channel information. However, the above beam sweeping approach is typically restricted to the millimeter/terahertz communication scenario with sharp beams. And deep learning has been considered in this area. As opposed to (Liu et al. 2020, 2021; Elbir et al. 2020; Gao et al. 2020; Huang et al. 2020; Feng et al. 2020) that use deep learning to either estimate channels or optimize phase shifts given channels, the recent work (Jiang et al. 2021) proposes learning to directly map the received pilot signals to the passive beamforming vector via deep neural networks. It is argued in (Jiang et al. 2021) that such unified learning policy is capable of extracting more pertinent information from the raw data. Differing from these neural net-based approaches, the method called RFocus in (Arun and Balakrishnan 2020) uses the statistics of the received signal. The present chapter also pursues a statistical approach and shows that the proposed blind beamforming method can strike provable better performance than RFocus in (Arun and Balakrishnan 2020). Aside from the theoretical justifications, we further demonstrate this novel blind beamforming method through field tests in a commercial 5G network.

Throughout the chapter, we use the Bachmann–Landau notation extensively: $f(n) = O(g(n))$ if there exists some $c > 0$ such that $|f(n)| \leq cg(n)$ for n sufficiently large; $f(n) = o(g(n))$ if there exists some $c > 0$ such that $|f(n)| < cg(n)$ for n sufficiently large; $f(n) = \Omega(g(n))$ if there exists some $c > 0$ such that $f(n) \geq cg(n)$ for n sufficiently large; $f(n) = \Theta(g(n))$ if $f(n) = O(g(n))$ and $f(n) = \Omega(g(n))$ both hold.

12.2 System Model

For ease of discussion, the majority of the present chapter focuses on the single-user single-antenna transmission; the extension to the general case with multiple users and multiple antennas is postponed till Section 12.5.3. Assume that the IRS consists of a total of N reflective elements. Let $h_n^{\mathrm{I}} \in \mathbb{C}$ be the channel from the transmitter to the nth reflective element, and $h_n^{\mathrm{II}} \in \mathbb{C}$ the channel from the nth reflective element to the receiver. The cascaded channel h_n associated with the nth reflective element is then obtained as

$$h_n = h_n^{\mathrm{I}} h_n^{\mathrm{II}}, \quad \text{for } n = 1, \dots, N. \tag{12.1}$$

Let $h_0 \in \mathbb{C}$ be the superposition of the rest channels from the transmitter to the receiver (including the direct channel as well as those reflected channels not due to the IRS). We frequently represent the above channels in a polar form, i.e.,

$$h_n = \beta_n e^{j\alpha_n}, \quad \text{for } n = 0, \dots, N, \tag{12.2}$$

where the channel magnitude $\beta_n \in (0,1)$ and the channel phase $\alpha_n \in (0, 2\pi]$. Moreover, use $\theta_n \in (0, 2\pi]$ to denote the phase shift induced by the nth reflective element in its corresponding channel h_n, and θ the passive beamforming vector $(\theta_1, \ldots, \theta_N)$. Thus, for the transmit signal $X \in \mathbb{C}$, the received signal $Y \in \mathbb{C}$ is given by

$$Y = \left(h_0 + \sum_{n=1}^{N} h_n e^{j\theta_n} \right) X + Z, \tag{12.3}$$

where a complex Gaussian random variable $Z \sim \mathcal{CN}(0, \sigma^2)$ models the additive background noise. Further, for practical reasons, we assume that each θ_n is selected from a prescribed discrete set

$$\Phi_K = \{\omega, 2\omega, \ldots, K\omega\}, \tag{12.4}$$

where K is the number of phase shift choices and the distance ω is given by

$$\omega = \frac{2\pi}{K}. \tag{12.5}$$

For the field tests as shown in Section 12.6, we adopt $K = 4$ and $\omega = \pi/2$, namely the quadrature phase shifting.

Assuming a mean transmit power of P, i.e., $\mathbb{E}[|X|^2] = P$, the signal-to-noise ratio (SNR) can be computed as

$$\text{SNR} = \frac{\mathbb{E}[|Y - Z|^2]}{\mathbb{E}[|Z|^2]} = \frac{P}{\sigma^2} \left| \beta_0 e^{j\alpha_0} + \sum_{n=1}^{N} \beta_n e^{j(\alpha_n + \theta_n)} \right|^2. \tag{12.6}$$

Notice that we are only interested in how much the SNR could be improved by configuring the IRS properly, rather than the specific value of SNR. Toward this end, we define the baseline case without IRS to be

$$\text{SNR}_0 = \frac{P\beta_0^2}{\sigma^2}, \tag{12.7}$$

and consider the *SNR boost* $f(\theta)$ as follows:

$$f(\theta) := \frac{\text{SNR}}{\text{SNR}_0} = \frac{1}{\beta_0^2} \left| \beta_0 e^{j\alpha_0} + \sum_{n=1}^{N} \beta_n e^{j(\alpha_n + \theta_n)} \right|^2. \tag{12.8}$$

The IRS beamforming task is now formulated as a combinatorial optimization problem of maximizing the SNR boost over the discrete phase shifts:

$$\underset{\theta}{\text{maximize}} \quad f(\theta) \tag{12.9a}$$

$$\text{subject to} \quad \theta_n \in \Phi_K \text{ for } n = 1, \ldots, N. \tag{12.9b}$$

We remark that the channel information $\{h_0, \ldots, h_N\}$ is not available in the above problem setup. A common approach in the literature is to first estimate channels and subsequently optimize θ explicitly in (12.9). Nevertheless, as elaborated in Section 12.3, channel acquisition can be quite costly in practice. A *blind beamforming* approach is then proposed to optimize θ in the absence of channel information. As the main contribution of this chapter, we illustrate a somewhat surprising result that a statistical blind beamforming method is capable of reaching the global optimum without knowing the channels.

12.3 Random-Max Sampling (RMS)

The simplest method for blind beamforming is just to try out a bunch of random samples of θ and then choose the best, referred to as the *random-max sampling (RMS)* method. Specifically, we generate a total of T samples of θ at random. For the tth sample denoted as $\theta_t = (\theta_{1t}, \ldots, \theta_{Nt})$, each entry θ_{nt} is drawn uniformly and independently from the discrete set Φ_K. The received signal corresponding to the tth random sample is given by

$$
Y_t = \left(h_0 + \sum_{n=1}^N h_n e^{j\theta_{nt}} \right) X_t + Z_t. \tag{12.10}
$$

The RMS method simply decides θ according to the received signal power, i.e.,

$$
\theta^{\mathrm{RMS}} = \theta_{t^\star} \quad \text{where} \quad t^\star = \arg\max_{1 \le t \le T} |Y_t|^2. \tag{12.11}
$$

Clearly, the sample size T plays a key role in the performance of RMS. Before proceeding to the performance analysis, we first define the *average per element reflection power gain* to be

$$
\overline{\beta}^2 = \frac{1}{N} \sum_{n=1}^N \beta_n^2. \tag{12.12}
$$

The following theorem shows how the SNR boost by RMS grows with the sample size T and the IRS size N.

Theorem 12.1 *Consider T i.i.d. random samples of θ uniformly drawn over Φ_K. The expected SNR boost achieved by the RMS method in (12.11) has the following order bounds:*

$$
\mathbb{E}\left[f(\theta^{\mathrm{RMS}}) \right] = \frac{\overline{\beta}^2}{\beta_0^2} \cdot \Theta(N \log T) \quad \text{if } T = o\left(\sqrt{N} \right), \tag{12.13}
$$

$$
\mathbb{E}\left[f(\theta^{\mathrm{RMS}}) \right] = \frac{\overline{\beta}^2}{\beta_0^2} \cdot O(N \log T) \quad \text{in general,} \tag{12.14}
$$

where the expectation is taken over random samples of θ.

The proof of the above theorem is beyond the scope of this chapter. We refer the interested readers to (Ren et al. 2021) for the mathematical details.

But how far is this scaling rate of RMS from the optimum? To answer this question, we construct an upper bound on the SNR boost by assuming that the channels are already known and also that each phase shift θ_n can be arbitrarily chosen on $(0, 2\pi]$. It can be easily seen that under the above two assumptions the maximum SNR boost is achieved when every phase-shifted reflected channel $h_n e^{j\theta_n}$ is aligned with the direct channel h_0 exactly. In other words, each θ_n is optimally determined as

$$\theta_n = \alpha_0 - \alpha_n. \tag{12.15}$$

Plugging the above θ_n into (12.8) yields an upper bound on the achievable SNR boost as stated in the following theorem.

Theorem 12.2 *The SNR boost is bounded from above as*

$$f(\theta) \leq \frac{\left(\sum_{n=0}^{N} \beta_n\right)^2}{\beta_0^2}, \tag{12.16}$$

and thus it is at most quadratic in the number of reflective elements, i.e.,

$$f(\theta) = \frac{\overline{\beta}^2}{\beta_0^2} \cdot O(N^2). \tag{12.17}$$

Consequently, the highest SNR boost we can expect for RMS (and also for any other algorithms) is quadratic in N. According to Theorem 12.1, it takes an exponential number of samples for RMS to reach this upper bound, i.e., $T = \Omega(2^N)$. Intuitively, RMS can figure out the optimal θ only after it has almost exhausted the entire solution space of Φ_K^N. But is it possible to figure out the optimal solution by using merely a polynomial number of samples? The next Section 12.4 aims to answer the above question.

12.4 Conditional Sample Mean (CSM)

In this section, we propose a novel statistical method that is guaranteed to achieve the quadratic SNR boost much more efficiently than RMS. We still generate T random samples of θ in a uniform and independent fashion, only that T just needs to be polynomially large as shown in the end of this section.

We begin by defining $\mathcal{Q}_{nk} \subseteq \{1, \dots, T\}$ to be the subset of the indices of all those random samples with $\theta_{nt} = k\omega$, i.e.,

$$\mathcal{Q}_{nk} = \{t : \theta_{nt} = k\omega\}, \quad \forall n = 1, \dots, N \text{ and } \forall k = 1, \dots, K. \tag{12.18}$$

A conditional sample mean of the received signal power $|Y|^2$ is computed within each subset Q_{nk}, denoted by

$$\hat{\mathbb{E}}[|Y|^2 \mid \theta_n = k\omega] = \frac{1}{|Q_{nk}|} \sum_{t \in Q_{nk}} |Y_t|^2. \tag{12.19}$$

The main idea of the *conditional sample mean (CSM)* method is to choose each θ_n such that the corresponding conditional sample mean is maximized:

$$\theta_n^{\text{CSM}} = \arg\max_{\varphi \in \Phi_K} \hat{\mathbb{E}}[|Y|^2 \mid \theta_n = \varphi]. \tag{12.20}$$

The above method is summarized in Algorithm 12.1.

The intuition behind CSM follows. We use $\hat{\mathbb{E}}[|Y|^2|\theta_n = k\omega]$ to measure the average goodness of a particular choice $\theta_n = k\omega$ when the rest θ_m's are still pending. With all the phase shifts determined in this way simultaneously, a kind of equilibrium among the N reflective elements can be reached. We examine this equilibrium in the following theorem.

Algorithm 12.1 Conditional sample mean (CSM)

1: **input:** Φ_K, N, T;
2: **for** $t = 1, 2, \ldots, T$ **do**
3:   generate $\theta_t = (\theta_{1t}, \ldots, \theta_{Nt})$ i.i.d. based on Φ_K;
4:   measure received signal power $|Y_t|^2$ with θ_t;
5: **end for**
6: **for** $n = 1, 2, \ldots, N$ **do**
7:   **for** $k = 1, 2, \ldots, K$ **do**
8:     compute $\hat{\mathbb{E}}[|Y|^2|\theta_n = k\omega]$ according to (12.19);
9:   **end for**
10:   decide θ_n^{CSM} according to (12.20);
11: **end for**
12: **output:** $\theta^{\text{CSM}} = (\theta_1^{\text{CSM}}, \ldots, \theta_N^{\text{CSM}})$.

Theorem 12.3 *Consider T i.i.d. random samples of θ uniformly drawn over Φ_K. When $K \geq 3$, the expected SNR boost achieved by the CSM method in Algorithm 12.1 has a tight order bound:*

$$\mathbb{E}[f(\theta^{\text{CSM}})] = \frac{\bar{\beta}^2}{\beta_0^2} \cdot \Theta(N^2) \quad \text{if} \quad T = \Omega(N^2(\log N)^3), \tag{12.21}$$

where the expectation is taken over random samples of θ.

Two observations in the above scaling law analysis of CSM are worth noting. First, CSM requires only a polynomial number of random samples to attain the quadratic SNR boost in N, which makes CSM much more practical than the standard method of RMS. Second, the quadratic scaling rate of CSM holds only when $K \geq 3$; we will delve into this issue in Section 12.5.1.

12.5 Some Comments on CSM

12.5.1 Connection to Closest Point Projection

In Section 12.3 we assume that the channel information is already known and also that each phase shift can be arbitrarily chosen on $(0, 2\pi]$ as $K \to \infty$, in order to derive an upper bound on the SNR boost. The resulting optimal θ_n is shown in (12.15). What if we only make the first assumption (i.e., channel information available) whereas the number of phase shift choices K is still finite?

A simple method under the above new setting is to project the relaxed continuous solution, which is given in (12.15), to the closest point within the discrete set Φ_K, referred to as the *closest point projection (CPP)* method. If the projection is performed in the Euclidean distance sense, then each θ_n is determined as

$$\theta_n^{\mathrm{CPP}} = \arg \min_{\varphi \in \Phi_K} \left| \mathrm{Arg} \left(h_0 e^{-j(\varphi + \alpha_n)} \right) \right|, \tag{12.22}$$

CPP can be alternatively interpreted as choosing each θ_n such that $h_n e^{j\theta_n}$ is rotated to the closest possible position to h_0 on the complex plane. We are interested in this CPP method because of its close connection to CSM, as specified in the following theorem.

Theorem 12.4 *The CSM method without channel information is equivalent to the CPP method with channel information if the sample size is sufficiently large, i.e.,*

$$\theta^{\mathrm{CSM}} = \theta^{\mathrm{CPP}} \quad as \ n \to \infty. \tag{12.23}$$

Proof: When the phase shift of the nth reflective element is fixed at $k\omega$, the received signal is given by

$$Y = \left(h_0 + h_n e^{jk\omega} \right) X + \left(\sum_{m=1, m \neq n}^{N} h_m e^{j\theta_m} \right) X + Z. \tag{12.24}$$

Furthermore, assuming that the rest phase shifts are randomly and independently chosen from the discrete set Φ_K, and also that X is i.i.d. with $\mathbb{E}[|X|^2] = P$

and Z is drawn i.i.d. from $\mathcal{CN}(0, \sigma^2)$, the conditional expectation of the received signal power can be computed as

$$
\mathbb{E}\left[|Y|^2 \mid \theta_n = k\omega\right] = \mathbb{E}_{\theta_m, X, Z}\left|\left(h_0 + h_n e^{jk\omega}\right)X + \left(\sum_{m=1, m \neq n}^{N} h_m e^{j\theta_m}\right)X + Z\right|^2
$$

$$
= \left|h_0 + h_n e^{jk\omega}\right|^2 P + \sum_{m=1, m \neq n}^{N} \beta_m^2 P + \sigma^2
$$

$$
= 2\beta_0 \beta_n P \cos(k\omega - \alpha_0 + \alpha_n) + \sum_{m=0}^{N} \beta_m^2 P + \sigma^2. \tag{12.25}
$$

By the law of large numbers, the conditional sample mean $\hat{\mathbb{E}}[|Y|^2 \mid \theta_n = k\omega]$ in (12.19) approaches the above value when the sample size increases. Consequently, as $T \to \infty$, the CSM method in (12.20) boils down to

$$
\theta^{\mathrm{CSM}} = \arg\max_{\theta_n \in \Phi_K} \mathbb{E}\left[|Y|^2 \mid \theta_n = k\omega\right] \tag{12.26}
$$

$$
= \arg\max_{\theta_n \in \Phi_K} \cos(\theta_n - \alpha_0 + \alpha_n). \tag{12.27}
$$

Clearly, the solution of (12.27) is to make θ_n close to $\alpha_0 - \alpha_n$ as much as possible, namely the CPP method. The equivalence between CPP and CSM as $T \to \infty$ is thus established. □

We further show that CPP yields a constant approximation ratio factor for any fixed K.

Theorem 12.5 *The CPP method in (12.22) satisfies*

$$
\cos^2(\pi/K) \cdot f^\star \leq f(\theta^{\mathrm{CPP}}) \leq f^\star, \tag{12.28}
$$

where $f^\star$ is the maximum SNR boost.

Proof: The right inequality is evident. We focus on showing the left inequality in what follows:

$$
f(\theta^{\mathrm{CPP}}) = \frac{1}{\beta_0^2} \cdot \left|\beta_0 e^{j\alpha_0} + \sum_{n=1}^{N} \beta_n e^{j(\theta_n^{\mathrm{CPP}} + \alpha_n)}\right|^2 \tag{12.29a}
$$

$$
= \frac{1}{\beta_0^2} \cdot \left|\beta_0 + \sum_{n=1}^{N} \beta_n e^{j(\theta_n^{\mathrm{CPP}} - \alpha_0 + \alpha_n)}\right|^2 \tag{12.29b}
$$

$$
\geq \frac{1}{\beta_0^2} \cdot \left|\beta_0 + \sum_{n=1}^{N} \beta_n \cos\left(\theta_n^{\mathrm{CPP}} - \alpha_0 + \alpha_n\right)\right|^2 \tag{12.29c}
$$

$$\geq \frac{1}{\beta_0^2} \cdot \left(\beta_0 + \sum_{n=1}^{N} \beta_n \cos(\pi/K) \right)^2 \qquad (12.29\text{d})$$

$$\geq \frac{\cos^2(\omega/2)}{\beta_0^2} \cdot \left(\sum_{n=0}^{N} \beta_n \right)^2 \qquad (12.29\text{e})$$

$$\geq \cos^2(\omega/2) \cdot f^\star, \qquad (12.29\text{f})$$

where (12.29c) follows since each $\beta_n e^{j(\theta_n^{\text{CPP}} - \alpha_0 + \alpha_n)} = \beta_n \cos(\theta_n^{\text{CPP}} - \alpha_0 + \alpha_n) + j\beta_n \sin(\theta_n^{\text{CPP}} - \alpha_0 + \alpha_n)$ and the removal of the sin component does not decrease the absolute square, (12.29d) follows by the fact that $|\theta_n^{\text{CPP}} - \alpha_0 + \alpha_n| \leq \pi/K$ under the closest point projection, and (12.29f) follows by the upper bound in Theorem 12.2. The proof is then completed. □

Incorporating the above result into Theorem 12.4 partly verifies the scaling law of CSM in Theorem 12.3, as stated in the following corollary.

Corollary 12.1 *For CPP, the approximation ratio $\cos^2(\pi/K) \geq 0.5$ if $K \geq 3$, whereas $\cos^2(\pi/K) = 0$ if $K = 2$. Consequently, as $T \to \infty$, CSM optimally reaches a quadratic SNR boost in N if $K \geq 3$, whereas its SNR boost cannot be bounded from below if $K = 2$.*

Figure 12.1 gives a concrete example to illustrate the failure of CSM when $K = 2$. Notice that the above corollary only claims that CSM would be equivalent to CPP when the sample size T is sufficiently large. It entails considerable efforts to verify the specific threshold on T as stated in Theorem 12.3; the interested readers are referred to (Ren et al. 2021) for the complete proof of Theorem 12.3.

Furthermore, an enhanced CSM can yield a quadratic SNR boost in the number of reflective elements N for any $K \geq 2$. The main idea is to improve

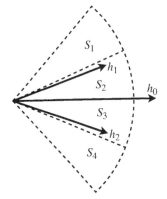

Figure 12.1 Consider four sectors $\{S_1, S_2, S_3, S_4\}$; each sector spans an angle of π/K. For $N = 2$, assume that h_0 is located right between S_2 and S_3, and that h_1 is inside S_2 but arbitrarily close to S_1, while h_2 and h_1 are symmetric about h_0. When $K = 4$ and $|h_1| = |h_2|$, it follows that h_1 and h_2 cancel out each other under CPP or CSM, so IRS does not boost SNR.

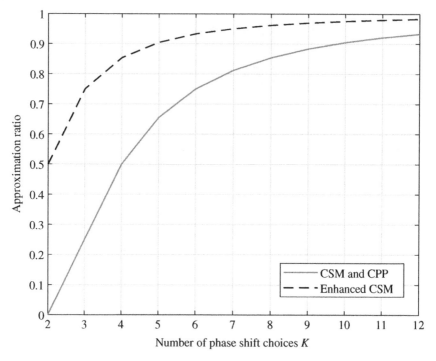

Figure 12.2 Consider four sectors $\{S_1, S_2, S_3, S_4\}$; each sector spans an angle of π/K. For $N = 2$, assume that h_0 is located right between S_2 and S_3, and that h_1 is inside S_2 but arbitrarily close to S_1, while h_2 and h_1 are symmetric about h_0. When $K = 4$ and $|h_1| = |h_2|$, it follows that h_1 and h_2 cancel out each other under CPP or CSM, so IRS does not boost SNR.

the approximation ratio factor of CPP, as shown in Fig. 12.2. Observe that the enhanced CSM yields an approximation ratio factor of 0.5 even at $K = 2$, so it guarantees a quadratic SNR boost for the binary beamforming case. More details can be found in (Ren et al. 2021).

12.5.2 Connection to Phase Retrieval

Although CSM does not perform channel estimation explicitly, we can somehow retrieve the phase information of the channels $\{h_0, h_1, \ldots, h_N\}$ from the beamforming decision θ by CSM.

When all the θ_n's are uniformly and independently distributed over Φ_K, the expectation of the received signal power is given by

$$\mathbb{E}[|Y|^2] = \beta_0^2 P + \sum_{m=1}^{N} \beta_m^2 P + \sigma^2. \tag{12.30}$$

When a particular θ_n is fixed at $k\omega$ and the rest θ_m's are randomized, the resulting conditional expectation of the received signal power is given by

$$\mathbb{E}\left[|Y|^2 \mid \theta_n = k\omega\right] = P\left|h_0 + h_n e^{jk\omega}\right|^2 + \sum_{m\neq n} \beta_m^2 P + \sigma^2. \tag{12.31}$$

We use J_{nk} to denote the difference between the above two expectations, which can be computed as

$$J_{nk} = \mathbb{E}[|Y|^2 \mid \theta_n = k\omega] - \mathbb{E}[|Y|^2] \tag{12.32a}$$

$$= 2\beta_0 \beta_n P \cos(k\omega - \alpha_0 + \alpha_n). \tag{12.32b}$$

Observe that the value of J_{nk} depends on the phase difference

$$\Delta_n = \alpha_0 - \alpha_n \tag{12.33}$$

between the direct channel h_0 and the reflected channel h_n.

Moreover, the above expectation difference can be evaluated empirically based on the random samples, i.e.,

$$\hat{J}_{nk} = \frac{1}{|Q_{nk}|} \sum_{t\in Q_{nk}} |Y_t|^2 - \frac{1}{T} \sum_{t=1}^{T} |Y_t|^2. \tag{12.34}$$

The main idea of phase retrieval is to recover the phase difference Δ_n through minimizing the gap between J_{nk} and $\hat{J}_{nk}$. For instance, if we consider a *square-max* loss function

$$\mathcal{L}_n(\Delta_n) = \left|\max_{1\leq k\leq K} \{J_{nk}\} - \max_{1\leq k\leq K} \{\hat{J}_{nk}\}\right|^2, \tag{12.35}$$

then the phase retrieval problem is formulated as

$$\underset{\{\Delta_n\}}{\text{minimize}} \quad \sum_{n=1}^{N} \mathcal{L}_n(\Delta_n) \tag{12.36a}$$

$$\text{subject to} \quad 0 \leq \Delta_n < 2\pi, \quad \text{for } n = 1, \ldots, N. \tag{12.36b}$$

The above problem can be optimally solved as

$$\Delta_n = k_0 \omega \quad \text{where} \quad k_0 = \arg\max_{1\leq k\leq K} \hat{\mathbb{E}}[|Y|^2 \mid \theta_n = k\omega]. \tag{12.37}$$

An analogy can be immediately seen between the above solution and the CSM method in (12.20), both of which seek the phase shift $\theta_n \in \Phi_K$ to maximize the conditional sample mean $\hat{\mathbb{E}}[|Y|^2 \mid \theta_n = k\omega]$. Thus, CSM boils down to recovering the phase difference Δ_n according to the loss function in (12.35).

But we could have used a different loss function in the phase retrieval. If $\mathcal{L}_n(\Delta_n)$ in problem (12.36) is replaced with the following *sum-of-squares* loss function

$$\mathcal{L}'_n(\Delta_n) = \sum_{k=1}^{K} \left|J_{nk} - \hat{J}_{nk}\right|^2, \tag{12.38}$$

the solution of Δ_n becomes

$$
\Delta_n = \begin{cases} -\arctan \dfrac{F_n}{E_n} + \dfrac{\pi}{2} & \text{if } E_n \geq 0, \\[3mm] -\arctan \dfrac{F_n}{E_n} - \dfrac{\pi}{2} & \text{if } E_n < 0, \end{cases}
\tag{12.39}
$$

where

$$
E_n = \sum_{k=1}^{K} \hat{J}_{nk} \sin(k\omega) \quad \text{and} \quad F_n = \sum_{k=1}^{K} \hat{J}_{nk} \cos(k\omega).
\tag{12.40}
$$

If the closest point projection is performed based on the above Δ_n, i.e., $\theta_n = \arg\min_{\varphi \in \Phi_K} |\varphi - \Delta_n|$, we would arrive at another version of CSM. Furthermore, it turns out that every choice of loss function in the phase retrieval problem (12.36) can be recognized as a variation of the CSM method.

12.5.3 CSM for General Utility Functions

The CSM method can be further extended to a general utility function in order to account for multiple users and multiple antennas. For each random sample $t = 1, \ldots, T$, we now measure a utility value $U_t \in \mathbb{R}$ at the receiver side. For instance, if we have multiple users and aim at a max-min fairness, the utility could be set to the minimum SNR among the users.

Recall the conditional sample subset $\mathcal{Q}_{nk}$ in (12.18). We now compute the conditional sample mean of U within each subset $\mathcal{Q}_{nk}$, i.e.,

$$
\hat{\mathbb{E}}[U \mid \theta_n = k\omega] = \frac{1}{|\mathcal{Q}_{nk}|} \sum_{t \in \mathcal{Q}_{nk}} U_t.
\tag{12.41}
$$

Following Algorithm 12.1, we decide each θ_n according to the respective conditional sample mean, i.e.,

$$
\theta_n^{\text{CSM}} = \arg\max_{\varphi \in \Phi_K} \hat{\mathbb{E}}[U \mid \theta_n = \varphi].
\tag{12.42}
$$

Deciding the optimal utility function for the generalized CSM remains an open problem.

12.6 Field Tests

In this section, we demonstrate the performance of the blind beamforming approach through prototype tests in the real-world wireless environment. Our tests are carried out in a public downlink network over a 200 MHz wide spectrum band centered at 2.6 GHz. It is worth pointing out that our blind beamforming

Figure 12.3 A panoramic view of the field test site. The base station is located on a 20-meter-high terrace while the user terminal is located inside an underground parking lot. The IRS is placed at the entrance of the parking lot. The IRS is approximately 250 m away from the base station, and the user terminals are approximately 40 m away from the IRS.

(a)   (b)

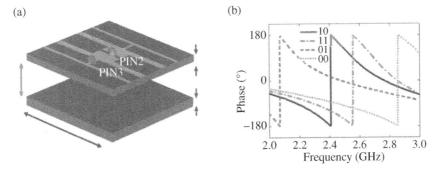

Figure 12.4 The ON-OFF state of a PIN diode results in two distinct resonance frequencies in the series RLC circuit, which correspond to two phase shifts. Further, with a pair of PIN diodes integrated into each reflective element (part a), we can realize four phase shifts by controlling the respective ON-OFF states of the two PIN diodes (part b).

method does not require any collaboration from the service provider side. Thus, the IRS can be deployed and then configured in a plug-and-play fashion (Fig. 12.3).

The hardware realization of each reflective element is illustrated in Fig. 12.4. A quadrature beamforming with $\theta_n \in \{0, \pi/2, \pi, 3\pi/2\}$ is implemented by using a pair of PIN diodes at each reflective element. Moreover, as shown in Fig. 12.5, the IRS is formed by 16 "reflecting tiles" – a tiny IRS prototype that is 50×50 cm large – arranged in a 4×4 array. Each reflecting tile consists of 16 reflective elements, so the assembled large IRS consists of 256 reflective elements in total. There are four phase shift choices $\{0, \pi/2, \pi, 3\pi/2\}$ for each individual reflective element.

As shown in Figs. 12.3 and 12.6, the base station is located on a 20-meter-high terrace while the user terminal is located in an underground parking lot.

Figure 12.5 The IRS is formed by a 4×4 array of reflecting tiles. Each reflecting tile is 50×50 cm large and consists of 16 reflective elements.

Figure 12.6 A satellite image of the field test site. The base station and the IRS are outdoor while the user terminals are indoor.

There is no line-of-sight propagation from the base station to the user terminal. The IRS is placed outdoors near the entrance of the parking lot. The distance from the base station to the IRS is approximately 250 m; the distance from the IRS to the user terminal is approximately 40 m. It is worth remarking that the wireless environment is highly volatile in our case because of the busy traffic in the parking lot, as can be observed from Fig. 12.7.

We use the sample size $T = 2560$ (i.e., $T = 10N$) for both RMS and CSM. Moreover, we include a baseline method called "OFF," which simply fixes the phase shift θ_n at the initial state without beamforming.

We start with the single-input single-output (SISO) transmission, aiming to improve the SNR boost. There are two measurements: Reference Signal Received

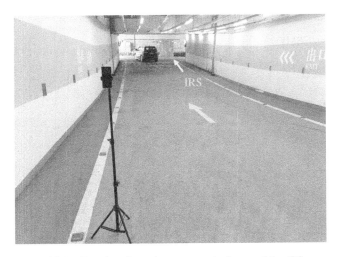

Figure 12.7   The view from the user terminal toward the IRS.

Power (RSRP) and Signal-to-Interference-plus-Noise Ratio (SINR). Notice that the SNR cannot be measured directly because of co-channel interference. Following the definition of the SNR boost, we let the RSRP (or SINR) boost be the ratio between the achieved RSRP (or SINR) and the baseline RSRP (or SINR) without IRS. Figure 12.9 shows the RSRP boosts achieved by the various methods. It can be seen that CSM outperforms the other methods significantly. CSM gives an approximately 5 dB improvement upon CSM and OFF. As shown in the figure, although CSM encounters two sharp drops, which are due to the shadowing effect caused by vehicles, its overall performance is still more consistent over time than RMS and OFF. Observe from Fig. 12.9 that the RSRP boost by OFF is mostly below 0 dB; the reason is that the reflected signals without proper beamforming can result in a destructive superposition. Observe also that RMS yields the worst RSRP performance, even 4 dB lower than not using IRS. This surprising result indicates that, in a complicated wireless environment with interference and noise, the beamformer decision based on the best single sample is not reliable.

We further compare the SINR boosts of the various methods in Fig. 12.8. It can be seen that the SINR boosts and the RSRP boosts have similar profiles (Fig. 12.9). The average RSRP boosts and the average SINR boosts are summarized in Table 12.1. According to Table 12.1, the SINR gain is smaller than the RSRP gain. One reason for this gain reduction is that IRS incurs additional reflected interference. Nevertheless, the constructive effect on the desired signals outweighs that on the interfering signals. As a result, CSM can still bring considerable performance gains as compared to the benchmark methods and not using channel state information (CSI).

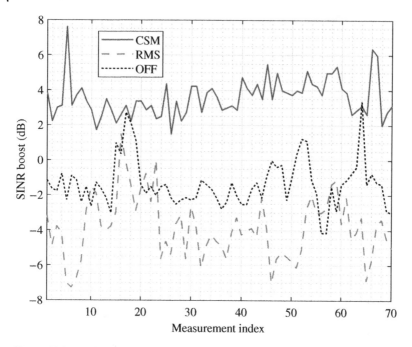

Figure 12.8 SINR boost for SISO transmission.

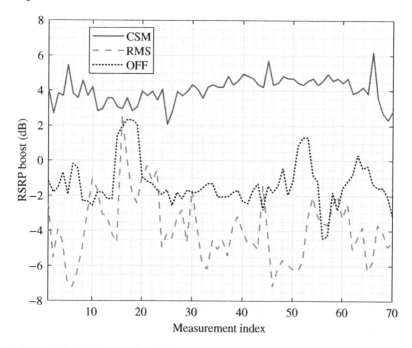

Figure 12.9 RSRP boost for SISO transmission.

Table 12.1 Performance of the different algorithms.

Algorithm	SISO RSRP boost (dB)	SISO SINR boost (dB)	MIMO SE increment (bps/Hz)
CSM	4.02	3.57	2.02
RMS	−3.93	−3.84	1.97
OFF	−1.69	−1.69	0.77

Moreover, we consider the multiple-input multiple-output (MIMO) transmission. In our case, the base station has 64 transmit antennas while the user terminal has four receive antennas, so at most four data streams are supported. Because the base station is a black box to us, how the transmit precoding is performed is unknown. We use the generalized CSM in Section 12.5.3 and let the utility be the *Spectral Efficiency (SE)*. Thus, we measure the SE in bps/Hz at the user terminal for each random sample. Figure 12.10 shows the SE increments by the various algorithms against the baseline SE without IRS. Observe that all the algorithms can bring improvements, although OFF occasionally gives negative effects. The figure shows that RMS becomes more robust in the MIMO case. Actually,

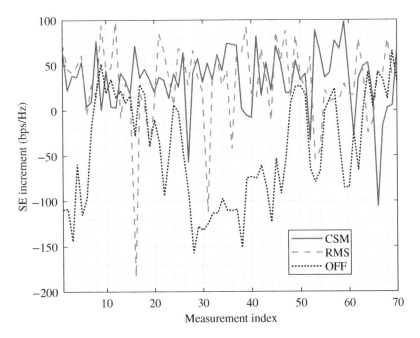

Figure 12.10 SE increment for MIMO transmission.

RMS is sometimes even better than CSM, but it still has inferior performance on average. The average SE increment results summarized in Table 12.1 agree with what we observe from Fig. 12.10.

12.7 Conclusion

In this chapter, we consider passive beamforming for IRS without any channel information, because channel acquisition can be costly and technically difficult in practice. We begin with the standard method of RMS – which simply tries out different beamformer samples at random and chooses the best, but it requires an exponentially large number of samples to achieve a quadratic SNR boost in the number of reflective elements. In contrast, the proposed statistical blind beamforming method called CSM is capable of achieving the quadratic SNR boost by using merely a polynomial number of samples. We then examine CSM from the closest point projection and the phase retrieval points of view. Furthermore, CSM can be enhanced to reach a higher approximation ratio and can be extended to the multiuser multi-antenna scenario by means of utility function. Finally, we demonstrate the effectiveness of CSM in improving the data transmission in a commercial 5G network.

Bibliography

Arun, V. and Balakrishnan, H. (2020). RFocus: beamforming using thousands of passive antennas. *USENIX Symposium on Networked Systems Design and Implementation (NSDI)*, 1047–1061, February 2020.

Elbir, A.M., Papazafeiropoulos, A., Kourtessis, P., and Chatzinotas, S. (2020). Deep channel learning for large intelligent surfaces aided mm-Wave massive MIMO systems. *IEEE Wireless Communications Letters* 9 (9): 1447–1451.

Feng, K., Wang, Q., Li, X., and Wen, C.-K. (2020). Deep reinforcement learning based intelligent reflecting surface optimization for MISO communication systems. *IEEE Wireless Communications Letters* 9 (5): 745–749.

Gao, J., Zhong, C., Chen, X. et al. (2020). Unsupervised learning for passive beamforming. *IEEE Communications Letters* 24 (5): 1052–1056.

Huang, C., Mo, R., and Yuen, C. (2020). Reconfigurable intelligent surface assisted multiuser MISO systems exploiting deep reinforcement learning. *IEEE Journal on Selected Areas in Communications* 38 (8): 1839–1850.

Jiang, T., Cheng, H.V., and Yu, W. (2021). Learning to reflect and to beamform for intelligent reflecting surface with implicit channel estimation. *IEEE Journal on Selected Areas in Communications* 39 (6): 1913–1945.

Liu, S., Gao, Z., Zhang, J. et al. (2020). Deep denoising neural network assisted compressive channel estimation for mmWave intelligent reflecting surfaces. *IEEE Transactions on Vehicular Technology* 69 (8): 9223–9228.

Liu, C., Liu, X., Ng, D.W.K., and Yuan, J. (2021). Deep residual network empowered channel estimation for IRS-assisted multi-user communication systems. *IEEE International Conference on Communications*, June 2021.

Luo, Z.-Q., Ma, W.-K., So, A.M. et al. (2010). Semidefinite relaxation of quadratic optimization problems. *IEEE Signal Processing Magazine* 27 (3): 20–34.

Nadeem, Q.-U.-A., Zappone, A., and Chaaban, A. (2021). Intelligent reflecting surface enabled random rotations scheme for the MISO broadcast channel. *IEEE Transactions on Wireless Communications* 20 (8): 5226–5242.

Ning, B., Chen, Z., Chen, W. et al. (2021). Terahertz multi-user massive MIMO with intelligent reflecting surface: beam training and hybrid beamforming. *IEEE Transactions on Vehicular Technology* 70 (2): 1376–1393.

Psomas, C. and Krikidis, I. (2021). Low-complexity random rotation-based schemes for intelligent reflecting surfaces. *IEEE Transactions on Wireless Communications* 20 (8): 5212–5225.

Ren, S., Shen, K., Zhang, Y. et al. (2021). Configuring intelligent reflecting surface with performance guarantees: blind beamforming. *IEEE Transactions on Wireless Communications* 22 (5): 3355–3370. https://kaimingshen.github.io/doc/BlindBeamforming.pdf.

Shen, K. and Yu, W. (2018). Fractional programming for communication systems—Part I: power control and beamforming. *IEEE Transactions on Signal Processing* 66 (10): 2616–2630.

Shi, Q., Razaviyayn, M., Luo, Z.-Q., and He, C. (2011). An iteratively weighted mmse approach to distributed sum-utility maximization for a MIMO interfering broadcast channel. *IEEE Transactions on Signal Processing* 59 (9): 4331–4340.

Wang, W. and Zhang, W. (2021). Joint beam training and positioning for intelligent reflecting surface assisted millimeter wave communications. *IEEE Transactions on Communications* 20 (10): 6282–6297.

Wang, P., Fang, J., Zhang, W., and Li, H. (2022). Fast beam training and alignment for IRS-assisted millimeter wave/Terahertz systems. *IEEE Transactions on Wireless Communications* 21 (4): 2710–2724.

You, C., Zheng, B., and Zhang, R. (2020). Fast beam training for IRS-assisted multi-user communications. *IEEE Wireless Communications Letters* 9 (11): 1845–1849.

13

RIS in Wireless Information and Power Transfer

Yang Zhao and Bruno Clerckx

Department of Electrical and Electronics Engineering, Imperial College London, London, UK

13.1 Introduction

13.1.1 WPT and WIPT

Koomey's law (Koomey et al. 2011) predicts the computing energy efficiency roughly doubles every 19 months and the amount of battery needed for the same operation decreases to 1% in a decade. Meanwhile, the number of connected devices is estimated to reach 41.2 billion by the end of 2025, with around 75% related to IOT (Vailshery 2022). The upsurge in the number of low-power wireless devices not only challenges spectrum allocation and network coordination, but also calls for a sustainable energy solution. Most existing autonomous devices are powered by batteries that need frequent recharging/replacement, whose maintenance cost has become a major limiting factor for large-scale wireless networks. Although new charging techniques, such as solar, piezoelectric, and near-field (nonradiative) electromagnetic coupling, have been extensively studied and applied in the past decades, their applications in IOT remain questionable due to the bulky energy converter, unstable source dependency, and limited operation range. One promising solution is far-field (radiative) wireless power supply via electromagnetic waves, which is further classified into WEH and radiative WPT.[1] In the case of WEH, low-power nodes capture ambient signals (e.g., radio, television, and Wi-Fi) from legacy transmitters in a space-to-point manner, and thus require an omnidirectional antenna pattern, wideband/multiband capability, and polarization-insensitive characteristics (Zhou et al. 2022). On the other hand, the aim of WPT is to jointly design the dedicated signal and system architecture

---

1 In the following part of this chapter, WPT refers to radiative WPT.

Intelligent Surfaces Empowered 6G Wireless Network, First Edition.
Edited by Qingqing Wu, Trung Q. Duong, Derrick Wing Kwan Ng, Robert Schober, and Rui Zhang.

to boost end-to-end power efficiency (Clerckx et al. 2019). High directivity and narrow bandwidth are generally involved in such a point-to-point scenario for broader coverage and higher receive power.

WPT brings numerous opportunities to future wireless networks. First, it eliminates wire connections, avoids physical contacts, and shrinks the size of batteries. This results in reduced dimension, weight, and manufacturing cost for low-power electronic devices. Second, the radiated energy can be delivered on demand in a adjustable, sustainable, and reliable manner, in contrast to other intermittent energy sources. Finally and most importantly, radio waves can carry power and information simultaneously, and WPT can be jointly designed with WIT to make the most of radiation, spectrum, and infrastructures, i.e., WIPT. A unified WIPT system can integrate power supply and communications to provide an ultimate platform for trillions of low-power devices to charge, operate, and connect in a ubiquitous manner. Such a capability can also address the need for cost-effective, energy-efficient, eco-friendly, and environment-insensitive IOT. However, various cross-topic challenges from communication theory, information theory, circuit theory, RF design, signal processing, protocol design, optimization, prototyping, and experimentation need to be resolved for the paradigm shift toward WIPT.

13.1.2 RIS

Since the power density of radiative field decreases drastically with the propagation distance (a.k.a. inverse-square law in RF engineering and path loss in wireless communications), the low RF-to-RF efficiency is considered the main bottleneck for WPT and WIPT. Reconfigurable intelligent surface (RIS) is a promising technique to engineer the wireless channel for enhanced spectrum and energy efficiencies. A RIS is a smart programmable reflector consisting of numerous low-power and low-cost elements, whose amplitude and/or phase responses can be adjusted in real-time to realize signal enhancement, interference suppression, NLOS bypassing, and scattering enrichment. It recycles and redistributes RF signals to boost the WIPT performance with no additional hardware cost for the target devices. Thanks to the reflecting characteristics, a passive RIS[2] operates without power-hungry RF chains (i.e., power amplifier, phase shifter, attenuator, mixer, and analog-to-digital converter) and introduces no additional thermal noise on the reflected signal. The architecture mainly consists of three stacked layers and a smart controller:

---

2 In the following part of this chapter, RIS refers to passive RIS.

- *Outer layer*: Numerous non-resonant reflective patches with sub-wavelength dimension and spacing, fabricated from metamaterial units (i.e., metallic or dielectric materials with tunable effective permittivity or permeability) (Zhou et al. 2022) or scattering antennas (Liang et al. 2022);
- *Middle layer*: A ground copper plate reflecting residual signals to avoid leakage;
- *Inner layer*: A circuit board adjusting the amplitude and phase properties of reflective patches via analog (e.g., varactor diodes and liquid crystal) or digital components (e.g., PIN diodes and electromechanical switches) (Abbasi et al. 2021);
- *Controller*: An integrated microcontroller or FPGA coordinating with the network and controlling the circuit components.

Such a novel adaptive channel reconfiguration technique also comes with appealing features like compact, lightweight, mountable, scalable, and full-duplex, which may be seamlessly integrated into (and inherently supported by) WIPT networks.

13.1.3 RIS in WPT and WIPT

The main advantages of introducing RIS to WIPT are summarized below:

- *Range*: RIS operates as an adaptive reflecting and focusing mirror/lens to confine and concentrate electromagnetic waves for extended WIPT coverage;
- *Efficiency*: Non-reflected and reflected signals can add constructively at desired directions to boost the receive power and SNR;
- *Propagation*: Proper deployment of RIS (e.g., at static LOS location) creates additional propagation paths for improved reliability;
- *Accessibility*: Dedicated WIPT zones can be established near the RIS with lower hardware cost and energy consumption;
- *Safety*: Energy signals are split into direct and reflective components to mitigate the potential high directional beams toward users;
- *Diversity*: The wireless channel can be flexibly evolved to balance information and power performance and meet specific network demands.

In the following context, we first introduce the fundamentals and challenges of WPT/WIPT, then reveals how RIS unlocks wireless channel control to enhance end-to-end power efficiency and information-energy tradeoff. We aim to provide a tutorial overview of existing RIS-aided WPT/WIPT literature and discuss some state-of-the-art simulation and experiment results. Topics covered in this chapter include RF circuits, waveform design, active/passive beamforming, harvester combining, channel acquisition, prototypes, and experiments.

13.2 RIS-Aided WPT

13.2.1 WPT Architecture

A closed-loop WPT system architecture is illustrated in Fig. 13.1. The RF signal is generated and radiated by the ET, propagated through a wireless channel, captured by the antenna(s) at the ER, converted to DC power by rectifier(s), then passed to the power management unit. Upon successful harvesting, the DC power is either passed directly to the device, or stored in a battery/super capacitor for future operations. When a reverse communication link is available, it is possible to acquire CSI at the ET for signal optimization. Such a closed-loop WPT can provide complete system control to maximize the end-to-end power transfer efficiency

$$e = \frac{P^r_{dc}}{P^t_{dc}} = \underbrace{\frac{P^t_{rf}}{P^t_{dc}}}_{e_1} \underbrace{\frac{P^r_{rf}}{P^t_{rf}}}_{e_2} \underbrace{\frac{P^r_{dc}}{P^r_{rf}}}_{e_3} \underbrace{\frac{P^s_{dc}}{P^r_{dc}}}_{e_4} , \qquad (13.1)$$

where P^t_{dc}, P^t_{rf}, P^r_{rf}, P^r_{dc}, P^s_{dc} are power levels in Fig. 13.1 and e_1, e_2, e_3, e_4, respectively, denote the DC-to-RF, RF-to-RF, RF-to-DC, and DC-to-DC conversion efficiencies, respectively. Extensive efforts have been contributed from RF, wireless communications, and power electronic communities to boost specific energy conversion efficiencies via efficient power amplifier, adaptive signal, and channel, appropriate rectenna (combination of antenna and rectifier), and effective DC-to-DC converter designs. However, optimizing e_1–e_4 independently does not necessarily maximize the end-to-end power efficiency e due to the coupling between modules and the characteristics of power electronics. In particular, the RF-based energy harvesting circuits involve nonlinear devices such as diodes and capacitors, and the modeling and analysis are subject to practical constraints like diode threshold and reverse-breakdown voltages, devices parasitics, impedance matching, and harmonic generation (Valenta and Durgin 2014).

Figure 13.2(a) and (b) illustrate the equivalent circuit of a lossless antenna and a single-diode half-wave rectifier, respectively. For a typical rectenna, the power

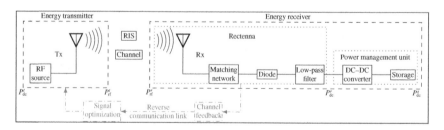

Figure 13.1 Block diagram of a closed-loop adaptive RIS-aided WPT architecture.

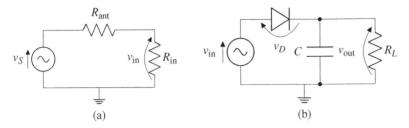

Figure 13.2 Antenna equivalent circuit and a single-diode half-wave rectifier. (a) Antenna equivalent circuit and (b) single-diode half-wave rectifier.

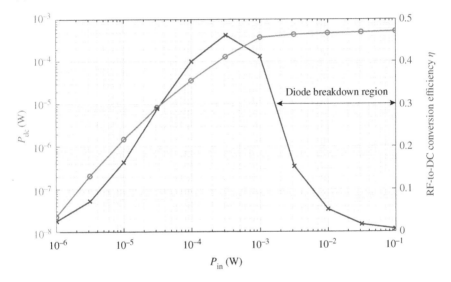

Figure 13.3 Power response and RF-to-DC conversion efficiency of a single-diode half-wave rectifier (Clerckx 2018). The input signal is a continuous wave at 5.18 GHz and rectifier is designed for −20 dBm input power.

response and RF-to-DC efficiency e_3 were obtained by circuit simulation and the results are shown in Fig. 13.3 (Clerckx 2018). Generally speaking, the behavior of any rectenna can be separated into three operation regions:

- *Linear region*: The RF-to-DC conversion efficiency e_3 is constant and the harvester output power is proportional to its input power;
- *Nonlinear (transition) region*: e_3 is significantly higher in this region thanks to the appropriate rectifier input power level;
- *Saturation region*: The harvested power reaches the maximum and the diode works in the reverse breakdown region.

Most existing WPT literature assume the rectenna works in the linear/nonlinear region since the power of radiated signal can be relatively weak in a few meters.

Based on the (truncated) Taylor expansion of the diode characteristic function, Clerckx (Clerckx and Bayguzina 2016) derived a tractable signal model for general unsaturated rectennas, where the output DC voltage v_{out} across the load R_{ant} is a function of the input signal $y_{rf}(t)$:

$$v_{out} = \sum_{i \text{ even}} \beta_i \mathbb{E}[y_{rf}(t)^i], \tag{13.2}$$

where $\beta_i = R_{ant}^{i/2}/i!(nv_t)^{i-1}$, v_t is the thermal voltage, and n is the diode ideality factor. Since (13.2) is a convex function of $y_{rf}(t)^2$, Jensen's inequality suggests

$$\mathbb{E}[y_{rf}(t)^i] \geq (\mathbb{E}[y_{rf}(t)^2])^{\frac{i}{2}} = (P_{rf}^r)^{\frac{i}{2}}. \tag{13.3}$$

(13.2) and (13.3) together suggest

$$v_{out} \geq \sum_{i \text{ even}} \beta_i (P_{rf}^r)^{\frac{i}{2}}. \tag{13.4}$$

The harvester behavior in both linear and nonlinear regions can be modeled by (13.2). When the receive RF power is extremely low (e.g., long-range WPT), v_{out} mainly captures the $i = 2$ term and the rectenna output DC power is proportional to its input RF power. In contrast, when P_{rf}^r can reach the desired rectifier input power level (but below the saturation level), the contribution from $i \geq 4$ terms becomes dominant and the i/o relationship is no longer linear. (13.4) demonstrates that the conventional strategy that maximizes the receive power P_{rf}^r without considering its high-order statistics only maximizes a loose lower bound of P_{dc}^r, and accounting the nonlinear harvester behavior in signal and system design is essential for practical WPT.

On the other hand, (Boshkovska et al. 2015) proposed a nonlinear parametric saturation model for a pre-defined rectenna circuit and input waveform based on curve fitting over measurement results. The harvested DC power is expressed as

$$P_{dc}^r = \frac{\Psi_{dc} - P_{sat}\Omega}{1 - \Omega}, \quad \Psi_{dc} = \frac{P_{sat}}{1 + \exp\left(-a(P_{rf}^r - b)\right)}, \quad \Omega = \frac{1}{1 + \exp(ab)}, \tag{13.5}$$

where the constant P_{sat} is the maximal harvested power when the receive power saturates the rectifier, and the parameters a, b model the nonlinear charging rate with respect to input power and the minimum turn-on voltage of the rectifier, respectively. This model describes the saturation behavior of the rectenna.

The exact boundaries between different harvester regions depend on the circuit layouts, component parameters, and rectenna input signal (Zeng et al. 2017). In particular, signals with a higher PAPR usually encounter the nonlinear and saturation effects at lower rectenna input power levels. For example, the nonlinear region is typically $[-20, 0]$ dBm for a CW (i.e., single sinusoidal) and $[-30, -10]$

dBm for a multisine signal (Del Prete et al. 2016). Overall, the RF-to-DC conversion efficiency e_3 relates to the *power* and *shape* of the rectenna input signal, and thus also depends on the transmit waveform and wireless channel. This motivates adaptive multisine waveform designs (Clerckx and Bayguzina 2016; Zeng et al. 2017; Huang and Clerckx 2017; Shen and Clerckx 2021) to leverage the nonlinear harvester region for enhanced energy efficiency product $e_2 e_3$. Such a coupling effect calls for a joint design of the entire WPT system, including ET, ER, and possibly RIS, to improve the end-to-end power transfer efficiency.

Wireless power and communication systems share many resources like frequency spectrum, propagation medium and RF infrastructures. However, adapting communication techniques such as network coordination, channel acquisition, and multi-antenna cooperation in WPT is nontrivial due to the differences in signal model, system architecture, objective function, and receiver input sensitivity, and signal processing capability. Next, we address relevant design issues and provide an overview of RIS-aided WPT literature.

13.2.2 Waveform and Beamforming

Despite the nonlinear rectenna behavior, when a *single* energy branch (over frequency and/or spatial domains) is fed to the ER, maximizing the receive power is equivalent to maximizing the harvested power. Based on the linear harvester model (Wu et al. 2022) considered weighted sum-receive power maximization for single-carrier single-antenna ERs via joint active and passive beamforming design. Let M, L, K be the number of transmit antennas, reflecting elements and users, respectively, $\theta_l \in [0, 2\pi)$ be the phase shift of RIS element l, $\boldsymbol{w}_k$ be the active beamformer for user k, and $\boldsymbol{\phi}^H \triangleq [e^{j\theta_1}, \ldots, e^{j\theta_L}]$ be the passive beamformer[3] at the RIS. Denote the direct ET-ER k channel as $\boldsymbol{h}_{\mathrm{D},k}^H \in \mathbb{C}^{1 \times M}$, the forward ET-RIS forward channel as $\boldsymbol{H}_{\mathrm{F}} \in \mathbb{C}^{L \times M}$, and the backward RIS-ER k channel as $\boldsymbol{h}_{\mathrm{B},k}^H \in \mathbb{C}^{1 \times L}$. The composite ET-ER k channel can be expressed as $\boldsymbol{h}_k^H \triangleq \boldsymbol{h}_{\mathrm{D},k}^H + \boldsymbol{\phi}^H \mathrm{diag}(\boldsymbol{h}_{\mathrm{B},k}^H) \boldsymbol{H}_{\mathrm{F}}$, and the weighted sum-receive power maximization problem is formulated as

$$\max_{\{\boldsymbol{w}_k\}, \boldsymbol{\phi}} \quad \sum_k \alpha_k \boldsymbol{w}_k^H \boldsymbol{h}_k \boldsymbol{h}_k^H \boldsymbol{w}_k \tag{13.6a}$$

$$\text{s.t.} \quad \sum_k \|\boldsymbol{w}_k\|^2 \leq P \tag{13.6b}$$

$$|\boldsymbol{\phi}| = \mathbf{1}, \tag{13.6c}$$

where α_k is the relative weight of ER k, and P is the average transmit power budget. The authors proposed an AO approach to decouple the non-convex problem, where the optimal active beamformer was obtained by eigen decomposition and

---

3 For simplicity, we assume lossless reflection at all RIS elements throughout this chapter.

the phase shift was updated by SCA and SDR. It was concluded that sending only one common energy beam is sufficient for problem (13.6), and deploying a RIS around ERs can significantly increase the receive power and operation range.

On the other hand, capturing and combining energy from frequency and spatial domains can significantly boost the end-to-end power efficiency. When the power signal occupies multiple frequency bands, the *coupling effect* in the frequency domain creates more balanced DC components after rectification (Clerckx and Bayguzina 2016). For RIS-aided multiuser multi-carrier WPT, Feng (Feng et al. 2022) proposed an adaptive multisine waveform and passive beamforming design to maximize the weighted sum-harvested power, based on the nonlinear harvester model (13.2) truncated to the fourth order. With N subcarriers, the transmit multisine waveform at time t is given as

$$x(t) = \Re\left\{ \sum_n s_n e^{j2\pi f_n t} \right\}, \tag{13.7}$$

where s_n, f_n denote the complex weight and carrier frequency of the nth sinusoidal tone, respectively. When uncorrelated phase-shift control over different frequencies is available at the RIS,[4] the composite ET-ER k channel at subband n is $h_{k,n} \triangleq \boldsymbol{\phi}_n^H \mathrm{diag}(\boldsymbol{h}_{\mathrm{B},k,n}^H)\boldsymbol{h}_{\mathrm{F},n}$, and the signal received by user k at time t is[5]

$$y_k(t) = \Re\left\{ \sum_n h_{k,n} s_n e^{j2\pi f_n t} \right\}. \tag{13.8}$$

For uniformly spaced frequency bands, the rectenna output DC voltage (13.2) can be explicitly written as

$$\begin{aligned}
v_{\mathrm{out},k} = \ &\frac{1}{2}\beta_2 \sum_n (h_{k,n}s_n)(h_{k,n}s_n)^* \\
&+ \frac{3}{8}\beta_4 \sum_{\substack{n_1,n_2,n_3,n_4 \\ n_1+n_2=n_3+n_4}} (h_{k,n_1}s_{n_1})(h_{k,n_2}s_{n_2})(h_{k,n_3}s_{n_3})^*(h_{k,n_4}s_{n_4})^*,
\end{aligned} \tag{13.9}$$

and the weighted sum-harvested power maximization problem is

$$\max_{\{s_n,\boldsymbol{\phi}_n\}} \quad \sum_k \alpha_k v_{\mathrm{out},k} \tag{13.10a}$$

$$\text{s.t.} \quad \|s\|^2/2 \leq P \tag{13.10b}$$

$$\quad |\boldsymbol{\phi}_n| = 1, \quad \forall n, \tag{13.10c}$$

---

4 In practice, reflection coefficients at different frequencies are correlated due to their dependency on the element input impedance. The analysis of frequency-selective/uncorrelated RIS provides an upper bound for the passive beamforming gain.
5 It is assumed the noise power is too small to be harvested.

where $s \triangleq [s_1, \dots, s_n]^T$ and (13.10b) is the average power constraint of multisine. The passive beamforming design subproblem of (13.10a) generalizes that of (13.6) by accounting for harvester nonlinearity and allowing uncorrelated phase shift control at different frequencies. Problem (13.10a) was also solved by AO, where both waveform and passive beamformer were obtained by SCA and SDR.

Figure 13.4 shows the average single-user output DC current versus the number of subbands N and RIS elements L, where "FF" means frequency-flat (i.e., correlated) and "FS" means frequency-selective (i.e., uncorrelated) RIS (Feng et al. 2022). We observe that increasing N and L both enhances the harvested power at the user, where the former produces more balanced DC components and the latter boosts the received signal power (thus further exploits harvester nonlinearity). It was summarized that:

- The proposed design leverages the beamforming gain, frequency-diversity gain, and rectifier nonlinearity to boost the end-to-end power efficiency;
- RIS significantly improves the harvested power, and uncorrelated phase shift at different frequencies can further increase the passive beamforming gain;

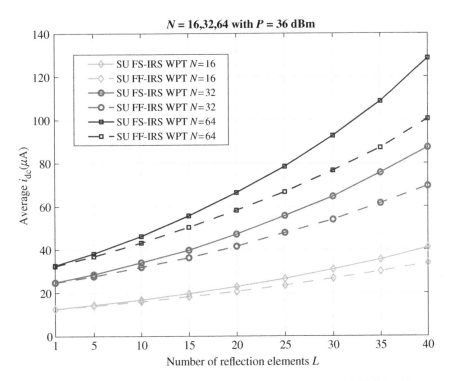

Figure 13.4 Average single-user output DC current versus N and L at 5.18 GHz with 10 MHz bandwidth and 15 m ET-ER distance. Source: (Feng et al. 2022)/IEEE.

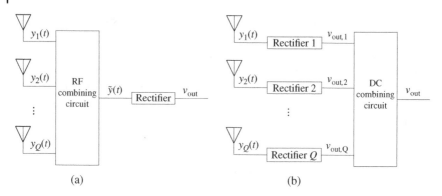

Figure 13.5 Combining techniques for multiantenna ERs. (a) RF combing and (b) DC combining.

- Both waveform power allocation and RIS phase shift design experience a trade-off between favoring multiple frequencies (to exploit rectifier nonlinearity) and favoring a single subband (to exploit channel frequency selectivity).

So far we considered single antenna at the WPT receiver. When the power signal can be captured by multiple receive antennas, there exist two combining techniques as illustrated in Fig. 13.5:

- *RF combining*: The signal components are coherently combined in the spatial domain and then fed into one rectifier;
- *DC combining*: The signal components are immediately rectified by individual rectifiers and then accumulated in the power domain.

RF combining can be realized with an equal-gain combiner and one phase shifter attached to each receive antenna (i.e., analog receive beamforming). The concept is somewhat similar to ideal RIS, but the combination happens within the receiver circuit rather than propagation environment. In RF combining, the rectifier requires the phase information of all signal branches for a constructive superposition. It has to be designed for higher RF input power. In DC combing, phase estimation is unnecessary and the rectifiers are designed for lower input power (Shen and Clerckx 2021). For RIS-aided single-user multiple-input multiple-output (MIMO) WPT with DC combining, (Yue et al. 2022) proposed a joint active and passive beamforming design to maximize the harvested DC power for CW, QAM, and Gaussian waveforms based on the saturation harvester model (13.5). The authors concluded Gaussian waveform is preferred at low harvester input power while CW is preferred at higher input power.

13.2.3 Channel Acquisition

Adaptive waveform and beamforming designs require the CSI of direct ET-ER and cascaded ET-RIS-ER links. However, accurate and efficient channel estimation can be very challenging due to the large number of extra channels introduced by the RIS, as well as the potential lack of estimators/oscillators at both RIS and ER. Next, we introduce several common CSI acquisition approaches for single- and multiuser RIS-aided WPT.

13.2.3.1 Direct Channel

Figure 13.6 illustrates four-channel acquisition schemes at the ET (Clerckx et al. 2022):

- *Forward-link training*: The ET generates pilots, while the ER estimates the downlink CSI and feeds back to the ET;
- *Reverse-link training*: The ER generates pilots, while the ET estimates the uplink CSI and exploits channel reciprocity for the downlink CSI;
- *Codebook-based training (power probing) with limited feedback (Huang and Clerckx 2018)*: The ET sequentially sends every codeword (e.g., waveform and active beam) within a codebook, and the ER returns the index that achieves the largest P_{dc}^r;
- *Backscatter-based training (Mishra and Larsson 2019)*: The ET generates pilots, estimates the round-trip channel from the backscatter signal, then extracts the downlink CSI by channel reciprocity.

Forward and backward link training involve power-hungry operations (signal processing or pilot generation) at the wirelessly powered devices and can be unaffordable in practice. In contrast, power probing can exploit the waveform and beamforming gain with a few bits of CSI feedback (Huang and Clerckx 2018), and the backscatter-based training completely shifts the pilot generation and channel estimation burden to the ET.

13.2.3.2 RIS-Related Channels

To estimate the cascaded CSI in RIS-aided WPT systems, it is possible to combine the forward/reverse-link training with some channel acquisition approaches in RIS-aided WIT systems:

- *Group-based training (Zheng and Zhang 2020)*: Adjacent RIS elements with high channel correlation are grouped into a group sharing one common reflection coefficient;
- *Hierarchical training (You et al. 2020)*: The group size reduces over time blocks and the channel for each element is progressively refined;
- *Reference-based estimation (Wang et al. 2020)*: The estimated CSI for a reference user is used to reduce the training overhead of other users, based on their correlation;

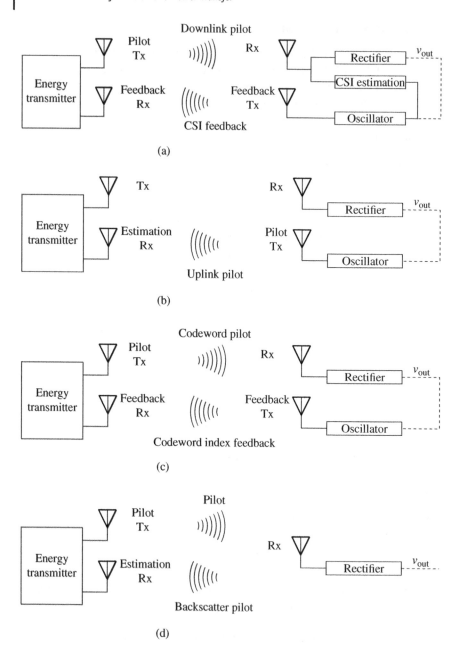

Figure 13.6 Direct CSI acquisition schemes at the ET. (a) Forward-link training, (b) reverse-link training, (c) codebook-based training, (d) backscatter-based training.

- *Anchor-assisted training (Guan et al. 2020)*: Active anchors help to estimate the common ET-RIS channel of all users and reduce the individual training cost.

It is noteworthy the RIS controller is an active device attached to the RIS through physical connections, and the channel in between is typically LOS. Inspired by this, (Wu et al. 2022) proposed a codebook-based active and passive beam training for practical RIS-aided WPT systems. In the active beam training phase, the RIS controller transmits consecutive pilots, while the ET estimates the ET-controller and ET-RIS-controller channels. The active beamforming vector is then trained to maximize the receive power at the controller. In the passive beam training phase, the ET transmits pilots with the trained beam, the RIS traverses every passive beam in the codebook, and the ER feeds back the best codeword index to the controller.

13.2.4 Prototype and Experiments

We now introduce two recent experimental setups and prototypes for RIS-aided WPT. A medium-range 5.8 GHz RIS-aided single-user MIMO WPT prototype was built in (Tran et al. 2022b). Its modules are presented in Fig. 13.7(a)–(c).

The ET is equipped with an 8×8 phased planar array over the stacked power divider, phase control, and antenna amplifier boards. The ER employs a 4×4 planar array with DC combining for reduced training overheads. The RIS incorporates 16×16 one-bit PIN diodes resonating at 5.8 GHz over an FPGA control board for serial-to-parallel conversion. Due to the difficulties of accurate signal-phase acquisition at the ER and frequent reflection pattern switching at the RIS, the authors proposed a multi-tile beam scanning method. It combines the codebook-based and the group-based active and passive beam training to maximize the receive power at the ER. Fig. 13.7(d) shows the experiment setup, where the total transmit power is 2.1 W and the horizontal distance between the RIS and ET/ER is 2 m. The proposed beam training method can achieve a 20 dB receive power gain compared to the case where all RIS elements are turned off.

The work was then extended to the multiuser scenario, where three multifocus techniques were proposed to formulate multiple beams in the radiative near-field (Fresnel region) with adjustable power levels (Tran et al. 2022a). In particular, *pattern addition* superposes the required reflection coefficients at various beam directions, *random unit cell interleaving* allocates different unit cells to different beams, and *tile division* assigns each RIS tile to the ER with the highest channel gain. In the simulation setup, a 16×32 RIS is divided into 32 tiles, the distances between devices are around 2 m, and all other parameters are the same as above. Figure 13.8 compares the measured receive power regions achieved by those multifocus techniques. The power allocated to each beam can be flexibly

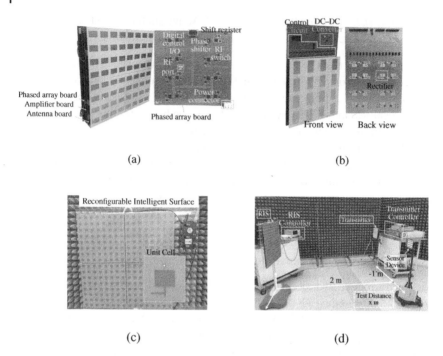

(a)

(b)

(c)

(d)

Figure 13.7 5.8 GHz RIS-aided MIMO WPT prototype testbed. (a) Phased array transmitter, (b) receiver, (c) RIS, and (d) experiment setup. Source: Nguyen Minh Tran et al. 2022/Reproduced from IEEE.

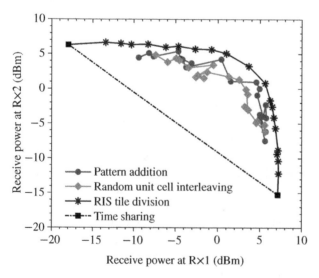

Figure 13.8 Measured receive power regions for different multifocus techniques. Source: (Tran et al. 2022a)/IEEE.

adjusted by the proposed multifocus schemes to improve the energy efficiency and tradeoff at multiple ERs.

13.3 RIS-Aided WIPT

13.3.1 WIPT Categories

Based on information and energy flows, WIPT can be classified into the three categories shown in Fig. 13.9:

- *SWIPT*: Energy and information are simultaneously delivered in the downlink to co-located and/or separated information receiver(s) and energy receiver(s);
- *WPCN*: Energy is delivered in the downlink to active device(s) that generate and transmit information-bearing signals in the uplink;
- *WPBC*: Energy is delivered in the downlink to passive node(s) that reflect part of incoming waves and embed information in the uplink.

Thanks to the backscatter properties, the power consumption of WPBC nodes can be several orders of magnitude lower than conventional active transmitters (Boyer and Roy 2014). This accounts for the great success of WPBC applications such as RFID and passive sensor networks in modern IOT networks. However, SWIPT/WPCN nodes typically employ complex signal processors/oscillators for reception/transmission and thus require a decent amount of power to be harvested. Next, we provide an overview of RIS-aided WIPT networks, with an emphasis on SWIPT.

13.3.2 RIS-Aided SWIPT

13.3.2.1 SWIPT Architecture

The random time-varying transmit signal of SWIPT carries information and power simultaneously. It can be demodulated and harvested by co-located or separated

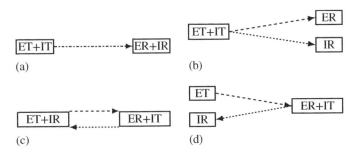

Figure 13.9 Information and energy flows in different WIPT schemes. Dashed and dotted lines denote energy and information flows, respectively. (a) SWIPT with co-located receivers, (b) SWIPT with separated receivers, (c) monostatic WPCN/WPBC, and (d) bistatic WPCN/WPBC.

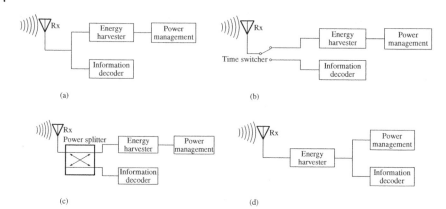

Figure 13.10 Single-antenna SWIPT receiver architectures. (a) Ideal receiver, (b) time switching receiver, (c) power splitting receiver, and (d) integrated receiver.

information and energy receivers. Figure 13.10 illustrates four single-antenna co-located SWIPT receiver architectures (Clerckx et al. 2022):

- *Ideal receiver*: The entire received signal is simultaneously used for information decoding and energy harvesting;
- *TS receiver*: Each transmission block is divided into orthogonal WIT and WPT phases, where the transmitter employs dedicated signals and the receiver activates the decoder and harvester, respectively;
- *PS receiver*: The transmit signals are jointly optimized for WIT and WPT, while the received signal is split in the power domain and fed into the decoder and harvester, respectively;
- *Integrated receiver (Kim and Clerckx 2022)*: The transmit signal is modulated in properties that can be preserved after rectification (e.g., pulse position), while the received signal is harvested and then decoded.

In SWIPT, the tradeoff between WIT and WPT is often characterized by the RE region. It consists of all pairs of achievable transmission rate and harvested power under certain design constraints. The ideal receiver simultaneously achieves the optimal WIT and WPT performance. TS and PS are two commonly considered practical receiving schemes. Different RE point are achieved by adjusting the transmit signals as well as the phase duration and power splitting ratio, respectively. In contrast, integrated receiver eliminates the need for RF chains and complicated architectures, but suffers information degradation and is more suitable for low-throughput applications.

13.3.2.2 Waveform and Beamforming
Similar to the WPT case, adaptive waveform and beamforming design can enlarge the achievable RE region of a SWIPT user, especially in the presence of a RIS.

For the multi-carrier scenario, it has been proved that a modulated Gaussian waveform (for WIT) superposed to an unmodulated multisine (for WPT) can significantly enhance the RE tradeoff for a single user (Clerckx 2018). In such cases, the hybrid transmit waveform at time t is expressed as

$$x(t) = \Re \left\{ \sum_n (w_{I,n}\tilde{x}_{I,n}(t) + w_{P,n})e^{j2\pi f_n t} \right\}, \tag{13.11}$$

where $w_{I,n}/w_{P,n}$ is the information/power precoder at subband n, and $\tilde{x}_{I,n} \sim \mathcal{CN}(0,1)$ is the information symbol at subband n. The received signal at the single-antenna SWIPT user is

$$y(t) = \Re \left\{ \sum_n \left(h_n^H(w_{I,n}\tilde{x}_{I,n}(t) + w_{P,n}) + \tilde{n}_n(t) \right) e^{j2\pi f_n t} \right\}, \tag{13.12}$$

where $h_n \triangleq \phi^H \operatorname{diag}(h_{B,n}^H)h_{F,n}$ is the equivalent channel[6] at subband n and $\tilde{n}_n(t)$ is the RF noise at subband n. Note that the modulated component can be used for harvesting if necessary, but the deterministic multisine component carries no information and creates no interference (by waveform cancelation or translated demodulation) (Clerckx 2018). Therefore, the achievable transmission rate is

$$R = \sum_n \log_2 \left(1 + \frac{(1-\rho)|h_n^H w_{I,n}|^2}{\sigma_n^2} \right), \tag{13.13}$$

where ρ is the power splitting ratio for WPT, and σ_n^2 is the variance of the total noise (at the RF-band and during RF-to-baseband conversion) on subband n. Due to the additional coupling between the modulated and multisine waveforms, the rectenna output DC voltage (13.2) is explicitly expressed as

$$
\begin{aligned}
v_{\text{out}} = {} & \frac{1}{2}\beta_2\rho \sum_n (h_n^H w_{I,n})(h_n^H w_{I,n})^* + (h_n^H w_{P,n})(h_n^H w_{P,n})^* \\
& + \frac{3}{8}\beta_4\rho^2 \Bigg[2\left(\sum_n (h_n^H w_{I,n})(h_n^H w_{I,n})^* \right)^2 \\
& + \sum_n 4(h_n^H w_{I,n})(h_n^H w_{I,n})^*(h_n^H w_{P,n})(h_n^H w_{P,n})^* \\
& + \sum_{\substack{n_1,n_2,n_3,n_4 \\ n_1+n_2=n_3+n_4}} (h_{n_1}^H w_{P,n_1})(h_{n_2}^H w_{P,n_2})(h_{n_3}^H w_{P,n_3})^*(h_{n_4}^H w_{P,n_4})^* \Bigg].
\end{aligned} \tag{13.14}
$$

---

6 Here we assume $\phi = \phi_n$ for all n, but the discussion also applies to the frequency-selective/uncorrelated RIS.

On top of this, we can jointly optimize the waveform, splitting ratio, active, and passive beamforming to characterize the RE boundary for a PS receiver by maximizing the output DC voltage subject to sum-rate, average transmit power, and RIS amplitude constraints (Zhao et al. 2022):

$$\max_{\{w_{I,n}, w_{P,n}\}, \phi, \rho} \quad v_{\text{out}} \tag{13.15a}$$

$$\text{s.t.} \quad R \geq \overline{R} \tag{13.15b}$$

$$\sum_n \left(\|w_{I,n}\|^2 + \|w_{P,n}\|^2 \right) / 2 \leq P \tag{13.15c}$$

$$|\phi| = 1 \tag{13.15d}$$

$$0 \leq \rho \leq 1, \tag{13.15e}$$

where $\overline{R}$ is the guaranteed transmission rate. The RE region can be characterized by successively varying $\overline{R}$ for the PS receiver, and time sharing between the WIT and WPT points for the TS receiver. Problem (13.15a) generalizes (13.6) and (13.10a) by introducing an information waveform at the hybrid transmitter and a power splitter at the SWIPT receiver. It was solved by the BCD method.

Figure 13.11 reveals the dissimilar preferences of WIT and WPT on passive beamforming. The WIT-optimized RIS provides a fair amplitude gain over all subchannels to improve the sum rate, reminiscent of water-filling power allocation at high SNR. In comparison, the WPT-optimized RIS further leverages the strong subbands to exploit the rectifier nonlinearity and enhance the harvested power.

The average RE regions for PS and TS receivers under different number of subbands are presented in Fig. 13.12. As N increases, each subband receives less power and the per-subband rate is reduced, while the frequency coupling effect creates more balanced DC terms to boost the harvested energy. Besides, PS is preferred at small N and TS is preferred at large N, and a combination of PS and TS is generally the best strategy. This is fundamentally different from the observation over the linear harvester model that PS always outperform TS. It was also concluded that:

- The flexible subchannel shaping capabilities of RIS can significantly improve the information-energy tradeoff;
- Active (resp. passive) beamforming enjoys an array gain of L (resp. L^2). The power scaling order is at least L^2 (resp. L^4) when the rectifier is not saturated;

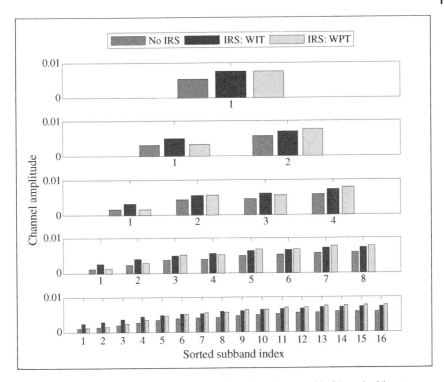

Figure 13.11 Sorted equivalent subchannel amplitude versus *N* with and without an 100-element RIS at 2.4 GHz with 10 MHz bandwidth and 12 m transmission distance. Soruce: (Zhao et al. 2022)/IEEE.

- For frequency-flat channels, the optimal active and passive beamforming for any RE point is also optimal for the whole RE region.

13.3.2.3 Channel Acquisition

The aforementioned channel estimation techniques for RIS-aided WIT and WPT can be straightforwardly applied to RIS-aided SWIPT with separated and/or co-located information and energy receivers. Besides, the beam training protocol for WPT was also extended to SWIPT (Wu et al. 2022). In the active beam training phase, both RIS controller and information receivers send pilots, while the hybrid transmitter first estimates all relevant channels, then optimizes one energy beam to the RIS controller and multiple information beams to the information receivers. In the passive beam training phase, the hybrid transmitter sends pilots, the RIS consecutively reflects every passive beam in the codebook, and the

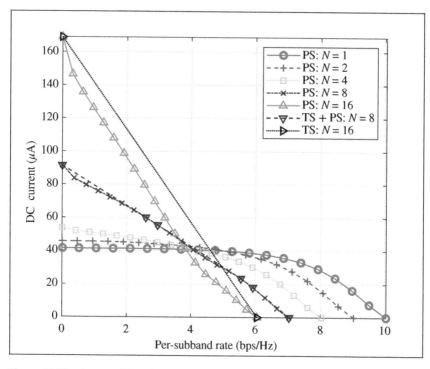

Figure 13.12 Average RE region versus N with a 20-element RIS at 2.4 GHz with 10 MHz bandwidth and 12 m transmission distance. Source: (Zhao et al. 2022)/IEEE.

energy receivers report the best codeword indexes to the controller. The final RIS reflection pattern is determined by a time-sharing of different passive beams.

13.3.3 RIS-Aided WPCN and WPBC

WPCN and WPBC both deliver power in the downlink and transmit information in the uplink. The former involves dedicated energy accumulation and information transmission phases in the time domain, while the latter performs absorption and reflection in the power domain. The behavior is similar to TS and PS, but now the information is carried in the uplink. Many previous discussions on RIS-aided SWIPT can therefore be extended to RIS-aided WPCN and WPBC. We briefly review some specific issues and applications in this subsection.

Adding a RIS to WPCN/WPBC can provide a two-fold benefit: In the downlink, the passive beamforming gain boosts the receive signal strength for a reduced charging time/enhanced reflection efficiency. In the uplink, it also improves the SNR at the hybrid AP. In the multiuser scenario, the resource allocation and beamforming design depend on practical multiple-access schemes. Since the charging time for active transmission can be relatively long, TDMA is usually considered as a benchmark for WPCN where one user transmits at a time. The other users can harvest power (Hua and Wu 2021), relay information (Zheng et al. 2020; Lyu et al. 2021), or simply be idle (Xu et al. 2021; Hua et al. 2022). At the cost of additional network coordination, dynamic RIS control during each slot provides a higher passive forming gain. Besides, power-domain NOMA for WPCN was also studied in (Wu et al. 2021) to maximize the sum throughput via joint time allocation and passive beamforming optimization. Under the linear harvester model, the optimal RIS phase shift in the downlink WPT and uplink WIT phases were demonstrated to be completely the same. A hybrid scheme of intra-cluster NOMA and inter-cluster TDMA can also increase the system throughput, reduce the successive interference cancelation complexity, and preserve the dynamic passive beamforming gain (Zhang et al. 2021). On the other hand, both WPBC and RIS operate over scatter/reflect principles. The former (resp. latter) exploits the reflection pattern for information modulation (resp. channel reconfiguration), absorbs part of incident waves for energy harvesting (resp. full reflection), and does not require CSI knowledge (resp. does need CSI). The integration of RIS and WPBC not only improves the data rate, operation range, and link reliability, but also provides more design flexibility for novel backscatter protocols. For example, AMBC recycles ambient signals from legacy transmitters (e.g., radio, television, Wi-Fi) to transmit without dedicated carrier emitter and frequency spectrum (Liu et al. 2013), and the detection performance is mainly restricted by the strong direct-link interference. In this case, RIS can enable destructive interference superposition at the AMBC receiver to improve the system reliability (Nemati et al. 2020; Idrees et al. 2022). A RIS-aided AMBC system using Wi-Fi signal was built and tested in (Fara et al. 2021), and the bit error rate advantage of passive beam training was experimentally validated.

13.4 Conclusion

In this chapter, we presented how RIS operates as a key enabling technology to complete the end-to-end (transmitter, channel, and receiver) design for wireless information and power transfer. We first revealed the potential of RIS to address the requirements and challenges of WIPT, then explained the motivation and importance of integrated RF, signal, and system design, with an emphasis on the nonlinear harvester gain and how it can be further leveraged by passive beamforming. An overview on RIS-aided WPT and WIPT was also provided, where we drew particular attention to RF circuits, waveform design, active/passive beamforming, harvester combining, channel acquisition, prototypes, and experiments. It was concluded that the powerful and flexible channel reconfiguration capability of RIS can significantly boost the end-to-end power efficiency for WPT, enhance the information-energy tradeoff for SWIPT, and provide more design flexibility for WPCN and WPBC.

Bibliography

Abbasi, Q.H., Abbas, H.T., Alomainy, A., and Imran, M.A. (2021). *Backscattering and RF Sensing for Future Wireless Communication*. Wiley. ISBN 9781119695721 https://doi.org/10.1002/9781119695721. https://onlinelibrary.wiley.com/doi/book/10.1002/9781119695721.

Boshkovska, E., Ng, D.W.K., Zlatanov, N., and Schober, R. (2015). Practical non-linear energy harvesting model and resource allocation for SWIPT systems. *IEEE Communications Letters* 19: 2082–2085. https://doi.org/10.1109/LCOMM.2015.2478460.

Boyer, C. and Roy, S. (2014). Backscatter communication and RFID: coding, energy, and MIMO analysis. *IEEE Transactions on Communications* 62: 770–785. https://doi.org/10.1109/TCOMM.2013.120713.130417.

Clerckx, B. (2018). Wireless information and power transfer: nonlinearity, waveform design, and rate-energy tradeoff. *IEEE Transactions on Signal Processing* 66: 847–862. https://doi.org/10.1109/TSP.2017.2775593.

Clerckx, B. and Bayguzina, E. (2016). Waveform design for wireless power transfer. *IEEE Transactions on Signal Processing* 64: 6313–6328. https://doi.org/10.1109/TSP.2016.2601284.

Clerckx, B., Zhang, R., Schober, R. et al. (2019). Fundamentals of wireless information and power transfer: from RF energy harvester models to signal and system designs. *IEEE Journal on Selected Areas in Communications* 37: 4–33. https://doi.org/10.1109/JSAC.2018.2872615.

Clerckx, B., Kim, J., Choi, K.W., and In Kim, D. (2022). Foundations of wireless information and power transfer: theory, prototypes, and experiments. *Proceedings of the IEEE* 110: 8–30. https://doi.org/10.1109/JPROC.2021.3132369.

Del Prete, M., Costanzo, A., Magno, M. et al. (2016). Optimum excitations for a dual-band microwatt wake-up radio. *IEEE Transactions on Microwave Theory and Techniques* 64: 4731–4739. https://doi.org/10.1109/TMTT.2016.2622699.

Fara, R., Phan-Huy, D.-T., Ratajczak, P. et al. (2021). Reconfigurable intelligent surface-assisted ambient backscatter communications–experimental assessment. *2021 IEEE International Conference on Communications Workshops (ICC Workshops)*, 1–7. IEEE. ISBN 978-1-7281-9441-7 https://doi.org/10.1109/ICCWorkshops50388.2021.9473842.

Feng, Z., Clerckx, B., and Zhao, Y. (2022). Waveform and beamforming design for intelligent reflecting surface aided wireless power transfer: single-user and multi-user solutions. *IEEE Transactions on Wireless Communications* 21 (7): 5346–5361. https://doi.org/10.1109/TWC.2021.3139440.

Guan, X., Wu, Q., and Zhang, R. (2020). Anchor-assisted intelligent reflecting surface channel estimation for multiuser communications. *GLOBECOM 2020 - 2020 IEEE Global Communications Conference*, volume 2020-January, 1–6. IEEE, December 2020. ISBN 978-1-7281-8298-8 https://doi.org/10.1109/GLOBECOM42002.2020 .9347985. https://ieeexplore.ieee.org/document/9347985/.

Hua, M. and Wu, Q. (2021). Joint dynamic passive beamforming and resource allocation for IRS-aided full-duplex WPCN. *IEEE Transactions on Wireless Communications* 21 (7): 4829–4843. https://doi.org/10.1109/twc.2021.3133491.

Hua, M., Wu, Q., and Poor, H.V. (2022). Power-efficient passive beamforming and resource allocation for IRS-aided WPCNs. *IEEE Transactions on Communications* 70: 3250–3265. https://doi.org/10.1109/TCOMM.2022.3161688.

Huang, Y. and Clerckx, B. (2017). Large-scale multiantenna multisine wireless power transfer. *IEEE Transactions on Signal Processing* 65: 5812–5827. https://doi.org/10 .1109/TSP.2017.2739112.

Huang, Y. and Clerckx, B. (2018). Waveform design for wireless power transfer with limited feedback. *IEEE Transactions on Wireless Communications* 17: 415–429. https://doi.org/10.1109/TWC.2017.2767578.

Idrees, S., Jia, X., Durrani, S., and Zhou, X. (2022). Design of intelligent reflecting surface (IRS)-boosted ambient backscatter systems. *IEEE Access* 10: 65000–65010. https://doi.org/10.1109/ACCESS.2022.3184017.

Kim, J. and Clerckx, B. (2022). Wireless information and power transfer for IoT: pulse position modulation, integrated receiver, and experimental validation. *IEEE Internet of Things Journal* 9: 12378–12394. https://doi.org/10.1109/JIOT.2021 .3135712.

Koomey, J., Berard, S., Sanchez, M., and Wong, H. (2011). Implications of historical trends in the electrical efficiency of computing. *IEEE Annals of the History of Computing* 33: 46–54. https://doi.org/10.1109/MAHC.2010.28.

Liang, Y.C., Zhang, Q., Wang, J. et al. (2022). Backscatter communication assisted by reconfigurable intelligent surfaces. *Proceedings of the IEEE* 110 (9): 1339–1357. https://doi.org/10.1109/JPROC.2022.3169622.

Liu, V., Parks, A., Talla, V. et al. (2013). Ambient backscatter: wireless communication out of thin air. *ACM SIGCOMM Computer Communication Review* 43: 39–50. https://doi.org/10.1145/2534169.2486015.

Lyu, B., Ramezani, P., Hoang, D.T. et al. (2021). Optimized energy and information relaying in self-sustainable IRS-empowered wpcn. *IEEE Transactions on Communications* 69: 619–633. https://doi.org/10.1109/TCOMM.2020.3028875.

Mishra, D. and Larsson, E.G. (2019). Multi-tag backscattering to MIMO reader: channel estimation and throughput fairness. *IEEE Transactions on Wireless Communications* 18: 5584–5599. https://doi.org/10.1109/TWC.2019.2937763.

Nemati, M., Ding, J., and Choi, J. (2020). Short-range ambient backscatter communication using reconfigurable intelligent surfaces. *IEEE Wireless Communications and Networking Conference, WCNC* (1–6 May 2020). ISSN 15253511. https://doi.org/10.1109/WCNC45663.2020.9120813.

Shen, S. and Clerckx, B. (2021). Beamforming optimization for MIMO wireless power transfer with nonlinear energy harvesting: RF combining versus DC combining. *IEEE Transactions on Wireless Communications* 20: 199–213. https://doi.org/10 .1109/TWC.2020.3024064.

Tran, N.M., Amri, M.M., Park, J.H. et al. (2022a). Multifocus techniques for reconfigurable intelligent surface-aided wireless power transfer: theory to experiment. *IEEE Internet of Things Journal* 9: 17157–17171. https://doi.org/10 .1109/JIOT.2022.3195948.

Tran, N.M., Amri, M.M., Park, J.H. et al. (2022b). Reconfigurable intelligent surface-aided wireless power transfer systems: analysis and implementation. *IEEE Internet of Things Journal* 9 (21): 21338–21356. https://doi.org/10.1109/JIOT.2022 .3179691.

Vailshery, L.S. (2022). IoT and non-IoT connections worldwide 2010–2025. https:// www.statista.com/statistics/1101442/iot-number-of-connected-devices-worldwide/ (accessed 10 November 2022).

Valenta, C.R. and Durgin, G.D. (2014). Harvesting wireless power: survey of energy-harvester conversion efficiency in far-field, wireless power transfer systems. *IEEE Microwave Magazine* 15 (4): 108–120. https://doi.org/10.1109/MMM.2014 .2309499.

Wang, Z., Liu, L., and Cui, S. (2020). Channel estimation for intelligent reflecting surface assisted multiuser communications: framework, algorithms, and analysis. *IEEE Transactions on Wireless Communications* 19: 6607–6620. https://doi.org/10 .1109/TWC.2020.3004330.

Wu, Q., Zhou, X., and Schober, R. (2021). IRS-assisted wireless powered NOMA: do we really need different phase shifts in DL and UL? *IEEE Wireless Communications Letters* 10: 1493–1497. https://doi.org/10.1109/LWC.2021.3072502.

Wu, Q., Guan, X., and Zhang, R. (2022). Intelligent reflecting surface-aided wireless energy and information transmission: an overview. *Proceedings of the IEEE* 110: 150–170. https://doi.org/10.1109/JPROC.2021.3121790.

Xu, S., Liu, J., and Zhang, J. (2021). Resisting undesired signal through IRS-based backscatter communication system. *IEEE Communications Letters* 25: 2743–2747. https://doi.org/10.1109/LCOMM.2021.3077093.

You, C., Zheng, B., and Zhang, R. (2020). Intelligent reflecting surface with discrete phase shifts: channel estimation and passive beamforming. *ICC 2020 - 2020 IEEE International Conference on Communications (ICC)*, 1–6. IEEE, June 2020. ISBN 978-1-7281-5089-5 https://doi.org/10.1109/ICC40277.2020.9149292.

Yue, Q., Hu, J., Yang, K., and Wong, K.-K. (2022). Intelligent reflecting surface aided wireless power transfer with a DC-combining based energy receiver and practical waveforms. *IEEE Transactions on Vehicular Technology* 71: 9751–9764. https://doi .org/10.1109/TVT.2022.3182379.

Zeng, Y., Clerckx, B., and Zhang, R. (2017). Communications and signals design for wireless power transmission. *IEEE Transactions on Communications* 65: 2264–2290. https://doi.org/10.1109/TCOMM.2017.2676103.

Zhang, D., Wu, Q., Cui, M. et al. (2021, 2021). Throughput maximization for IRS-assisted wireless powered hybrid NOMA and TDMA. *IEEE Wireless Communications Letters* 10: 1944–1948. https://doi.org/10.1109/LWC.2021 .3087495.

Zhao, Y., Clerckx, B., and Feng, Z. (2022). IRS-aided SWIPT: joint waveform, active and passive beamforming design under nonlinear harvester model. *IEEE Transactions on Communications* 70: 1345–1359. https://doi.org/10.1109/TCOMM .2021.3129931.

Zheng, B. and Zhang, R. (2020). Intelligent reflecting surface-enhanced OFDM: channel estimation and reflection optimization. *IEEE Wireless Communications Letters* 9: 518–522. https://doi.org/10.1109/LWC.2019.2961357.

Zheng, Y., Bi, S., Zhang, Y.J. et al. (2020). Intelligent reflecting surface enhanced user cooperation in wireless powered communication networks. *IEEE Wireless Communications Letters* 9: 901–905. https://doi.org/10.1109/LWC.2020.2974721.

Zhou, J., Zhang, P., Han, J. et al. (2022). Metamaterials and metasurfaces for wireless power transfer and energy harvesting. *Proceedings of the IEEE* 110: 31–55. https:// doi.org/10.1109/JPROC.2021.3127493.

14

Beamforming Design for Self-Sustainable IRS-Assisted MISO Downlink Systems

Shaokang Hu and Derrick Wing Kwan Ng

School of Electrical Engineering and Telecommunications, University of New South Wales, Sydney, NSW, Australia

14.1 Introduction

In the previous several decades, wireless communication technologies have typically evolved to the next generation every 10 years on average (Zhang et al. 2019). While the current fifth-generation (5G) networks aim to connect things by providing ultra-reliable low-latency communication (uRLLC) services and massive machine type communications (mMTCs), the future sixth-generation (6G) networks are anticipated to construct the digital society by 2030 via connecting intelligence, which has garnered significant attention from academia and industry (Zhang et al. 2019; Chen et al. 2020). Specifically, 6G intends to provide better communication services compared to the present 5G networks. For instance, ultra-high data rates with a peak data rate of at least 1 Tb/s are required (Boulogeorgos et al. 2018), which is 10 times than that of 5G. Also, energy efficiency improvement of 10 to 100 times higher than that of 5G is needed (Zhang et al. 2019), and global coverage is provided (Zhang et al. 2019). As such, numerous disruptive technologies have been proposed to address the upcoming requirements expected in 6G networks, e.g., massive multiple-input multiple-output (MIMO), millimeter wave (mmWave) communications, and ultra-dense networks (UDN), etc. Although these technologies have the potential to increase the capacity and coverage of wireless networks, they frequently result in the rapid large-scale deployment of active nodes and antennas, which escalates the associated hardware prices, energy consumption, and signal processing complexity (Wu et al. 2017; Shafi et al. 2017). To address this fundamental challenge, it is of utmost importance to design future wireless communication systems to provide global coverage, low-cost, and energy-efficient communications.

Intelligent Surfaces Empowered 6G Wireless Network, First Edition.
Edited by Qingqing Wu, Trung Q. Duong, Derrick Wing Kwan Ng, Robert Schober, and Rui Zhang.
© 2024 The Institute of Electrical and Electronics Engineers, Inc. Published 2024 by John Wiley & Sons, Inc.

Recent developments in intelligent reflecting surface (IRS)-assisted wireless communications (Yu et al. 2021a,b; Qi et al. 2019; Wu and Zhang 2019b) have drawn significant momentum in an effort to fulfill the severe criteria of 6G era (Zhang et al. 2019). In particular, an IRS is constituted by a fraction of low-cost passive reflection elements that can independently reflect the incident electromagnetic wave with controlled phase shifts. By intelligently adjusting the phase shift of each IRS element to the characteristics of communication channels, the desired receivers can integrate the reflected signals coherently. As a result, IRSs are able to reshape the signal propagation and establish a favorable communication environment for various purposes, including enhancing physical layer security, exploiting multiple access interference, and enhancing communication quality in terms of spectral efficiency and energy efficiency (Wu et al. 2021). For instance, (Wu and Zhang 2019b) confirmed that IRS-assisted communication systems can extend signal coverage in comparison to conventional systems that lack IRS. Also, the results of (Wu and Zhang 2019d) showed that adding an IRS to systems for simultaneous wireless information and power transmission (SWIPT) can greatly increase both the achievable system data rate and the total harvested power.

In spite of the fruitful results that have been published in the literature, e.g., (Wu and Zhang 2019b; Yu et al. 2021a; Wu et al. 2021; Liu et al. 2021; Cai et al. 2020; Abeywickrama et al. 2020; Pan et al. 2020; Zhang and Dai 2021), most studies assume, idealistically, that the IRS's energy consumption is negligible. However, the typical power consumption of each phase shifter is 1.5 to 7.8 mW for phase shifters with 3 to 6 bits of resolution, respectively (Méndez-Rial et al. 2016; Ribeiro et al. 2018; Huang et al. 2018, 2019). Taking into account 100 IRS elements, the operational power of IRS with 3-bit phase shifters is 0.15 W, which is about the same as the typical radiated power for wireless communication. In reality, IRSs are anticipated to be powered by batteries or the electrical grid (Wu and Zhang 2019c). However, powering IRSs using standard powerline networks would not only increase the implementation costs but also reduce IRS deployment options. Conversely, battery-powered IRS are typically outfitted with limited energy storage, resulting in a limited operating lifetime that congests communication networks. Furthermore, due to potential environmental dangers, replacing IRS's batteries manually may be costly or impossible. Therefore, the combination of energy harvesting and IRSs is a promising viable solution for providing self-sufficient and uninterrupted communication services. In practice, the availability of major traditional energy harvesting sources, such as wind, geothermal, and solar, is typically constrained by the weather and the locations of energy sources. Since wireless power transfer (WPT) can be controlled by advanced, it is more trustworthy and suited to acquiring energy from radio frequency (RF) in wireless communication systems than that from natural sources (Clerckx et al. 2018; Wei et al. 2021).

In this chapter, a WPT-enabled self-sustainable IRS is examined for enhancing the system's sum-rate and offering uninterrupted communication services. Specifically, this design supports some IRS elements to exploit the received power from RF chain to support the power consumption of the IRS such that the IRS requires no additional power source.

The remainder of this chapter is organized as follows: In Section 14.2, we introduce the adopted system model, including the self-sustainable IRS model, channel and signal models, and energy harvesting model. Sections 14.3 and 14.4 studied the beamforming design for self-sustainable IRS-assisted wireless communication system. Simulation results are provided in Section 14.5. In Sections 14.6 and 14.7, a brief summary and further extension of this chapter are provided, respectively.

Notations: We denote lowercase letter x as the scalars, boldface lowercase letter $\mathbf{x}$ as vectors, and boldface uppercase letter $\mathbf{X}$ as matrices. The space of $N \times M$ matrices with real and complex entries are represented as $\mathbb{R}^{N \times M}$ and $\mathbb{C}^{N \times M}$, respectively. An $N \times N$ identity matrix is denoted as $\mathbf{I}_N$. The Hermitian matrices are represented as $\mathbb{H}^N$. $|x|$ denotes the modulus of a complex-valued scalar. $\|\mathbf{x}\|$ represents an Euclidean norm of a vector. The transpose, conjugate transpose, rank, and trace of a matrix are denoted as $\mathbf{X}^T$, $\mathbf{X}^H$, Rank($\mathbf{X}$), and Tr($\mathbf{X}$), respectively. $\mathbb{E}\{\cdot\}$ is the function of statistical expectation. The conjugate of a complex-value scalar is denoted as x^*. A diagonal matrix whose diagonal elements are $\mathbf{x} \in \mathbb{C}^{N \times 1}$ is denoted by diag($\mathbf{x}$). j is the imaginary unit. For a continuous function $f(\mathbf{X})$, $\nabla_{\mathbf{X}} f(\cdot)$ represents the gradient of $f(\cdot)$ with respect to matrix $\mathbf{X}$. $\mathcal{CN}(\mu, \sigma^2)$ is the distribution of a circularly symmetric complex Gaussian (CSCG) random variable whose mean is μ and variance is σ^2. $\sim$ stands for "distributed as."

14.2 System Model

As shown in Fig. 14.1, to draw important insights into the optimal joint beamforming design, this chapter considers a wireless-powered IRS-assisted single-user downlink multiple-input single-output (MISO) system. Specifically, the system model consists of three nodes: a base station with M antennas, a self-sustainable IRS with N IRS elements, and a user with a single antenna. The user receives signals over two links, denoted by the BS-user link and the BS-IRS-user link, respectively.

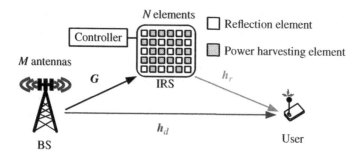

Figure 14.1 A self-sustainable IRS-assisted wireless communication system.

14.2.1 Self-Sustainable IRS Model

In this section, we introduce a realistic model of the IRS, which is equipped with discrete phase shifters and has the ability to harvest power[1] from the incident signals.

In Fig. 14.2, we show the block diagram of the self-sustainable IRS circuit. Specifically, the IRS is integrated with a controller that toggles the operation mode of the IRS elements between the reflection mode and the power harvesting mode. Positive-intrinsic-negative (PIN) diodes are built in each element's switch (Wu and Zhang 2019c). This makes it possible to switch between different modes. In particular, when an IRS element is in the reflection mode, all signal waveforms

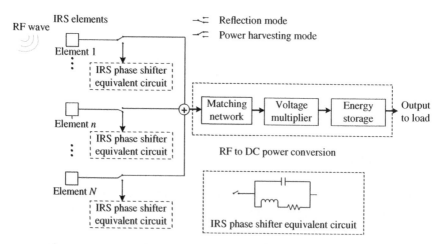

Figure 14.2 An energy harvesting block diagram of the self-sustainable IRS. Source: Hu et al. 2021/ IEEE.

---

1 The unit of energy consumption in this chapter is the Joule per second. Consequently, the terms "power" and "energy" are synonymous.

impinging the IRS element pass through the IRS phase shifter equivalent circuit, which alters the phase of the signals. As for the energy harvesting mode, the elements collect all the energy they receive and convert it into direct current by the RF to DC power conversion circuit, which provides the energy that the IRS needs. Once an IRS element adopts the reflection mode, all the received signals are reflected, and it is unable to harvest energy. Similarly, the elements operating in the power harvesting mode do not reflect any incoming signals. Note that the proposed algorithm in this chapter will optimize the mode selection strategy to find a good balance between the number of energy-harvesting IRS elements and reflection IRS elements.

Now, we construct a mathematical model for the self-sustainable IRS by introducing an IRS reflection matrix, $\mathbf{\Theta} = \mathbf{A}\mathbf{\Phi}$. In particular, $\mathbf{A} = \mathrm{diag}(\alpha_1, \ldots, \alpha_n, \ldots, \alpha_N)$, $\mathbf{A} \in \mathbb{B}^{N \times N}$, $\forall n \in \mathcal{N}$, is an IRS mode selection matrix, which captures the mode of each IRS elements. Here, $\alpha_n \in \{0, 1\}$ is an IRS mode selection variable which is defined as:

$$\alpha_n = \begin{cases} 1, & \text{Reflection mode at IRS element } n, \\ 0, & \text{Energy harvesting mode at IRS element } n. \end{cases} \tag{14.1}$$

On the other hand, $\mathbf{\Phi} = \mathrm{diag}(\beta_1 e^{j\theta_1}, \ldots, \beta_n e^{j\theta_n}, \ldots, \beta_N e^{j\theta_N})$, $\mathbf{\Phi} \in \mathbb{C}^{N \times N}$ is a diagonal matrix that represents reflection-coefficients of IRS elements. Here, $\theta_n \in [0, 2\pi)$ is phase shift. In practice, θ_n are often chosen from a finite set of discrete values ranging from 0 to 2π to facilitate practice circuit construction. Thus, we adopt a discrete phase shifter in each IRS element. In particular, the interval $[0, 2\pi)$ is uniformly quantized into $B = 2^b$ levels and b is the given constant bit resolution. As such, the phase shift θ_n belongs to the set $\mathcal{F}$, i.e., $\theta_n \in \mathcal{F} = \{0, \ldots, \Delta\theta, \ldots, \Delta\theta(B-1)\}$, $\forall n \in \mathcal{N}$, where $\Delta\theta = 2\pi/B$. Note that in reflection mode, the phase shift level is selected by adjusting the biassing voltage of each IRS element through a direct current feeding line (Wu and Zhang 2019c). As a result, each IRS reflection element's power consumption is proportional to its bit resolution (Méndez-Rial et al. 2016; Ribeiro et al. 2018; Huang et al. 2018, 2019). Hence, we represent $P_{\mathrm{IRS}}(b)$ as the amount of power required by the IRS reflection element with a resolution of b bits. Moreover, $\beta_n \in [0, 1]$, $\forall n \in \mathcal{N}$, is an amplitude coefficient. As is customary in the literature (Wu and Zhang 2019a,b,d; Huang et al. 2019), the reflection coefficient β_n is fixed at 1 for practical implementation of the IRS.

14.2.2 Channel and Signal Models

In this chapter, we adopt a quasi-static flat fading channel model for all the involved links. For the sake of simplicity, we assume that the channel state information (CSI) is precisely known at the BS for all the links, which may be realized by applying existing CSI estimation algorithms (Liu et al. 2021; Wang et al. 2020). The extension to the case of imperfect CSI can be done by following a similar approach as in (Hu et al. 2021). Matrix $G \in \mathbb{C}^{N \times M}$ denotes the baseband equivalent channels from the BS to the IRS. Matrix $\mathbf{h}_r \in \mathbb{C}^{N \times 1}$ represents the channel from the IRS to the user. Matrix $\mathbf{h}_d \in \mathbb{C}^{M \times 1}$ stands for the link from the BS to the user.

As shown in Eq. (14.2), we use $\mathbf{x}$ to denote the signal transmitted from the BS to the user.

$$\mathbf{x} = \mathbf{w}x. \tag{14.2}$$

In particular, $\mathbf{w} \in \mathbb{C}^{M \times 1}$ is the beamforming vector and $x \sim C\mathcal{N}(0, 1)$ denotes information symbol. Without loss of generality, we assume $\mathbb{E}\{|x|^2\} = 1$. The total transmit power from the BS is denoted by $P_{\max}$, i.e., $\mathbb{E}\{\mathbf{x}\} = \|\mathbf{w}\|^2 \leq P_{\max}$. Thus, the signal received at the user is

$$y = \left(\mathbf{h}_d^H + \mathbf{h}_r^H \mathbf{A}\boldsymbol{\Phi}\mathbf{G}\right)\mathbf{w}x + n, \tag{14.3}$$

where the background noise at the user is denoted as $n \sim C\mathcal{N}(0, \sigma^2)$ and σ^2 is the corresponding noise power.

Accordingly, the received signal-to-noise-ratio (SNR) and the achievable rate (bits/s/Hz) at the user are given by

$$\text{SNR} = \frac{1}{\sigma^2}|(\mathbf{h}_d^H + \mathbf{h}_r^H \mathbf{A}\boldsymbol{\Phi}\mathbf{G})\mathbf{w}|^2 \quad \text{and} \tag{14.4}$$

$$R = \log_2(1 + \text{SNR}), \tag{14.5}$$

respectively.

14.2.3 Power Harvesting Model at the IRS

This section discusses the power harvesting model at the IRS. First, the total received signals for power harvesting at the IRS are computed in (14.6).

$$\mathbf{y}_{\text{EH}}(\mathbf{A}, \mathbf{w}) = \mathbf{A}_{\text{EH}}(\mathbf{G}\mathbf{w}x + \mathbf{n}_a), \tag{14.6}$$

where the receiving thermal noise at the IRS is represented as $\mathbf{n}_a \in \mathbb{C}^{N \times 1} \sim C\mathcal{N}(\mathbf{0}, \sigma_a^2 \mathbf{I}_N)$ and the noise power per IRS element is denoted as σ_a^2. Because the reflection IRS element is represented by $\alpha_n = 1$, the IRS elements in the power harvesting mode can be represented by the binary matrix $\mathbf{A}_{\text{EH}} = \mathbf{I}_N - \mathbf{A}$. For simplicity, we adopt linear power harvesting model [2] in this chapter and thus the

---

2 Note that the linear power harvesting model proposed system model can be easily extend to a more practical non-linear power harvesting model as presented in (Hu et al. 2021).

total harvested power by the IRS is given by

$$P_{EH} = \eta_h \mathbb{E}\left(\|\mathbf{A}_{EH}(\mathbf{G}\mathbf{w}x + \mathbf{n}_a)\|^2\right), \tag{14.7}$$

where $0 \leq \eta_h \leq 1$ is the power harvesting efficiency of the conversion from the incoming RF signals into electrical energy.

14.3 Problem Formulation

This section proposes a resource allocation for a system sum-rate maximization problem in Eq. (14.8) while maintaining the self-sustainability of the IRS by jointly designing the three optimization variables, i.e., precoding vector $\mathbf{w}$ at the BS, the mode selection $\{\alpha_n\}_{n \in \mathcal{N}}$, and the discrete phase shifter $\{\theta_n\}_{n \in \mathcal{N}}$ adopted at IRS:

$$
\begin{aligned}
&\underset{\mathbf{w},\, \alpha_n,\, \theta_n}{\text{maximize}} \quad \log_2(1 + \text{SNR}) \\
&\text{s.t.} \ \ \text{C1:} \ \|\mathbf{w}\|^2 \leq P_{\max}, \\
&\quad\ \ \text{C2:} \ \theta_n \in \mathcal{F}, \forall n \in \mathcal{N}, \\
&\quad\ \ \text{C3:} \ \sum_{n=1}^{N} \alpha_n P_{\text{IRS}}(b) \leq P_{EH}, \\
&\quad\ \ \text{C4:} \ \alpha_n \in \{0,1\}, \forall n.
\end{aligned}
\tag{14.8}
$$

Constraint C1 limits the maximum radiated power such that it is below $P_{\max}$. Constraint C2 ensures that the phase shift of the b-bit resolution of IRS reflecting element is chosen from the set $\mathcal{F}$. Constraint C3 assures that the total amount of power harvested from the BS, P_{EH}, is sufficient to support the total amount of power consumed by the IRS. The binary constraint C4 makes sure that each IRS element can only work in one of the two modes: reflection or power harvesting.

14.4 Solution

Due to the couplings between variables $\mathbf{w}$, θ_n, and α_n in the objective function and constraint C3, the discrete phase shift constraint C2, and the binary variable α_n in constraint C4, the formulated problem in (14.8) is non-convex. Acquiring the globally optimal solution of (14.8) typically requires the brute-force search, which is computationally impossible for even moderately sized systems. As a middle ground, this section devises a suboptimal iterative algorithm that is easy to compute, i.e., with polynomial-time complexity. In particular, we first derive an optimal solution of the transmit beamforming $\mathbf{w}$ for any given IRS phase shifts, such that the original problem (14.8) is transformed to its equivalent form with variables α_n and θ_n. Then, we introduce some slack optimization variables and constraints that facilitates the application of apply successive convex approximation (SCA) to deal with the reminding non-convexities.

14.4.1 Problem Transformation

For any given $\{\alpha_n, \theta_n\}$, the optimal transmit beamforming solution of problem (14.8) can be found by adopting the maximum-ratio transmission (MRT), (Tse and Viswanath 2005), i.e., $\mathbf{w}^* = \sqrt{P}\frac{(\mathbf{h}_r^H \mathbf{A}\mathbf{\Phi}\mathbf{G} + \mathbf{h}_d^H)^H}{\|\mathbf{h}_r^H \mathbf{A}\mathbf{\Phi}\mathbf{G} + \mathbf{h}_d^H\|^2}$, where P denotes the transmit power at the BS. By substituting $\mathbf{w}^*$ into problem (14.8), we have

$$
\begin{aligned}
&\underset{P, \alpha_n, \theta_n}{\text{maximize}} \quad \frac{P}{\sigma^2}\|\mathbf{h}_d^H + \mathbf{h}_r^H \mathbf{A}\mathbf{\Phi}\mathbf{G}\|^2 \\
&\text{s.t.} \quad \overline{\text{C1}}: P \le P_{\max}, \\
&\quad\quad \text{C2, C3, C4.}
\end{aligned}
\tag{14.9}
$$

It can be easily verified that at the optimal solution, the transmit power at BS satisfies $P^* = P_{\max}$. As such, problem (14.9) is equivalent to the following problem:

$$
\begin{aligned}
&\underset{\alpha_n, \theta_n}{\text{maximize}} \quad \|\mathbf{h}_d^H + \mathbf{h}_r^H \mathbf{A}\mathbf{\Phi}\mathbf{G}\|^2 \\
&\text{s.t.} \quad \text{C2, C3, C4.}
\end{aligned}
\tag{14.10}
$$

14.4.2 Address the Coupling Variables and Binary Variables

However, due to non-convexities in problem (14.10), i.e., the coupling between variables $\{\theta_n, \alpha_n\}$ in the objective function and constraint C3, the discrete phase-shift constraint C2, and the binary variable α_n in constraint C4, the optimization problem need to be further transformed.

First, the coupling variables in the objective function are addressed. This makes it easier to design discrete IRS phase shifts. In order to accomplish this, we define an enhanced mode selection matrix $\tilde{\mathbf{A}} = \text{diag}(\tilde{\alpha}) = \text{diag}(\tilde{\alpha}_1, \ldots, \tilde{\alpha}_n, \ldots, \tilde{\alpha}_N)$. Different from α_n, $\tilde{\alpha}_n$ belongs to a generalized mode selection set

$\tilde{F} = \{0, e^{j0}, \dots, e^{j\Delta\theta}, \dots, e^{j\Delta\theta(B-1)}\}$, which combines the energy harvesting mode and B discrete phase shifts, i.e., the number of total modes is $B+1$. As a result, $\tilde{\alpha}_n = 0$ for the nth IRS element set to power harvesting mode. It is in reflection mode if $\tilde{\alpha}_n$ equals other values. As such, the binary constraint C4 merged with constraint C2, which makes constraints C3 and C4 in (14.8) equivalently transformed as:

$$C3: \sum_{n=1}^{N} |\tilde{\alpha}_n| P_{IRS}(b) \le P_{EH} \quad \text{and} \tag{14.11}$$

$$\overline{C4}: \tilde{\alpha}_n \in \tilde{F} = \{0, e^{j0}, e^{j\Delta\theta}, \dots, e^{j\Delta\theta(B-1)}\}, \forall n,$$

respectively. Next, we further introduce a binary mode selection optimization variable $s_{i,n}, \forall i \in I = \{1, \dots, B+1\}, n \in \mathcal{N}$, and a mode selection binary matrix $\mathbf{S} \in \mathbb{R}^{(B+1)\times N}$ to address the non-convex constraint $\overline{C4}$. In particular, $s_{i,n}$ is the element in matrix $\mathbf{S}$'s ith row and nth column. When $s_{i,n} = 1$, the i-th mode is chosen as the nth element. Otherwise, $s_{i,n} = 0$. Note that the binary integers in the first row of the mode selection matrix $\mathbf{S}$ indicate whether or not each element of the IRS is configured for power harvesting mode. Thus, by defining vector $\mathbf{s}_1 = [s_{1,1}, \dots, s_{1,n}, \dots, s_{1,N}]^T$ as the transpose of the first row of $\mathbf{S}$, we can rewrite $\mathbf{A}_{EH}$ in (14.6) as a function of $\mathbf{s}_1$, which is given by

$$\mathbf{A}_{EH} = \text{diag}(\mathbf{s}_1). \tag{14.12}$$

Now, constraint C3 and $\overline{C4}$ in (14.11) are equivalently transformed as:

$$\overline{C3}: (N - \sum_{n=1}^{N}(s_{1,n}))P_{IRS}(b) \le \eta_h \left(P_{max} \, \text{Tr}(\mathbf{G}\mathbf{G}^H \, \text{diag}(\mathbf{s}_1)) + \sigma_a^2 \sum_{n=1}^{N} s_{1,n} \right), \tag{14.13}$$

$$C4a: \sum_{i\in I} s_{i,n} \le 1, \forall n, \tag{14.14}$$

$$C4b: \tilde{\alpha}_n = \sum_{i\in I} s_{i,n} f_i, \forall n, \tag{14.15}$$

$$C4c: s_{i,n} \in \{0,1\}, \forall i, n, \tag{14.16}$$

where f_i is the ith element of set $\tilde{F}$ defined in (14.11).

According to (14.13), there is a nontrivial trade-off between the system performance and the number of IRS elements in power harvesting mode. In order to realize the self-sustainability of the IRS, a portion of the IRS elements are leveraged for power harvesting. Yet, this means that there are fewer IRS elements to reflect signals and improve the system sum-rate.

To convert the binary variable $s_{i,n}$ into continuous variables, constraint C4c is equivalently transformed as the following two constraints:

$$\overline{C4c}: s_{i,n} - s_{i,n}^2 \leq 0, \forall i, n, \text{ and} \tag{14.17}$$

$$C4d: 0 \leq s_{i,n} \leq 1, \forall i, n. \tag{14.18}$$

Now, the problem in (14.10) can be equivalent transformed as

$$\underset{S,v,t}{\text{maximize}} \quad t \tag{14.19}$$

$$\text{s.t.} \quad \overline{C3}, C4a, C4b, \overline{C4c}, C4d,$$

$$C5: t \leq \mathbf{v}^H \mathbf{LL}^H \mathbf{v} + \mathbf{vLh}_d + \mathbf{h}_d^H \mathbf{L}^H \mathbf{v} + \mathbf{h}_d^H \mathbf{h}_d .$$

Here, t is a slack optimization variable, $\mathbf{L} = \text{diag}(\mathbf{h}_r^H)\mathbf{G}$, and $\mathbf{v} = [\tilde{\alpha}_1, \ldots, \tilde{\alpha}_n, \ldots \tilde{\alpha}_N]^H$. It is worth noting that the inequality constraints C5 are always satisfied with equality at the optimal solution of (14.19).

14.4.3 Successive Convex Approximation

Since constraints $\overline{C4c}$ and C5 are the functions in difference of convex (D.C.) form, we exploit the SCA to deal with their non-convexity. For any feasible point $s_{i,n}^{(t)}$ and $\mathbf{v}^{(t)}$, where (t) is defended as the iteration index for proposed algorithm in Fig. 14.3, a lower bound function of $s_{i,n}^2$ and $\mathbf{v}^H \mathbf{LL}^H \mathbf{v}$ are given by their first-order Taylor expansion:

$$s_{i,n}^2 \geq (s_{i,n}^{(t)})^2 + 2s_{i,n}^{(t)}(s_{i,n} - s_{i,n}^{(t)}), \forall i, n, \tag{14.20}$$

$$\mathbf{v}^H \mathbf{LL}^H \mathbf{v} \geq (\mathbf{v}^{(t)})^H \mathbf{L}_k \mathbf{W}_k \mathbf{L}_k^H \mathbf{v}^{(t)} + 2(\mathbf{v}^{(t)})^H \mathbf{L}_k \mathbf{W}_k^H \mathbf{L}_k^H (\mathbf{v} - \mathbf{v}^{(t)}), \tag{14.21}$$

respectively.

As such, subsets of constraints $\overline{C4c}$ and C5 are given by

$$\overline{\overline{C4c}}: s_{i,n} \leq \left(s_{i,n}^{(t)}\right)^2 + 2s_{i,n}^{(t)}\left(s_{i,n} - s_{i,n}^{(t)}\right), \forall i, n, \tag{14.22}$$

$$\overline{C5}: t \leq \left(\mathbf{v}^{(t)}\right)^H \mathbf{L}_k \mathbf{W}_k \mathbf{L}_k^H \mathbf{v}^{(t)} + 2\left(\mathbf{v}^{(t)}\right)^H \mathbf{L}_k \mathbf{W}_k^H \mathbf{L}_k^H (\mathbf{v} - \mathbf{v}^{(t)}) \tag{14.23}$$

$$+ \mathbf{vLh}_d + \mathbf{h}_d^H \mathbf{L}^H \mathbf{v} + \mathbf{h}_d^H \mathbf{h}_d.$$

Then, a sub-optimal solution of the problem in (14.19) can be found by solving the following optimization problem:

$$\underset{S,v,t}{\text{maximize}} \quad t \tag{14.24}$$

$$\text{s.t.} \quad \overline{C3}, C4a, C4b, \overline{\overline{C4c}}, C4d, \overline{C5}.$$

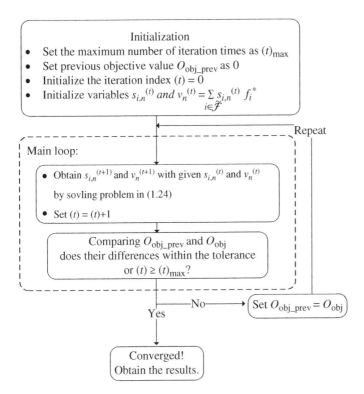

Figure 14.3 The flow chart of the proposed iterative algorithm.

Due to the use of SCA, solving problem (14.24) gives a lower bound of problem (14.19). By iteratively updating the feasible solutions $\{S, v\}$, the obtained performance lower bound is tightened by using a convex programming solver to solve the problem in (14.24). The proposed algorithm for handling (14.24) is shown in Fig. 14.3. The computational complexity of each iteration of the proposed algorithm is $\mathcal{O}\left((N_{\mathcal{O}}M_{\mathcal{O}}^3 + M_{\mathcal{O}}^2 N_{\mathcal{O}}^2 + N_{\mathcal{O}}^3) \times \sqrt{M_{\mathcal{O}}} \log \frac{1}{\rho}\right)$ (Pólik and Terlaky 2010), where $\mathcal{O}$ represents the big-O notation and ρ denotes the threshold of convergence tolerance of the proposed scheme. $N_{\mathcal{O}} = BN + 2N + 1$ and $M_{\mathcal{O}} = 5N + 2BN + 1$ are the number of variables and inequalities of optimization problem in (14.24), respectively. It can be seen that the proposed algorithm has a computational complexity of polynomial time, which makes it a fast algorithm to implement (Cormen et al. 2009).

14.5 Numerical Results

This section evaluates the system performance of the proposed wireless-powered IRS scheme by using simulation. The setup is depicted in Fig. 14.4. In particular,

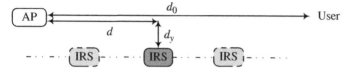

Figure 14.4 Simulation setup.

the distance between the BS and the user is $d_0 = 60\,\text{m}$. The IRS is placed between the BS and the user at a distance of $d_y = 1\,\text{m}$ in the vertical direction and d in the horizontal direction from the BS. The distance-dependent path loss model (Goldsmith 2005) is applied with a 10 m reference distance. Due to the relatively large distance and arbitrary scattering of the BS-user channel, we set the path loss exponent for the BS-user link at $\alpha_{\text{AU}} = 3.5$. Because the IRS is typically used to establish a LoS channel with the BS, we set the path loss exponents of the BS-IRS and IRS-user links to $\alpha_{\text{AI}} = \alpha_{\text{IU}} = 2.2$. The BS-user link, BS-IRS link, and IRS-user link small scale fading coefficients are generated as independent and identically distributed (i.i.d.) Rican random variables with Rician factors $\beta_{\text{AU}} = 0$, $\beta_{\text{AI}} = 2$, and $\beta_{\text{IU}} = 2$, respectively. We consider that the thermal noise and quantization noise account for the signal processing noise in each receiver. Specifically, the user adopts a 12-bit uniform quantizer at the receiver for quantizing the received data. Consequently, the thermal noise power is set to $-110\,\text{dBm}$ and the quantization noise power is $-47\,\text{dBm}$ (Wepman 1995). Additional essential parameters are given in Table 14.1.

We also examine the system performance of two alternative schemes for comparison: (i) An IRS-assisted system with an idealized IRS reaches a performance upper bound, e.g., all the IRS elements are used to reflect signals while

Table 14.1 Simulation parameters.

The number of antennas at the IRS	$N = 200$
Antenna gains at the BS,	20 dBi, 0 dBi,
the IRS, and receivers	and 0 dBi (Dai et al. 2020)
The system bandwidth	200 kHz
The carrier center frequency	2.4 GHz
The maximum power budget at the BS	$P_{\text{max}} = 38\,\text{dBm}$
The phase shifter bit resolution of an IRS element	$b = 3$ bits
Power consumption of	$P_{\text{IRS}}(b) = 1.5\,\text{mW}$ for
each IRS reflection element	$b = 3$ bits (Huang et al. 2019)
The power harvesting efficiency of IRS elements	$\eta_h = 0.8$

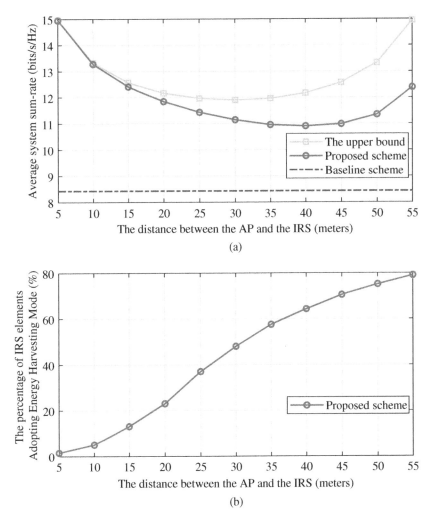

Figure 14.5 (a) Average system sum-rate (bits/s/Hz) versus the horizontal distance of BS-IRS link. (b) The percentage of IRS elements adopting power harvesting mode versus the horizontal distance of BS-IRS link.

consuming no power; (ii) The baseline scheme is set for the scenario in which the IRS is not deployed.

The average system sum-rate versus horizontal distance for various designs is depicted in Fig. 14.5(a). Both the upper bound design and the proposed method yield a significantly greater sum-rate than the baseline scheme, as observed. In fact, the IRS offers a second path gain, i.e., $\mathbf{h}_r^H \mathbf{\Theta} \mathbf{G}$, that contains the same important signals as the direct channel to the user. Also, the proposed scheme takes the advantage of and optimizes this extra path gain to improve the performance of the system. Moreover, when the IRS is positioned between the BS and the user, the average system sum-rate is the lowest for all the schemes with an IRS. When neither the BS nor the user is in close proximity to the IRS, both the path from the BS to the IRS and the path from the IRS to the user would experience significant signal loss. This makes it harder for the IRS to beamform the reflected signals toward the desired user. As shown in Fig. 14.5(a), the performance upper bound can be approached by positioning the IRS near the BS as it can facilitate effective energy harvesting at the IRS. In contrast, when the IRS deviates further from the BS, i.e., d increases, the sum-rate discrepancy between the proposed scheme and the upper bound develops significantly. This is because as d increases, each IRS element would, on average, captures less amount of energy. As predicted by constraint $\overline{C3}$ in (14.13), the increasing number of IRS elements that are converted to the power harvesting mode preserves the IRS's sustainability, resulting in a lower proportion of IRS elements for boosting the system's performance by signal reflection. This can also be seen in Fig. 14.5(b), which depicts the proportion of power-harvesting

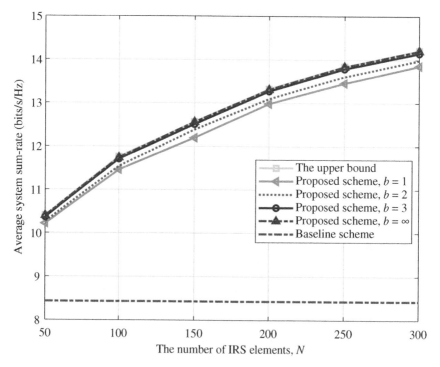

Figure 14.6 The impact of the total number of IRS elements on the average system sum-rate (bits/s/Hz).

IRS elements in relation to the horizontal distance of the BS-IRS link. The proportion of IRS elements that adopt power harvesting mode increases as d rises.

Figure 14.6 illustrates the variation of the average system sum-rate with various numbers of IRS elements N. It is observed that the average system sum-rate of the proposed method rises as the number of IRS elements N increases. Moreover, even when the self-sustainability of the IRS is taken into consideration, the proposed method can achieve a large sum-rate gain compared with the baseline scheme. In fact, the increased number of reflecting IRS elements offers greater spatial degrees of freedom allowing for further flexibility in beamforming to enhance the channel quality of the end-to-end BS-IRS-user link and increase the sum-rate of the system. In addition, Fig. 14.6 compares the system performance with various IRS phase shifters' bit resolutions, b. Here, the upper bound in Fig. 14.6 is the previously described upper bound scheme with b equal to infinity. When the bit resolution of $b = 2, 3$ bits, the system sum-rate approaches that of the upper bound. Specifically, raising bit resolution beyond 2 bits would only yield a negligible gain on the system's sum-rate. Indeed, the IRS-user links in the Rician fading channels are dominated by LoS components. Consequently, a small bit resolution in phase shifts is adequate for the beamformer to align the necessary signals with the dominated links. As such, for designs with minimum complexity, the bit resolution of each IRS element can therefore be set as low as 2 bits.

14.6 Summary

This chapter investigated the application of a self-sustainable IRS to a single-user MISO downlink communication system. To maximize the system sum-rate, the joint design of the beamformer at the BS and the phase shifts and power harvesting schedule at the IRS was formulated as a non-convex optimization problem. We proposed an iterative method for achieving a suboptimal design solution that is computationally efficient. The simulation results revealed that the suggested approach provides a substantial performance improvement over the typical MISO system without IRS. In addition, our results revealed a nontrivial trade-off between self-sustainability and the system's sum-rate. Last but not least, we demonstrated that a modest number of bit-resolution phase shifters at IRS can provide a significant average system sum-rate equivalent to the ideal case with continuous phase shifters.

14.7 Further Extension

To draw important insights on how self-sustainable IRS impact the communication system, this chapter considers a single-user case such that the optimal transmit beamforming can be derived analytically. In fact, the considered framework can be easily extended to a multiuser MISO downlink system with self-sustainable IRS (Hu et al. 2020). In particular, it has been shown in (Hu et al. 2020) that the self-sustainable IRS can improve the system performance in a multiuser case. Another possible extension of this study in self-sustainable IRS-assisted communication systems is to deal with potential wiretapping. Specifically, transmitters in SWIPT systems typically boost the signal strength of information-carrying signals to assist efficient energy harvesting at the desired receivers. Due to the increasing likelihood of information leakage, the sensitivity to eavesdropping has consequently grown. Therefore, communication security is a crucial concern for wireless IRS-assisted systems. To achieve secure communications, the authors of (Hu et al. 2021) examined a secure and robust multiuser MISO downlink system with self-sustaining IRS.

Bibliography

Abeywickrama, S., Zhang, R., Wu, Q., and Yuen, C. (2020). Intelligent reflecting surface: practical phase shift model and beamforming optimization. *IEEE Transactions on Communications* 68 (9): 5849–5863.

Boulogeorgos, A.-A.A., Alexiou, A., Merkle, T. et al. (2018). Terahertz technologies to deliver optical network quality of experience in wireless systems beyond 5G. *IEEE Communications Magazine* 56 (6): 144–151.

Cai, Y., Wei, Z., Hu, S. et al. (2020). Resource allocation for power-efficient IRS-assisted UAV communications. *Proceedings of the IEEE International Conference on Communications Workshops (ICC Workshops)*, 1–7.

Chen, X., Ng, D.W.K., Yu, W. et al. (2020). Massive access for 5G and beyond. *IEEE Journal on Selected Areas in Communications* 39 (3): 615–637.

Clerckx, B., Zhang, R., Schober, R. et al. (2018). Fundamentals of wireless information and power transfer: from RF energy harvester models to signal and system designs. *IEEE Journal on Selected Areas in Communications* 37 (1): 4–33.

Cormen, T.H., Leiserson, C.E., Rivest, R.L., and Stein, C. (2009). *Introduction to Algorithms*. MIT Press.

Dai, L., Wang, B., Wang, M. et al. (2020). Reconfigurable intelligent surface-based wireless communications: antenna design, prototyping, and experimental results. *IEEE Access* 8: 45913–45923.

Goldsmith, A. (2005). *Wireless Communications*. Cambridge University Press.

Hu, S., Wei, Z., Cai, Y. et al. (2020). Sum-rate maximization for multiuser MISO downlink systems with self-sustainable IRS. *IEEE Global Communications Conference 2020*, 1–7.

Hu, S., Wei, Z., Cai, Y. et al. (2021). Robust and secure sum-rate maximization for multiuser MISO downlink systems with self-sustainable IRS. *IEEE Transactions on Communications* 69 (10): 7032–7049.

Huang, C., Alexandropoulos, G.C., Zappone, A. et al. (2018). Energy efficient multi-user MISO communication using low resolution large intelligent surfaces. *2018 IEEE Globecom Communications Workshops*, 1–6.

Huang, C., Zappone, A., Alexandropoulos, G.C. et al. (2019). Reconfigurable intelligent surfaces for energy efficiency in wireless communication. *IEEE Transactions on Wireless Communications* 18 (8): 4157–4170.

Liu, C., Liu, X., Ng, D.W.K., and Yuan, J. (2021). Deep residual learning for channel estimation in intelligent reflecting surface-assisted multi-user communications. *IEEE Transactions on Wireless Communications* 21 (2): 898–912.

Méndez-Rial, R., Rusu, C., González-Prelcic, N. et al. (2016). Hybrid MIMO architectures for millimeter wave communications: phase shifters or switches? *IEEE Access* 4: 247–267.

Pan, C., Ren, H., Wang, K. et al. (2020). Multicell MIMO communications relying on intelligent reflecting surfaces. *IEEE Transactions on Wireless Communications* 19 (8): 5218–5233.

Pólik, I. and Terlaky, T. (2010). Interior point methods for nonlinear optimization. In: *Nonlinear Optimization*, Lecture Notes in Mathematics, vol. 1989 (ed. G. Di Pillo and F. Schoen), 215–276. Berlin, Heidelberg: Springer-Verlag.

Qi, Q., Chen, X., and Ng, D.W.K. (2019). Robust beamforming for NOMA-based cellular massive IoT with SWIPT. *IEEE Transactions on Signal Processing* 68: 211–224.

Ribeiro, L.N., Schwarz, S., Rupp, M., and de Almeida, A.L. (2018). Energy efficiency of mmWave massive MIMO precoding with low-resolution DACs. *IEEE Journal of Selected Topics in Signal Processing* 12 (2): 298–312.

Shafi, M., Molisch, A.F., Smith, P.J. et al. (2017). 5G: a tutorial overview of standards, trials, challenges, deployment, and practice. *IEEE Journal on Selected Areas in Communications* 35 (6): 1201–1221.

Tse, D. and Viswanath, P. (2005). *Fundamentals of Wireless Communication*. Cambridge University Press.

Wang, Z., Liu, L., and Cui, S. (2020). Channel estimation for intelligent reflecting surface assisted multiuser communications. *2020 IEEE Wireless Communications and Networking Conference (WCNC)*, 1–6.

Wei, Z., Yu, X., Ng, D.W.K., and Schober, R. (2021). Resource allocation for simultaneous wireless information and power transfer systems: a tutorial overview. *Proceedings of the IEEE* 110 (1): 127–149.

Wepman, J.A. (1995). Analog-to-digital converters and their applications in radio receivers. *IEEE Communications Magazine* 33 (5): 39–45.

Wu, Q. and Zhang, R. (2019a). Beamforming optimization for intelligent reflecting surface with discrete phase shifts. *Proceedings of the IEEE International Conference on Acoustics, Speech and Signal Process,* 7830–7833.

Wu, Q. and Zhang, R. (2019b). Intelligent reflecting surface enhanced wireless network via joint active and passive beamforming. *IEEE Transactions on Wireless Communications* 18 (11): 5394–5409.

Wu, Q. and Zhang, R. (2019c). Towards smart and reconfigurable environment: intelligent reflecting surface aided wireless network. *IEEE Communications Magazine* 58 (1): 106–112.

Wu, Q. and Zhang, R. (2019d). Weighted sum power maximization for intelligent reflecting surface aided SWIPT. *IEEE Wireless Communications Letters* 9 (5): 586–590.

Wu, Q., Li, G.Y., Chen, W. et al. (2017). An overview of sustainable green 5G networks. *IEEE Wireless Communications* 24 (4): 72–80.

Wu, Q., Zhang, S., Zheng, B. et al. (2021). Intelligent reflecting surface-aided wireless communications: a tutorial. *IEEE Transactions on Communications* 69 (5): 3313–3351.

Yu, X., Jamali, V., Xu, D. et al. (2021a). Smart and reconfigurable wireless communications: from IRS modeling to algorithm design. *IEEE Wireless Communications* 28 (6): 118–125.

Yu, X., Xu, D., Ng, D.W.K., and Schober, R. (2021b). IRS-assisted green communication systems: provable convergence and robust optimization. *IEEE Transactions on Communications* 69 (9): 6313–6329.

Zhang, Z. and Dai, L. (2021). A joint precoding framework for wideband reconfigurable intelligent surface-aided cell-free network. *IEEE Transactions on Signal Processing* 69: 4085–4101.

Zhang, Z., Xiao, Y., Ma, Z. et al. (2019). 6G wireless networks: vision, requirements, architecture, and key technologies. *IEEE Vehicular Technology Magazine* 14 (3): 28–41.

15

Optical Intelligent Reflecting Surfaces

Hedieh Ajam and Robert Schober

Department of Electrical, Electronics, and Communication Engineering, Institute for Digital Communications, Friedrich-Alexander-University Erlangen Nürnberg, Erlangen, Germany

15.1 Introduction

Free space optical (FSO) systems due to their directional narrow laser beams, easy-to-install transceivers, license-free bandwidth, and high data rate are suitable candidates for enhancing the next generation of wireless communication networks (Saad et al. 2020). However, FSO links are impaired by atmospheric turbulence, beam divergence, and misalignment errors in long-distance deployments and require line-of-sight (LOS) link connections. Optical intelligent reflecting surfaces (IRSs) are appealing solutions to circumvent the LOS requirement of FSO links.

IRSs can be implemented at optical frequencies using mirror- and metasurface-based technologies. Mirror-based IRSs include typical mirrors and micro-mirror arrays which are limited to specular reflection (Jamali et al. 2021). To provide a desired reflection angle, they can employ mechanical motors such as rotary gimbals or micro-electro-mechanical systems (MEMS). On the other hand, metasurface-based IRSs are designed to locally manipulate the incident electromagnetic wave using subwavelength unit cells. They also provide higher functionality to change the amplitude, phase, and polarization of the incident beam.

The optimization of the phase shifts of the IRS unit cells highly depends on the phase and power density distribution of the incident beam. In FSO systems, the laser beams incident on the optical IRS has a Gaussian power intensity profile and a nonlinear phase profile, whereas the plane waves in radio frequency (RF) systems have a uniform power intensity profile and a linear phase profile.

Intelligent Surfaces Empowered 6G Wireless Network, First Edition.
Edited by Qingqing Wu, Trung Q. Duong, Derrick Wing Kwan Ng, Robert Schober, and Rui Zhang.

For large IRSs, the unit cells outside the beam footprint receive negligible power and cannot improve the beamforming gain and the end-to-end performance (Ajam et al. 2022b). Moreover, due to highly concentrated power even small pointing errors can severely deteriorate the quality of the IRS-assisted FSO link. Furthermore, IRSs at optical frequencies have large electrical size, i.e., the IRS dimension in meter divided by the optical wavelength is large. As an important consequence, optical IRSs have higher flexibility for beam shaping. Furthermore, different analysis techniques, such as geometric optics, which models the wave propagation as rays, are applicable at optical frequencies. Given these substantial differences, the results available for IRS-assisted RF systems are not applicable to IRS-assisted FSO systems.

Thus, in this chapter, we develop an analytical channel model for IRS-assisted FSO links using scattering theory and geometric optics. The former provides a more general model which is applicable to linear and quadratic phase shift profiles across an IRS with flexible size and we mathematically characterize the range of intermediate and far-field distances where the proposed channel model is valid (Ajam et al. 2021). The latter can be used to derive an equivalent mirror-based model for IRSs, which only reflects, the beam toward a desired direction and whose size is larger than the beam footprint incident on the IRS (Najafi et al. 2021b). Furthermore, we discuss three different protocols for sharing a single IRS in a multi-link FSO system, namely, protocols based on time division (TD), IRS division (IRSD), and IRS homogenization (IRSH) (Ajam et al. 2022a). Finally, we provide simulation results to validate the analytical channel model and to compare the performance of different IRS-sharing protocols.

The remainder of this chapter is organized as follows: In Section 15.2, we introduce the adopted IRS-assisted system and channel models. Section 15.3 presents the theoretical modeling for optical IRSs using scattering theory and geometric optics. In Section 15.4, we design the IRS for a point-to-point FSO link and propose protocols for sharing the IRS among multiple FSO links. The theoretical models and the performance of the considered system are evaluated via simulations in Section 15.5. Conclusions and future research directions are provided in Section 15.6.

Notations: Boldface lower-case and upper-case letters denote vectors and matrices, respectively. Superscript $(\cdot)^{\mathrm{T}}$ denotes the transpose operator. $x \sim \mathcal{N}(\mu, \sigma^2)$ represents a Gaussian random variable with mean μ and variance σ^2. $\mathbf{I}_n$ is the $n \times n$ identity matrix, j denotes the imaginary unit, and $(\cdot)^*$ represent the complex conjugate of a complex number. Moreover, $\mathrm{erf}(\cdot)$ and $\mathrm{erfi}(\cdot)$ are the error function and the imaginary error function, respectively. Furthermore, $\mathbf{T}_v = \mathrm{diag}\{\sin(v), 1\}$, $\mathbf{V}_v = \begin{pmatrix} \cos(v) & -\sin(v) \\ \sin(v) & \cos(v) \end{pmatrix}$. Here, $\mathrm{diag}\{x_1, \dots, x_N\}$ denotes a diagonal matrix with $x_i, i \in \{1, \dots, N\}$, as main diagonal entries.

15.2 System and Channel Model

We consider an FSO transmitter equipped with a laser source (LS) connected via an optical IRS to a receiver equipped with a photo detector (PD) and a lens, see Fig. 15.1.

15.2.1 IRS Model

The IRS is installed on a building wall which we define as the xy-plane and the center of the IRS is at the origin of the xyz-coordinate system, see Fig. 15.1. The size of the IRS is $L_x \times L_y$ comprising a large number of subwavelength unit cells. As the size of the IRS is much larger than the optical wavelength, i.e., $L_x, L_y \gg \lambda$, the IRS can be modeled as a continuous surface with a continuous phase-shift profile centered at point $\mathbf{r}_t$ and denoted by $\Phi_{\mathrm{irs}}(\mathbf{r}, \mathbf{r}_t)$, see (Najafi et al. 2021a), where $\mathbf{r} = [x, y]^{\mathrm{T}}$ denotes a point in the xy-plane. As shown in Fig. 15.1, the LS is located at distance d_ℓ, from the center of its beam footprint on the IRS along the beam axis and its direction is denoted by $\mathbf{\Psi}_\ell = (\theta_\ell, \phi_\ell)$, where θ_ℓ is the angle between the xy-plane and the beam axis, and ϕ_ℓ is the angle between the projection of the beam axis on the xy-plane and the x-axis. Moreover, the laser beam footprint on the IRS plane is centered at point $\mathbf{r}_{\ell 0} = [x_{\ell 0}, y_{\ell 0}]^{\mathrm{T}}$. Furthermore, we assume the PD is equipped with a circular lens of radius a. The PD is located at distance d_p from the IRS along the normal vector of the lens plane and it has direction $\mathbf{\Psi}_p = (\theta_p, \phi_p)$, where θ_p is the angle between the xy-plane and the normal vector, and ϕ_p is the angle between the projection of the normal vector on the xy-plane and the x-axis. Without loss of

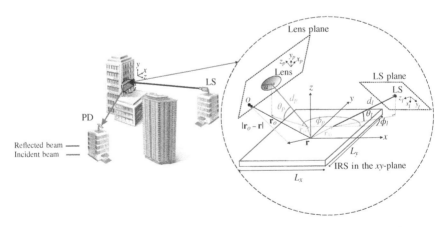

Figure 15.1 An intelligent reflecting surface (IRS)-assisted free space optical (FSO) communication system where the direct link between the transmitter laser source (LS) and the receiver photo detector (PD) is obstructed by a building.

generality, we assume $\phi_\ell = 0$. The normal vector of the lens plane intersects the IRS plane at point $\mathbf{r}_{p0} = [x_{p0}, y_{p0}]^T$, which we refer to as the lens center on the IRS.

15.2.2 Transmitter and Receiver Model

The transmitter is equipped with an LS which emits a monochromatic Gaussian wave. Assuming that the waist of the Gaussian laser beam, w_0, is larger than the wavelength, λ, the paraxial approximation is valid, and thus, the propagated beam in free space along the z_ℓ-axis is given by (Andrews and Phillips 2005)

$$E_\ell\left(\mathbf{r}_\ell\right) = \frac{E_0 w_0}{w(z_\ell, w_0)} \exp\left(-\frac{x_\ell^2 + y_\ell^2}{w^2(z_\ell, w_0)} - j\psi_\ell^G(\mathbf{r}_\ell, z_\ell, w_0)\right) \text{ with phase}$$

$$\psi_\ell^G(\mathbf{r}_\ell, z_\ell, w_0) = k\left(z_\ell + \frac{x_\ell^2 + y_\ell^2}{2R(z_\ell, w_0)}\right) - \tan^{-1}\left(\frac{z_\ell}{z_0}\right), \tag{15.1}$$

where $\mathbf{r}_\ell = [x_\ell, y_\ell]^T$ is a point in the $x_\ell y_\ell z_\ell$-coordinate system which has its origin at the LS. The z_ℓ-axis of this coordinate system is along the beam axis, its y_ℓ-axis is parallel to the intersection line of the LS plane and the IRS plane, and its x_ℓ-axis is orthogonal to the y_ℓ- and z_ℓ-axes. Here, E_0 is the electric field at the origin of the $x_\ell y_\ell z_\ell$-coordinate system, $k = \frac{2\pi}{\lambda}$ is the wave number, $w(z_\ell, w_0) = w_0\left[1 + \left(\frac{z_\ell}{z_0}\right)^2\right]^{1/2}$ is the beam width at distance z_ℓ, $R(z_\ell, w_0) = z_\ell\left[1 + \left(\frac{z_0}{z_\ell}\right)^2\right]$ is the radius of the curvature of the beam's wavefront, and $z_0 = \frac{\pi w_0^2}{\lambda}$ is the Rayleigh range. Here, the total power emitted by the LS is given by $P_\ell = \frac{\pi}{4\eta}|E_0|^2 w_0^2$ and the power density of the Gaussian beam is given by $I_\ell\left(\mathbf{r}_\ell\right) = \frac{1}{2\eta P_\ell}|E_\ell\left(\mathbf{r}_\ell\right)|^2$.

Moreover, we assume an intensity modulation and direct detection (IM/DD) FSO system where the transmitter modulates the data onto the intensity of the laser beam using on-off keying (OOK) modulation and the receiver is equipped with a PD which directly detects the intensity of the received photocurrent. The receiver model is highly dependent on the photo-detection process. Assuming a non-ideal PD, where the background induced shot noise and the thermal noise are dominant (i.e., signal-induced shot noise is negligible), the combined noise impairing the PD can be modeled as an additive white Gaussian noise (AWGN).

15.2.3 Channel Model

In general, FSO channels are affected by geometric and misalignment losses (GML), atmospheric losses, and atmospheric turbulence-induced fading. Thus, the IRS-assisted FSO channel gain h between the LS and the PD can be modeled as follows

$$h = h_p h_{\text{gml}} h_a,$$ (15.2)

where h_a represents the random atmospheric turbulence-induced fading, $h_p = 10^{-\frac{\kappa}{10}(d_{\text{e2e}})}$ is the atmospheric loss, which depends on the end-to-end distance $d_{\text{e2e}} = d_\ell + d_p$ and attenuation coefficient, κ, and h_{gml} characterizes the deterministic GML. Here, we assume that h_a is a Gamma–Gamma distributed variable, i.e., $h_a \sim \mathcal{GG}(\alpha, \beta)$, with small and large scale turbulence parameters α and β, see (Uysal et al. 2006). Gamma–Gamma distributed fading can model moderate and strong atmospheric turbulence.

Moreover, the GML factor, h_{gml}, comprises the deterministic geometric loss due to the divergence of the laser beam along the transmission path and the random misalignment loss due to transceiver sway. Here, we consider a deterministic misalignment loss and ignore its random fluctuations by considering the misalignment vector **u** between the reflected beam footprint center and the center of the receiver lens. Thus, we obtain the GML as the fraction of the power of the LS that is reflected by the IRS and collected by the PD, i.e.,

$$h_{\text{gml}} = \frac{1}{P_\ell} \iint\limits_{\mathcal{A}_p} I_{\text{lens}}(\mathbf{r}_p) \, d\mathcal{A}_p,$$ (15.3)

where $\mathcal{A}_p$ denotes the area of the lens of the PD, $I_{\text{lens}}(\mathbf{r}_p)$ is the power intensity of the beam emitted by the LS and reflected by the IRS in the plane of the lens, and $\mathbf{r}_p = [x_p, y_p]^T$ denotes a point in the lens plane. The origin of the $x_p y_p z_p$-coordinate system is the center of the lens and the z_p-axis is parallel to the normal vector of the lens plane, see Fig. 15.1. We assume that the y_p-axis is parallel to the intersection line of the lens plane and the IRS plane and the x_p-axis is perpendicular to the y_p- and z_p-axes.

The power density of the reflected beam is given as follows (Saleh and Teich 1991)

$$I_{\text{lens}}(\mathbf{r}_p) = \frac{1}{2\eta P_\ell} |E_{\text{lens}}(\mathbf{r}_p)|^2,$$ (15.4)

where η is the free-space impedance and $E_{\text{lens}}(\mathbf{r}_p)$ is the electric field emitted by the LS, reflected by the IRS, and collected by the lens.

15.3 Communication Theoretical Modeling of Optical IRSs

In this section, we characterize the optical IRS-assisted links by modeling the power density of the reflected beam in (15.4) using scattering theory (Ajam et al. 2022a) and geometric optics (Najafi et al. 2021b).

15.3.1 Scattering Theory

In principle, the propagation of a wave can be modeled using Maxwell's equations. However, if (i) the diffracting aperture (IRS) is large compared to the wavelength, i.e., $L_x, L_y \gg \lambda$, and (ii) the diffracted fields are not observed too close to the IRS given the large lens-to-IRS distance, i.e., $d_p \gg \lambda$, scalar diffraction theory can also be applied to model the IRS-assisted links. Given the size of the IRS and the use of FSO systems for long-distance communications, these conditions are met in practice. Moreover, employing wave optics with scalar fields allows us to explicitly capture the impact of arbitrary IRS phase shift profiles, the size of the IRS, and the size of the lens, which is not possible if geometric optics is applied.

In the following, using the electric field of the LS in (15.1), $E_\ell (\mathbf{r}_\ell)$, first, we determine the electric field incident on the IRS, $E_{\text{in}}(\mathbf{r})$. Then, we specify the electric field and power density reflected from the IRS, denoted by $E_{\text{lens}} (\mathbf{r}_p)$ and $I_{\text{lens}} (\mathbf{r}_p)$, respectively. Finally, we determine the point-to-point GML factor between the LS and the PD, h_{gml}.

15.3.1.1 Incident Beam on the IRS

Assuming that $\hat{d}_\ell \gg L_x, L_y$, the electric field emitted by the LS incident on the IRS plane, denoted by $E_{\text{in}}(\mathbf{r})$, is given by

$$E_{\text{in}}(\mathbf{r}) = \frac{E_0 w_0 \zeta_{\text{in}}}{w(\hat{d}_\ell)} \exp\left(-\frac{\hat{x}^2}{w_x^2(\hat{d}_\ell)} - \frac{\hat{y}^2}{w_y^2(\hat{d}_\ell)} - j\psi_{\text{in}}(\mathbf{r}) \right) \quad \text{with phase} \quad (15.5)$$

$$\psi_{\text{in}}(\mathbf{r}) = k\left(\hat{d}_\ell - x\cos(\theta_\ell) + \frac{\hat{x}^2}{2R_x(\hat{d}_\ell)} + \frac{\hat{y}^2}{2R_y(\hat{d}_\ell)} \right) - \tan^{-1}\left(\frac{\hat{d}_\ell}{z_0} \right), \quad (15.6)$$

where $\zeta_{\text{in}} = \sqrt{|\sin(\theta_\ell)|}$, $\hat{d}_\ell = d_\ell + x_{\ell 0} \cos(\theta_\ell)$, $\hat{\mathbf{r}} = [\hat{x}, \hat{y}]^{\text{T}} = [x - x_{\ell 0}, y - y_{\ell 0}]^{\text{T}}$, $w_x(\hat{d}_\ell) = \frac{w(\hat{d}_\ell)}{\sin(\theta_\ell)}$, $w_y(\hat{d}_\ell) = w(\hat{d}_\ell)$, $R_x(\hat{d}_\ell) = \frac{R(\hat{d}_\ell)}{\sin^2(\theta_\ell)}$, and $R_y(\hat{d}_\ell) = R(\hat{d}_\ell)$.

Eq. (15.6) describes an elliptical Gaussian beam on the IRS with beam widths w_x and w_y along the x- and y-axes, respectively, which has power density

$$I_{\text{in}}(\mathbf{r}, d_\ell, w_0) = \frac{2\zeta_{\text{in}}^2}{\pi w^2(\hat{d}_\ell)} \exp\left(-2\hat{\mathbf{r}}^{\text{T}} \mathbf{S}_{w_0}^{\hat{d}_\ell} \mathbf{T}_{\theta_\ell}^2 \, \hat{\mathbf{r}} \right), \quad (15.7)$$

where $\mathbf{S}_{w_0}^d = \text{diag}\left\{ \frac{1}{w^2(d, w_0)}, \frac{1}{w^2(d, w_0)} \right\}$.

15.3.1.2 Huygens–Fresnel Principle

In this section, as a scalar field analysis method, we apply the Huygens–Fresnel principle for deriving the beam reflected by the IRS. This principle states that every point on the wavefront of the beam can be considered as a secondary source emitting a spherical wave and, at any position, the new wavefront is determined by the sum of these secondary waves. Given this principle, the complex amplitude of the electric field reflected by the IRS, denoted by $E_{irs}(\mathbf{r}_o)$, at an arbitrary observation point $\mathcal{O}$ located at $\mathbf{r}_o = [x_o, y_o, z_o]^T$, see Fig. 15.1, is given by (Goodman 2005, Eq. (3-51))

$$E_{lens}(\mathbf{r}_o) = \frac{1}{j\lambda} \iint\limits_{(x,y)\in\Sigma_q} E_{in}(\mathbf{r}) S(\mathbf{r}, \mathbf{r}_o) T(\mathbf{r}) dx dy, \qquad (15.8)$$

where $S(\mathbf{r}, \mathbf{r}_o) = \frac{\exp(-jk\|\mathbf{r}_o - \mathbf{r}\|)}{\|\mathbf{r}_o - \mathbf{r}\|}$ represents a spherical wave, $T(\mathbf{r}) = \zeta e^{-j\Phi_{irs}(\mathbf{r}, \mathbf{r}_t)}$ is the IRS response, and Σ_{irs} is the IRS area. Here, ζ denotes the efficiency factor ($0 \leq \zeta \leq 1$), which accounts for the portion of incident power propagated toward the lens and $\Phi_{irs}(\mathbf{r}, \mathbf{r}_t)$ is the phase-shift profile of the IRS centered at $\mathbf{r}_t$. In (15.8), the total surface of the IRS is divided into infinitesimally small areas $dx dy$, and the light wave scattered by each small area is modeled as a secondary source emitting a spherical wave, $S(\mathbf{r}, \mathbf{r}_o)$. The complex amplitudes of the secondary sources are proportional to the incident electric field, $E_{in}(\mathbf{r})$, and an additional phase-shift term, $e^{-j\Phi_{irs}(\mathbf{r}, \mathbf{r}_t)}$, is introduced by the IRS. The phases of the spherical sources, $k\|\mathbf{r}_o - \mathbf{r}\|$, play an important role in our analysis, and to find a closed-form solution for the integral in (15.8), we approximate $\|\mathbf{r}_o - \mathbf{r}\|$ in the following.

15.3.1.3 Intermediate-Field Versus Far-Field

First, using $\|\mathbf{r}_o - \mathbf{r}\| = [(x - x_o)^2 + (y - y_o)^2 + z_o^2]^{1/2}$, we establish

$$\frac{\|\mathbf{r}_o - \mathbf{r}\|^2}{\|\mathbf{r}_o\|^2} = 1 + \frac{x^2 + y^2}{\|\mathbf{r}_o\|^2} - 2\frac{xx_o + yy_o}{\|\mathbf{r}_o\|^2}. \qquad (15.9)$$

Applying the Taylor series expansion with $(1 + x)^{1/2} = 1 + \frac{1}{2}x - \frac{1}{8}x^2 + \cdots$, see (Gradshteyn and Ryzhik 1994), we obtain

$$\|\mathbf{r}_o - \mathbf{r}\| = \|\mathbf{r}_o\| \underbrace{- \frac{xx_o + yy_o}{\|\mathbf{r}_o\|}}_{=t_1} + \underbrace{\frac{x^2 + y^2}{2\|\mathbf{r}_o\|} - \frac{x^2 x_o^2 + y^2 y_o^2}{2\|\mathbf{r}_o\|^3}}_{=t_2}$$

$$\underbrace{- \frac{(x^2 + y^2)^2}{8\|\mathbf{r}_o\|^3} + \frac{(x^2 + y^2)(xx_o + yy_o)}{2\|\mathbf{r}_o\|^3}}_{=t_3} + \cdots. \qquad (15.10)$$

For the commonly used far-field approximation (Fraunhofer region), it is assumed that the secondary waves reflected by the IRS surface experience only a linear phase shift with respect to (w.r.t.) each other (Hecht 2017). In other words, only t_1 in (15.10) is taken into account, and t_2 and all higher-order terms are neglected. For the far-field approximation to hold, the impact of t_2 in the argument of the exponential term, $k\|\mathbf{r}_o - \mathbf{r}\|$, should be much smaller than one period of the complex exponential, and thus,

$$k\frac{x^2 + y^2}{2\|\mathbf{r}_0\|} \ll 2\pi. \tag{15.11}$$

The range of the relevant values for x and y in (15.8) is bounded by the beam widths of the incident electric field $2w_x$ and $2w_y$ (where the power of the incident beam drops by $\frac{1}{e^4}$ compared to the peak value) and the size of the IRS L_x and L_y, i.e., $x_e = \min\left(\frac{L_x}{2}, w_x\right) \geq |x|$ and $y_e = \min\left(\frac{L_y}{2}, w_y\right) \geq |y|$. Assuming that the lens radius, a, is smaller than the IRS-lens distance, i.e., $a \ll d_p$, for any observation point $\mathcal{O}$ on the lens area, we can substitute $\|\mathbf{r}_0\| \approx d_p$. Thus, we substitute x_e and y_e for x and y in (15.11), respectively, and define the minimum far-field distance d_f as follows

$$d_f = \frac{x_e^2 + y_e^2}{2\lambda}, \tag{15.12}$$

such that for distances $d_p \gg d_f$, the approximation of (15.10) in (15.8) by term t_1 only is appropriate. However, depending on the values of x_e and y_e, and the IRS-lens distance, d_p, this condition might not hold in practice. For example, let us consider a typical IRS size $L_x = L_y = 50$ cm and a LS located at $d_\ell = 600$ m and $\mathbf{\Psi}_\ell = \left(\frac{\pi}{2}, 0\right)$ emitting a laser beam with $\lambda = 1550$ nm and $w_0 = 2.5$ mm. Since the incident beamwidth on the IRS is $w_x = w_y = 0.12$ m, then, the far-field distance according to (15.12) is $d_f = 9.1$ km. This means that the far-field analysis is valid for IRS-to-lenses distances larger than d_f, which exceeds the typical values for FSO systems. In order to obtain a model that is also valid for shorter distances, we have to consider both the linear term t_1 and the quadratic term t_2 in (15.10). Using a similar method as in (15.11), we define intermediate distances where the largest term in t_3 is smaller than a wavelength, i.e.,

$$k\frac{(x^2 + y^2)(xx_0 + yy_0)}{2\|\mathbf{r}_0\|^3} \ll 2\pi. \tag{15.13}$$

Substituting x_e and y_e for x and y and setting $x_0 = y_0 = \frac{\|\mathbf{r}_0\|}{2} \approx \frac{d_p}{2}$, we define the minimum intermediate distances d_n as follows

$$d_n = \left[\frac{(x_e^2 + y_e^2)(x_e + y_e)}{4\lambda}\right]^{1/2}. \tag{15.14}$$

For the previous example, we obtain $d_n = 32.7$ m which is much smaller than the practical range of link distances of FSO systems. Thus, for practical link distances $d_p \gg d_n$, the approximation of (15.10) in (15.8) by terms t_1 and t_2 is appropriate. Unlike optical IRSs, RF IRSs with typical wavelength $\lambda = 0.1$ m usually operate in the far-field regime, as $d_f = 0.14$ m. To take into account this difference, we analyze optical IRSs for the intermediate and far-field ranges (known as Fresnel range, see (Hecht 2017)) in the following.

15.3.1.4 Received Power Density

As mentioned previously, the Gaussian beam incident on the IRS in (15.6) has a nonlinear phase profile. In order to redirect the beam in a desired direction, the IRS must compensate the phase of the incident beam and apply an additional phase shift to redirect the beam. In the following, we assume a quadratic phase shift (QP) profile across the IRS as an approximation of more general nonlinear phase-shift profiles and obtain a closed-form solution for the reflected electric field in (15.8).

Theorem 15.1 *Assume a QP profile centered at point* $\mathbf{r}_t$*, i.e.,* $\Phi_{\text{irs}}^{\text{quad}}(\mathbf{r}, \mathbf{r}_t) = k\left(\Phi_0 + \Phi_x(x - x_t) + \Phi_y(y - y_t) + \Phi_{x^2}(x - x_t)^2 + \Phi_{y^2}(y - y_t)^2\right)$*, where* Φ_0*,* Φ_x*,* Φ_{x^2}*,* Φ_y*, and* Φ_{y^2} *are constants. Then, the electric field emitted by the LS at position* (d_ℓ, Ψ_ℓ)*, reflected by the IRS, and received at the lens located at* (d_p, Ψ_p) *for any intermediate distances* $d_\ell \gg L_x, L_y$ *and* $d_p \gg a, x_{\ell 0}, y_{\ell 0}, d_n$ *is given by*

$$E_{\text{lens}}\left(\mathbf{r}_p\right) = \frac{\pi C C_q}{4\sqrt{b_x b_y}} \exp\left(-\frac{k^2}{4b_x}X^2 - \frac{k^2}{4b_y}Y^2\right) D\left(L_x, L_y, \mathbf{r}_p\right), \qquad (15.15)$$

with

$$D(L_x, L_y, \mathbf{r}_p) = \left[\text{erf}\left(\sqrt{b_x}\frac{L_x}{2} + \frac{jk}{2\sqrt{b_x}}X\right) - \text{erf}\left(-\sqrt{b_x}\frac{L_x}{2} + \frac{jk}{2\sqrt{b_x}}X\right)\right]$$
$$\times \left[\text{erf}\left(\sqrt{b_y}\frac{L_y}{2} + \frac{jk}{2\sqrt{b_y}}Y\right) - \text{erf}\left(-\sqrt{b_y}\frac{L_y}{2} + \frac{jk}{2\sqrt{b_y}}Y\right)\right], \qquad (15.16)$$

where $X = A_0 + c_1 x_p + c_2 y_p + \Phi_x - 2x_t \Phi_{x^2}$ $A_0 = \frac{2jvx_{\ell 0}}{k}\sin^2(\theta_\ell) - \frac{x_{p0}}{d_p} + \varphi_x$*,* $Y = B_0 + c_3 x_p + c_4 y_p + \Phi_y - 2y_t \Phi_{y^2}$*,* $B_0 = \frac{2jvy_{\ell 0}}{k} - \frac{y_{p0}}{d_p} + \varphi_y$*,* $c_1 = -\frac{1}{d_p}\cos(\phi_p)\sin(\theta_p)$*,* $c_2 = \frac{1}{d_p}\sin(\phi_p)$*,* $c_3 = -\frac{1}{d_p}\sin(\phi_p)\sin(\theta_p)$*,* $c_4 = -\frac{1}{d_p}\cos(\phi_p)$*,* $c_5 = \frac{1}{d_p}\cos(\phi_p)\cos(\theta_p)$*,* $c_6 = \frac{1}{d_p}\sin(\phi_p)\cos(\theta_p)$*,* $\varphi_x = -\cos(\theta_\ell) - \cos(\theta_p)\cos(\phi_p)$*,* $\varphi_y = -\cos(\theta_p)\sin(\phi_p)$*,* $b_y = v + \frac{jk}{2d_p}\left(1 - c_6^2 d_p^2 - 2c_6 y_{p0}\right) + jk\Phi_{y^2}$*,* $b_x = v\sin^2(\theta_\ell) + \frac{jk}{2d_p}\left(1 - c_5^2 d_p^2 - 2c_5 x_{p0}\right) + jk\Phi_{x^2}$*,* $\delta = -\Phi_{x^2}(x_t)^2 - \Phi_{y^2}(y_t)^2 + \Phi_x x_t + \Phi_y y_t - \Phi_0$*,* $C = \frac{E_o w_0 \zeta_{\text{in}}}{j\lambda w(\hat{d}_\ell)d_p}$

$e^{-jk(\hat{d}_\ell + d_p) + j\tan^{-1}\left(\frac{\hat{d}_\ell}{z_0}\right) - v\sin^2(\theta_\ell)x_{\ell 0}^2 - vy_{\ell 0}^2}$*,* $C_q = \zeta e^{jk\delta}$*,* $v = \frac{1}{w^2(\hat{d}_\ell)} + \frac{jk}{2R(\hat{d}_\ell)}$*.*

Proof: We substitute $\|\mathbf{r} - \mathbf{r}_0\|$ in (15.8) by terms t_1 and t_2 in (15.10) and solve it by applying (Gradshteyn and Ryzhik 1994, Eq. (2.33-1)). Then, we exploit $[x_o, y_o]^{\text{T}} = \mathbf{V}_{-\phi_p}\mathbf{T}_{\theta_p}\mathbf{r}_p + \mathbf{d} + \mathbf{r}_{p0}$, where $\mathbf{d} = d_p[\cos(\phi_p)\cos(\theta_p), \sin(\phi_p)\cos(\theta_p)]^{\text{T}}$. Next, approximate $\|\mathbf{r}_0\| \approx d_p$, $z_o \approx d_p\sin(\theta_p)$, $\frac{x_o^2}{\|\mathbf{r}_0\|^3} = c_5^2 d_p + \frac{2}{d_p}c_5 x_{p0}$, and $\frac{y_o^2}{\|\mathbf{r}_0\|^3} = c_6^2 d_p + \frac{2}{d_p}c_6 y_{p0}$ and this completes the proof. $\qquad\square$

The power intensity received at the lens can be simplified to

$$I_{\text{lens}}^{\text{Huygens}}(\mathbf{r}_p) = \frac{\pi \zeta_{\text{in}}^2 \zeta^2}{8\lambda^2 w^2(\hat{d}_\ell) d_p^2 |b_x b_y|} |D(L_x, L_y, \mathbf{r}_p)|^2 \exp\left(-\frac{k^2 X^2}{4\tilde{b}_x} - \frac{k^2 Y^2}{4\tilde{b}_y}\right),$$

(15.17)

where $\tilde{b}_i = \frac{b_i b_i^*}{b_i + b_i^*}, i \in \{x, y\}$. Eq. (15.17) explicitly shows the impact of the positioning of the LS and the lens w.r.t. the IRS, the size of the IRS, and the phase-shift
configuration across the IRS on the electric field reflected by the IRS. The above
theorem is valid for both far-field and intermediate distances. The reflected power
density for linear phase shift (LP) profiles is included in the result in (15.17) as a
special case for $\Phi_{x^2} = \Phi_{y^2} = 0$.

Then, the GML factor represents the fraction of the power received by the lens
and can be obtained by approximating $\tilde{D}(L_x, L_y, \mathbf{r}_p) = D(L_x, L_y, [\frac{a}{2}, \frac{a}{2}])$ as follows

$$h_{\text{gml}} = \frac{C_h \sqrt{\pi}}{2\sqrt{\rho_x}} |C_q \tilde{D}(L_x, L_y)|^2 \exp\left(-\frac{k^2 \bar{A}^2}{\tilde{b}_x} - \frac{k^2 \bar{B}^2}{\tilde{b}_y}\right)$$

$$\int_{-\tilde{a}}^{\tilde{a}} \exp\left(-\rho_y y_p^2 - \varrho_y y_p + \frac{(\rho_{xy} y_p + \varrho_x)^2}{4\rho_x}\right)$$

$$\times \left[\text{erf}\left(\frac{\rho_x \sqrt{\pi}a + \rho_{xy} y_p + \varrho_x}{2\sqrt{\rho_x}}\right) - \text{erf}\left(\frac{-\rho_x \sqrt{\pi}a + \rho_{xy} y_p + \varrho_x}{2\sqrt{\rho_x}}\right)\right] dy_p,$$

(15.18)

where $\tilde{a} = \frac{\sqrt{\pi}a}{2}$, $C_h = \frac{\pi \zeta_{\text{in}}^2}{8|b_x||b_y|\lambda^2 d_p^2 w^2(\hat{d}_\ell)} e^{-\frac{2x_{\ell 0}^2 \sin(\theta_\ell)}{w^2(\hat{d}_\ell)} - \frac{2y_{\ell 0}^2}{w^2(\hat{d}_\ell)}}$, $\rho_x = \frac{k^2}{4}\left(\frac{c_1^2}{\tilde{b}_x} + \frac{c_3^2}{\tilde{b}_y}\right)$, $\rho_y =$
$\frac{k^2}{4}\left(\frac{c_2^2}{\tilde{b}_x} + \frac{c_4^2}{\tilde{b}_y}\right)$, $\rho_{xy} = \frac{k^2}{2}\left(\frac{c_1 c_2}{\tilde{b}_x} + \frac{c_3 c_4}{\tilde{b}_y}\right)$, $\bar{A} = A_0 + \Phi_x - 2x_t \Phi_{x^2}$, $\bar{B} = B_0 + \Phi_y - 2y_t \Phi_{y^2}$,
$\varrho_x = \frac{k^2}{2}\left(c_1 \frac{\bar{A}}{\tilde{b}_x} + c_3 \frac{\bar{B}}{\tilde{b}_y}\right)$, $\varrho_y = \frac{k^2}{2}\left(c_2 \frac{\bar{A}}{\tilde{b}_x} + c_4 \frac{\bar{B}}{\tilde{b}_y}\right)$.

15.3.2 Geometric Optics

Geometric optics models the waves reflected from the surface as rays and is
known to provide an accurate approximation of Huygens–Fresnel principle at
optical frequencies. In practice, at optical frequencies, the electrical size of the
IRS is very large and the reflected beam can often be modeled as a narrow beam.
This allows us to analyze an equivalent mirror-assisted system, instead of the
original IRS-assisted system facilitating the application of the image method from
geometric optics (Najafi et al. 2021b). We note that this model is only applicable
for a specific IRS phase-shift profile which will be provided in this section.

15.3.2.1 Equivalent Mirror-Assisted Analysis

In this section, we develop an equivalent mirror-assisted system for the IRS-assisted system such that the reflected electric fields of the mirror and of the IRS are identical. However, the location of the LS and the properties of the emitted Gaussian laser beam are different in the original and the equivalent systems.

To construct the equivalent system, first, we replace the IRS by a mirror. Then, the LS is replaced by a new LS, which is located in a different position such that it has distance d_ℓ from the mirror and its beam hits the mirror with incident angle $\mathbf{\Psi}_p$, see Fig. 15.2(b). Due to specular reflection of the mirror, the angle of reflection is also $\mathbf{\Psi}_p$. The new LS emits a rotated astigmatic Gaussian beam with waist radius $\mathbf{w}_0 = (w_{0x}, w_{0y})$, rotation angle φ, and electric field

$$E_\ell(\hat{\mathbf{r}}_\ell) = E_\ell \exp\left(-\hat{\mathbf{r}}_\ell^T \mathbf{V}_\varphi^T \mathbf{S}_{\mathbf{w}_0}^{\hat{z}_\ell} \mathbf{V}_\varphi \hat{\mathbf{r}}_\ell - j\psi_\ell^{AG}\left(\hat{\mathbf{r}}_\ell, \hat{z}_\ell, \mathbf{w}_0\right)\right), \text{ with phase}$$

$$\psi_\ell^{AG}\left(\hat{\mathbf{r}}_\ell, \hat{z}_\ell, \mathbf{w}_0\right) = k\hat{z}_\ell + k\left(\frac{1}{2}\hat{\mathbf{r}}_\ell^T \mathbf{V}_\varphi^T \mathbf{Q}_{\mathbf{w}_0}^{\hat{z}_\ell} \mathbf{V}_\varphi \hat{\mathbf{r}}_\ell\right) + \psi_0(\hat{z}_\ell, \mathbf{w}_0), \quad (15.19)$$

where $\hat{\mathbf{r}}_\ell = [\hat{x}_\ell, \hat{y}_\ell]^T$, $\psi_0(\hat{z}_\ell, \mathbf{w}_0) = \frac{1}{2}\left(\tan^{-1}\left(\frac{\hat{z}_\ell}{z_{R_x}}\right) + \tan^{-1}\left(\frac{\hat{z}_\ell}{z_{R_y}}\right)\right)$, $z_{R_i} = \frac{\pi w_{0i}^2}{\lambda}$, $\mathbf{S}_{\mathbf{w}_0}^d = \text{diag}\{\frac{1}{w^2(d,w_{0x})}, \frac{1}{w^2(d,w_{0y})}\}$, and $\mathbf{Q}_{\mathbf{w}_0}^d = \text{diag}\{\frac{1}{R(d,w_{0x})}, \frac{1}{R(d,w_{0y})}\}$. The power density of the rotated astigmatic Gaussian beam is given by

$$I_{AG}(\hat{\mathbf{r}}_\ell, \mathbf{w}_0, \varphi) = \frac{2}{\pi w(\hat{z}_\ell, w_{0x})w(\hat{z}_\ell, w_{0y})} \exp\left(-2\hat{\mathbf{r}}_\ell^T \mathbf{V}_\varphi^T \mathbf{S}_{\mathbf{w}_0}^{\hat{z}_\ell} \mathbf{V}_\varphi \hat{\mathbf{r}}_\ell\right). \quad (15.20)$$

Then, we obtain the incident power density of the new LS in (15.20) across the equivalent mirror as follows

$$I_{AG}^{irs}(d_\ell, \mathbf{w}_0, \varphi) = \frac{2\sin(\theta_p)}{\pi w(d, w_{0x})w(d, w_{0y})}$$
$$\times \exp\left(-2\mathbf{r}^T \mathbf{V}_{-\phi_p}^T \mathbf{T}_{\theta_p}^T \mathbf{V}_\varphi^T \mathbf{S}_{\mathbf{w}_0}^{d_\ell} \mathbf{V}_\varphi \mathbf{T}_{\theta_p} \mathbf{V}_{-\phi_p} \mathbf{r}\right). \quad (15.21)$$

To determine the parameters of the new LS, we consider that the power density distributions of the new LS across the mirror in (15.21) and that of the original LS

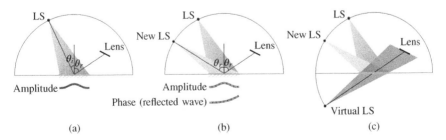

Amplitude

Phase (reflected wave)

(a) (b) (c)

Figure 15.2 Illustration of the proposed equivalent mirror-assisted system: (a) original IRS-assisted system, (b) equivalent mirror-assisted system, and (c) virtual image system.

across the IRS in (15.6) have to be identical. Thus, the parameters of the new LS, $\mathbf{w}_0$ and φ, are chosen as the solutions to equations $w(d_l, w_{0i}) = \frac{1}{\sqrt{\lambda_i}}$ and $T(\varphi) = [\mathbf{v}_x, \mathbf{v}_y]^T$. Here, λ_i and $\mathbf{v}_i$, $i \in \{x, y\}$, are respectively the eigenvalues and eigenvectors of matrix $\mathbf{A} = \mathbf{T}_{\theta_p}^{-T} \mathbf{V}_{\phi_p}^T \mathbf{S}_{w_0}^{d_e} \mathbf{T}_{\theta_e}^2 \mathbf{V}_{\phi_e} \mathbf{T}_{\theta_e}^{-1}$.

15.3.2.2 Received Power Density

To determine the received power density at the lens, first, we design the phase-shift profile, across the IRS such that the phases of the reflected beams in both systems are identical. This phase-shift profile is given by

$$\Phi_{\text{irs}}^{\text{equiv}} = \pi + \psi_\ell^{\text{AG}} \left(\mathbf{V}_\varphi \mathbf{T}_{\theta_p} \mathbf{V}_{-\phi_p} \mathbf{r}, d_\ell + x \cos\left(\theta_p\right) \cos\left(\phi_p\right) \right.$$
$$\left. + y \cos\left(\theta_p\right) \sin(\phi_p), \mathbf{w}_0 \right) - \psi_\ell^G \left(\sqrt{\mathbf{r}^T \mathbf{T}_{\theta_e}^2 \mathbf{r}}, d_\ell + x \sin\left(\theta_\ell\right), \mathbf{w}_0 \right).$$

$$(15.22)$$

This leads to the same reflected electric fields in both systems. Therefore, the angle of reflection in both systems is also identical and is given by $\mathbf{\Psi}_p$.

Using image theory from geometric optics, the system with the new LS and an infinite-size mirror in Fig. 15.2(b), can be transformed to an equivalent system with a virtual LS without the mirror, see Fig. 15.2(c). Thereby, the virtual LS is the image of the new LS w.r.t. the mirror. Thus, assuming a sufficiently large mirror in the equivalent system, the received beam at the lens is identical to the beam of the virtual source at distance d_{e2e} given by

$$I_{\text{lens}}(\mathbf{r}_p) = I_{\text{AG}}^{\text{irs}}(d_{\text{e2e}}, \mathbf{w}_0, \varphi).$$

$$(15.23)$$

For an IRS of finite size, the lines connecting the virtual LS and the boundaries of the mirror create a truncation region, outside of which the reflected optical power is negligible.

Given the power intensity in (15.23), the fraction of power collected by the lens is given by (Najafi et al. 2021b)

$$h_{\text{gml}} = G_0 \exp\left(-2\frac{\|\mathbf{u}\|^2}{W}\right),$$

$$(15.24)$$

where the vector of misalignment $\mathbf{u} = \mathbf{T}_{\theta_p}^{-1} \mathbf{V}_{-\phi_p}^T \tilde{\mathbf{u}}$, $\tilde{\mathbf{u}} = \mathbf{r}_{p0} - \mathbf{r}_{\ell 0}$, $G_0 = \frac{\pi a^2}{2\tau_1 \tau_2 w(d_{\text{e2e}}, w_{0x}) w(d_{\text{e2e}}, w_{0y})} \text{erf}(\tau_1) \text{erf}(\tau_2)$, $W = \frac{\pi a^2}{4\tau_1 \tau_2} \sqrt{\frac{\pi \text{erf}(\tau_1) \text{erf}(\tau_2)}{\tau_1 \tau_2 \exp(-(\tau_1^2 + \tau_2^2))}}$, and $\tau_\iota = a\sqrt{\frac{\pi \delta_\iota}{2}}$, $\iota \in \{1,2\}$. Here, δ_ι are the eigenvalues of matrix $\mathbf{V}_\varphi^T \mathbf{S}_{w_0}^{d_{\text{e2e}}} \mathbf{V}_\varphi$.

The above result is valid only for IRSs with the phase-shift profile in (15.22), although at distances $d_\ell \gg \max\left(\frac{\pi w_0^2}{2}, \frac{|\mathbf{r}^T\left(\mathbf{T}_{\theta_e}^2 - \mathbf{V}_{\phi_p} \mathbf{T}_{\theta_p}^2 \mathbf{V}_{-\phi_p}\right)\mathbf{r}|}{\lambda}\right)$, Eq. (15.22) can be approximated by the LP profile in (15.26). Comparing the analytical power density obtained with geometric optics in (15.23) and the power density obtained with scattering theory in (15.17) for the LP profile in (15.26) reveals the impact of diffraction. In (15.23), we ignore the size of the IRS and the power existing at the edges of the truncation area due to diffraction, although these factors are included in (15.17) by parameters $D(L_x, L_y, \mathbf{r}_p)$ and the beamwidths of the received beam $|b_x|$ and $|b_y|$. We can observe that by assuming a very large IRS size and ignoring the diffraction effect, the results in (15.23) and (15.17) are similar. Thus, the above geometric-based analysis is useful to provide a tractable model which can be extended to include the random fluctuations of misalignment errors facilitating a statistical analysis of the GML factor, see (Najafi et al. 2021b) for more details.

15.4 Design of Optical IRSs for FSO Systems

In this section, we propose IRS designs for single- and multi-link FSO systems.

15.4.1 IRS-Assisted Point-to-Point System

In this section, we consider the case where a single transmitter emits an FSO signal toward an IRS which redirects the beam to a single receiver. The signal intensity $y[t]$ received by the PD in time slot $t, t \in \mathcal{T}$, via a point-to-point IRS-assisted FSO link can be modeled as follows

$$y[t] = hs[t] + w[t], \quad t \in \mathcal{T}, \tag{15.25}$$

where $s[t]$ is the OOK modulated symbol transmitted by the LS with average transmit power P_ℓ, $h \in \mathcal{R}^+$ is the channel gain between the LS and the PD given in (15.2), and $w[t] \sim \mathcal{N}\left(0, \sigma_w^2\right)$ is the AWGN with zero mean and variance σ_w^2 impairing the PD.

In the following, we determine the phase-shift profile of the IRS, $\Phi_{\text{irs}}(\mathbf{r}, \mathbf{r}_t)$, needed to redirect the LS beam toward a desired PD and the corresponding efficiency factor, ζ, required for a passive-lossless IRS.

15.4.1.1 IRS Phase-Shift Profile $\Phi(\mathbf{r}, \mathbf{r}_t)$

The phase shift introduced by the IRS is exploited such that it compensates the phase difference between the incident and the desired beam. Moreover, the incident beam on the IRS has a nonlinear phase profile (15.6) and the phase profile of the secondary waves (15.8) includes linear or quadratic terms depending on the operating regime and the approximation used, cf. Section 15.3.1.3. Thus, the desired phase shift across the IRS is in general a nonlinear function. Assuming that the positions of the LS, $\left(d_\ell, \mathbf{\Psi}_\ell\right)$, and the PD, $\left(d_p, \mathbf{\Psi}_p\right)$, are known, we adopt

LP and QP profiles as approximations of general nonlinear phase-shift profiles as follows

$$\Phi_{irs}^{lin}\left(\mathbf{r},\mathbf{r}_t\right) = k\left(\Phi_0 + \Phi_x\left(x - x_t\right) + \Phi_y\left(y - y_t\right)\right), \tag{15.26}$$

$$\Phi_{irs}^{quad}\left(\mathbf{r},\mathbf{r}_t\right) = k\left(\Phi_0 + \Phi_x\left(x - x_t\right) + \Phi_y\left(y - y_t\right) \right.$$
$$\left. + \Phi_{q,x^2}\left(x - x_t\right)^2 + \Phi_{y^2}\left(y - y_t\right)^2\right), \tag{15.27}$$

where $\Phi_x = \cos(\theta_\ell)\cos(\phi_\ell) + \cos(\theta_p)\cos(\phi_p)$, $\Phi_y = \cos(\theta_\ell)\sin(\phi_\ell) + \cos(\theta_p)$ $\sin(\phi_p)$, $\Phi_{x^2} = \frac{1}{2d_p}\left(\cos^2(\theta_p)\cos^2(\phi_p) - 1\right) - \frac{\sin^2(\theta_\ell)}{2R(\hat{d}_\ell)} + \frac{\sin^2(\theta_p)}{4d_p}$, $\Phi_{y^2} = \frac{1}{2d_p}\left(\cos^2(\theta_p)\sin^2\right.$ $(\phi_p) - 1) - \frac{1}{2R(\hat{d}_\ell)} + \frac{1}{4d_p}$, and $\Phi_0 = d_p + \hat{d}_\ell$. The LP profile in (15.26) corresponds to the generalized Snell's law of reflection and is chosen such that the total accumulated phase is zero when the beam arrives at the lens center. Moreover, for the QP profile, the quadratic terms cancel the total accumulated phase due to the LS beam curvature and the IRS-to-lens distance. Furthermore, adding the term $\frac{1}{4d_p}$ causes a parabolic phase profile that focuses the beam at the lens center. The impact of both LP and QP profiles in (15.26) and (15.27) on the received power density in (15.17) can be analyzed using Huygens–Fresnel principle.

15.4.1.2 IRS Efficiency ζ

The efficiency factor ζ ensures that the power reflected from the IRS is smaller than or equal to the power incident on the IRS. We assume $\zeta = \zeta_0\bar{\zeta}$, where ζ_0 denotes the resistive loss of the IRS and $\bar{\zeta}$ is the passivity constant which ensures the incident power is completely reflected. To find $\bar{\zeta}$, we consider a large IRS size $L_x, L_y \to \infty$ and a lens with radius $a \to \infty$ to determine the maximum powers received by the IRS and the lens. Since we assumed the IRS is very large, we have $D(L_x, L_y, \mathbf{r}_p) = 4$ and $\mathbf{r}_{0\ell} = \mathbf{r}_{0p} = \mathbf{r}_t = \mathbf{0}$. Then, given the passivity of the IRS, the total average power on its surface should be zero, or equivalently, the reflected power and the incident power have to be equal.

Using (15.17), the received power for a very large lens P_{lens} is given as follows

$$P_{lens} = \int_{-\infty}^{\infty}\int_{-\infty}^{\infty} P_\ell I_{lens}^{Huygens}(\mathbf{r}_p)\,dx_p dy_p \overset{(a)}{=} \frac{P_\ell \zeta_0^2 \bar{\zeta}^2 \zeta_{in}^2}{|\sin(\theta_p)||\sin(\theta_\ell)|}, \tag{15.28}$$

where in (a) we apply (Gradshteyn and Ryzhik 1994, Eq. (3.323-2)).

Then, using (15.7) the incident power on the IRS, P_{in}, is also given by

$$P_{in} = \int_{-\infty}^{\infty}\int_{-\infty}^{\infty} P_\ell I_{in}(\mathbf{r})\,dx dy = \frac{\zeta_{in}^2 P_\ell}{|\sin(\theta_\ell)|}, \tag{15.29}$$

where $E_{in}(\mathbf{r})$ is given in (15.6). Then, (15.28) and (15.29) have to be equal and by assuming $\zeta_0 = 1$, we obtain the passivity factor $\overline{\zeta}$ of the IRS depends on $\mathbf{\Psi}_p$ and is given by

$$\overline{\zeta} = \sqrt{|\sin(\theta_p)|}.$$ (15.30)

15.4.2 IRS-Assisted Multi-Link System

In this section, we consider M FSO transmitter-receiver pairs connected via a single optical IRS. The mth LS and n-th PD are, respectively, located at $\left(d_{\ell m}, \mathbf{\Psi}_{\ell m}\right)$ and $\left(d_{pn}, \mathbf{\Psi}_{pn}\right), \forall m, n \in \mathcal{M}$, where $\mathcal{M} = \{1, \ldots, M\}$. The received signal at the nth PD comprises the desired signal intensity from the paired nth LS and the inter-link interference from the remaining $M - 1$ LSs. The signal intensity $y_n[t]$ received by the nth PD in time slot $t, t \in \mathcal{T}$, where $\mathcal{T} = \{1, \ldots, T\}$, can be modeled as follows

$$y_n[t] = h_{n,n}s_n[t] + \sum_{m=1, m \neq n}^{M} h_{m,n}s_m[t] + w_n[t], \quad n \in \mathcal{M}, t \in \mathcal{T},$$ (15.31)

where $s_m[t]$ is the OOK modulated symbol transmitted by the mth LS with average transmit power $P_{\ell m}$, $h_{m,n} \in \mathcal{R}^+$ is the channel gain between the mth LS and the n-th PD, and $w_n[t] \sim \mathcal{N}\left(0, \sigma_w^2\right)$ is the AWGN with zero mean and variance σ_w^2 impairing the nth PD.

To share the IRS among multiple links, the IRS is divided into $Q = Q_x Q_y$ tiles, each comprising a large number of subwavelength unit cells and a continuous phase-shift profile centered at $\mathbf{r}_t^q$ and denoted by $\Phi_q(\mathbf{r}, \mathbf{r}_t^q)$, where Q_x and Q_y are the numbers of tiles in x- and y-direction, respectively. Each tile is centered at $\mathbf{r}_q = [x_q, y_q]^T$ and has length l_x with tile spacing s_x in x-direction and width l_y with tile spacing s_y in y-direction. Thus, the IRS lengths in x- and y-direction are given by $L_i = Q_i l_i + (Q_i - 1)s_i$, $i \in \{x, y\}$. To employ the IRS for multi-link FSO transmission, we propose different IRS-sharing protocols in the following (Ajam et al. 2022a).

15.4.2.1 Time Division Protocol

For the TD protocol, in each time slot, one LS transmits, while the other LSs are inactive, thus, the number of time slots, T, is identical to the number of FSO links, M, i.e., $T = M$. In each time slot, the entire IRS surface is configured for the active LS-PD pair, i.e., only one tile is needed and $Q = 1$, see Fig. 15.3(a). Moreover, in order to maximize the received power, the phase-shift profile center, the incident beam footprint center, and the lens center on the IRS should coincide with the origin, i.e., $\mathbf{r}_t^q = \mathbf{r}_{\ell 0, m} = \mathbf{r}_{p 0, n} = \mathbf{0}$, $\forall m, n \in \mathcal{M}$, where $\mathbf{0}$ denotes the origin. For this protocol, the active PD receives maximum power as the IRS serves only one LS-PD pair at a time. However, the time sharing among LSs degrades the achievable data rate.

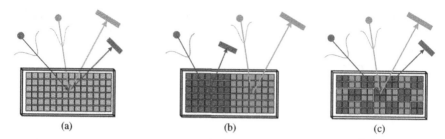

Figure 15.3 IRS-sharing protocols. The IRS is divided into $Q = 1, 2,$ and 21 tiles for the TD, IRSD, and IRSH protocols, respectively. (a) TD protocol, (b) IRSD protocol, and (c) IRSH protocol.

15.4.2.2 IRS Division Protocol

For this protocol, all LSs simultaneously illuminate the IRS, which is divided into $Q = M$ tiles, see Fig. 15.3(b). We assume that each LS-PD pair is assigned to a different tile such that the centers of the beam footprints, the lens centers, and the tile phase-shift profile centers coincide with the centers of the respective tiles, i.e., $\mathbf{r}_{\ell 0,m} = \mathbf{r}_{p0,n} = \mathbf{r}_t^q = \mathbf{r}_q, \forall m, n \in \mathcal{M}$. The IRSD protocol enables all LSs to transmit simultaneously which in turn increase the data rate compared to the TD protocol. Although, by partitioning the IRS among multiple FSO links, only part of the incident power is reflected by the respective tile toward the desired lens. Thereby, unless the incident beamwidth is smaller than the tile area, less power is received at the lens compared to the TD protocol. Moreover, misalignment errors may shift the beam footprint center or the lens center on the IRS toward a tile reserved for a different LS-PD pair. Thus, given the Gaussian beam intensity distribution, a portion of the power may be redirected in an undesired direction which in turn degrades the GML.

15.4.2.3 IRS Homogenization Protocol

Since perfect tracking of the tile, beam footprint, and lens centers may not always be feasible, we propose the homogenization of the IRS surface by dividing it into Q tiles such that the number of tiles is much larger than the number of LS-PD pairs, i.e., $Q \gg M$, see Fig. 15.3(c). In this protocol, the centers of the beam footprints, the lens centers, and the tile phase-shift profile centers coincide with the IRS center, i.e., $\mathbf{r}_{\ell 0,m} = \mathbf{r}_{p0,n} = \mathbf{r}_t^q = \mathbf{0}$. In this protocol, multiple small tiles are allocated to each LS-PD pair instead of only one tile for each pair in the IRSD protocol which helps to distribute the LS beam power across multiple tiles. Thus, if misalignment errors occur and the reflected beam footprint center on the lens is shifted, some tiles can still redirect the beam toward the desired PD. Although, this protocol is more resilient to misalignment errors, this comes at the cost of a power loss since only some of the tiles around the beam center are designed to redirect the beam toward the desired lens. Thus, without misalignment, the IRSD and TD protocols may achieve a higher performance as they redirect more power to the desired lenses.

15.5 Simulation Results

In this section, we validate the analytical results for modeling IRS-assisted links. Moreover, we compare the performance of multiple FSO links sharing an IRS. We consider LS 1 and LS 2 at distances $d_{\ell_i} = 1$ km and directions $\Psi_{\ell 1} = \left(\frac{\pi}{3}, 0\right), \Psi_{\ell 2} = \left(\frac{\pi}{4}, 0\right)$, respectively. Moreover, PD 1 and PD 2 are at distances $d_{p_i} = 3$ km, $\iota \in \{1,2\}$ and directions $\Psi_{p1} = \left(\frac{\pi}{3}, \pi\right)$, and $\Psi_{p2} = \left(\frac{\pi}{6}, \pi\right)$, respectively. For all figures, the parameter values provided in Table 15.1 are adopted, unless specified otherwise.

15.5.1 Validation of Channel Model

First, we consider a point-to-point IRS-assisted FSO system where LS 1 is connected via a single-tile IRS to PD 1. In Fig. 15.4, we validate the analytical GML factors calculated using Huygens–Fresnel principle in (15.18) and geometric optics in (15.24) with the numerical integration of the Huygens–Fresnel equations in (15.8), (15.4), and (15.3), where a squared-shape IRS with lengths $L_x = L_y = L$ m and two different beam waists $w_{01} = 2.5$ mm and $w_{01} = 7$ mm are assumed. We adopt the LP profile in (15.26) and the profile in (15.22) for the

Table 15.1 System parameters.

Parameters	Symbol	Value
FSO bandwidth	W_{FSO}	1 GHz
FSO wavelength	λ	1550 nm
Beam waist radius	w_{01}, w_{02}	0.25 mm
Electric fields at origin	E_{01}, E_{02}	$60 \frac{\text{kV}}{\text{m}}$
Noise spectral density	N_0	-114 dBm/MHz
Attenuation coefficient	κ	$0.43 \times 10^{-3} \frac{\text{dB}}{\text{m}}$
Gamma–Gamma parameters	(α, β)	$(2,2)$
Impedance of the propagation medium	η	377Ω
IRS size	$L_x \times L_y$	1 m × 0.5 m
Lenses radius	a	15 cm

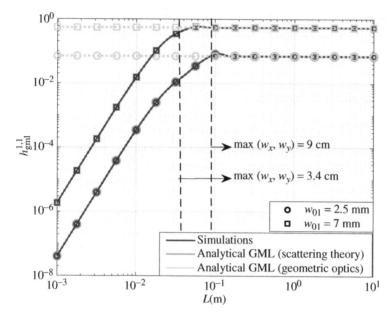

Figure 15.4 The GML factor versus the IRS length.

GML factor derived with Huygens–Fresnel (15.18) and geometric optics (15.24), respectively. However, given $d_\ell \gg \max(9.8\,\mu m, 0.016\,m)$, the proposed profile in (15.22) is practically identical to the LP profile, see Section 15.4.1.1. As can be observed, for larger IRS sizes, the GML increases as the IRS collects more power. Fig. 15.4 shows that the proposed analytical GML in (15.18) matches the Huygens–Fresnel simulation results for both beam waists, whereas the geometric optics approximation in (15.24) is only valid when the IRS length is much larger than the incident beam waist. As can be observed, for the larger beam waist, $w_{01} = 7\,mm$, the maximum incident beam width on the IRS $\max(w_x, w_y) = 3.4\,cm$ is smaller, see (15.6). Thus, the approximated GML based on geometric optics matches the simulation results at smaller values of L, since the equivalent mirror-based model is valid for very large IRSs, i.e., large enough to reflect most of the incident power.

15.5.2 Performance of Multi-Link IRS-Assisted FSO Systems

Figure 15.5 show the upper bounds on the outage probability of the link between LS 1 and PD 1 for required data rates $R_{th} = 1.3, 0.5$ Gbit/s. The performances of the three IRS-sharing protocols are compared for the LP profile and misalignment errors $\tilde{\mathbf{u}} = 0$ and $0.375\,m$. Here, by increasing angle θ_{p1}, IRS and lens become more

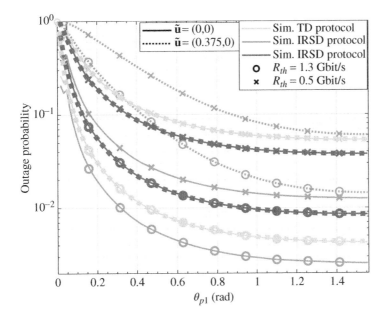

Figure 15.5 The outage probability versus the receiver lens angle.

parallel to each other which leads to a larger received power at PD 1 and a lower outage probability. For both considered threshold rates, in the absence of misalignment errors, the IRSD protocol performs better than the IRSH and TD protocols because of the larger received power at the lens and the simultaneous transmission of both LSs, respectively. Furthermore, for the larger data rate, i.e., $R_1 = 1.3$ Gbit/s, the IRSH protocol performs better than the TD protocol, where the LSs transmit only half of the time slots. For the smaller data rate, i.e., $R_1 = 0.5$ Gbit/s, the TD protocol performs better than the IRSH protocol because the benefit of a larger received power outweighs the rate loss associated with the TD protocol in this case. Furthermore, we observe that performance of the IRSD protocol degrades significantly in the presence of the misalignment error while the IRSH protocol is resilient to the misalignment error because of the homogenization of the IRS.

15.6 Future Extension

In this chapter, we developed an analytical channel model for optical IRS-assisted FSO links based on scattering theory and geometric optics, where we accounted for the non-uniform power distribution of Gaussian beams and a flexible phase-shift profile of the IRS. Moreover, we exploited the proposed model to study the performance of an IRS-assisted multi-link FSO system, where we proposed three IRS-sharing protocols. Our results revealed that, which of the proposed IRS-sharing protocols is preferable, depends on the position of the LSs and PDs, the target transmission rate, and on whether or not misalignment errors are present. Our channel model can be extended in future work to generalized atmospheric turbulence models with full phase control of the IRS. Further research is also required to model the impact of practical effects such as IRS phase-shift errors and wavefront distortions in FSO links.

Bibliography

Ajam, H., Najafi, M., Jamali, V., and Schober, R. (2021). Channel modeling for IRS-assisted FSO systems. *Proceedings of the IEEE WCNC*, 1–7.

Ajam, H., Najafi, M., Jamali, V. et al. (2022a). Modeling and design of IRS-assisted multi-link FSO systems. *IEEE Transactions on Communications* 70 (5): 3333–3349.

Ajam, H., Najafi, M., Jamali, V., and Schober, R. (2022b). Power scaling law for optical IRSs and comparison with optical relays. *Proceedings of the IEEE GLOBECOM*, 1527–1533.

Andrews, L.C. and Phillips, R.L. (2005). *Laser Beam Propagation through Random Media*. Bellingham, WA: SPIE Press.

Goodman, J.W. (2005). *Introduction to Fourier Optics*. Roberts & Co.

Gradshteyn, I.S. and Ryzhik, I.M. (1994). *Table of Integrals, Series, and Products*. San Diego, CA: Academic Press.

Hecht, E. (2017). *Optics*. Pearson.

Jamali, V., Ajam, H., Najafi, M. et al. (2021). Intelligent reflecting surface assisted free-space optical communications. *IEEE Communications Magazine* 59 (10): 57–63.

Najafi, M., Jamali, V., Schober, R., and Poor, H.V. (2021a). Physics-based modeling and scalable optimization of large intelligent reflecting surfaces. *IEEE Transactions on Communications* 69 (4): 2673–2691.

Najafi, M., Schmauss, B., and Schober, R. (2021b). Intelligent reflecting surfaces for free space optical communication systems. *IEEE Transactions on Communications* 69 (9): 6134–6151.

Saad, W., Bennis, M., and Chen, M. (2020). A vision of 6G wireless systems: applications, trends, technologies, and open research problems. *IEEE Network* 34 (3): 134–142.

Saleh, B.E.A. and Teich, M.C. (1991). *Fundamentals of Photonics*. New York: Wiley.

Uysal, M., Li, J., and Yus, M. (2006). Error rate performance analysis of coded free-space optical links over Gamma-Gamma atmospheric turbulence channels. *IEEE Transactions on Wireless Communications* 5 (6): 1229–1233.

Index

Intelligent Surfaces Empowered 6G Wireless Network, First Edition.
Edited by Qingqing Wu, Trung Q. Duong, Derrick Wing Kwan Ng, Robert Schober, and Rui Zhang.
© 2024 The Institute of Electrical and Electronics Engineers, Inc. Published 2024 by John Wiley & Sons, Inc.

Printed and bound by CPI Group (UK) Ltd, Croydon, CR0 4YY

16/04/2025

14658418-0002